Lubrication Fundamentals

Mechanical Engineering
A Series of Textbooks and Reference Books

editors

L. L. Faulkner
Department of Mechanical Engineering
The Ohio State University
Columbus, Ohio

S. B. Menkes
Department of Mechanical Engineering
The City College of the City University of New York
New York, New York

Other Volumes in Preparation

Lubrication Fundamentals

J. GEORGE WILLS

Mobil Oil Corporation

MARCEL DEKKER, INC.
New York and Basel

Wills, J. George
   Lubrication fundamentals.

   (Mechanical engineering: 3)
   Includes index.
   I. Lubrication and lubricants.
I. Title   II. Series.
TJ1075.W57   621.8'9   80-14095
ISBN 0-8247-6976-7

Marcel Dekker, Inc., 270 Madison Avenue, New York, New York 10016

10 9 8

Printed in the United States of America

Preface

Our modern world of machines moves on lubricating films, and everyone, whether knowingly or not, has an interest in lubrication. Sewing machines, vacuum cleaners, or clothes washers could not run without proper lubrication. Millions of automobiles, transporting people and goods, would never get out of the garage without automotive oils and greases. Complex equipment — to make cloth, process food from the field for the consumer, and produce electric energy for the many comforts we enjoy today — can operate only because there are lubricants formulated to satisfy the operating requirements.

Intricate machines spin and weave textiles; huge rolling mills turn out metal sheets; metal-working machines make intricate metal parts with extreme precision; special machines manufacture cement, rubber, and plastics — all these machines are continually being improved to reduce operating costs and speed up production.

The increased efficiency and complexity of modern machines often require improved lubricants to perform under much more severe operating conditions. The machine builder, the user, and the lubricant supplier cooperate in determining the requirements, and the researcher and the refiner work together to develop new and better lubricants to meet the changing needs of our ever more complex world.

This book was written for those who desire a fundamental knowledge of lubricants and lubrication, especially those concerned with the operation and maintenance of machines.

Chapter 1 presents an historical introduction to the field of lubrication. The next three chapters discuss lubricating oils and greases, and the new field of synthetic lubricants. Chapters 5 and 6 concern the machine elements that require lubrication and the methods of application. Chapters 7 through 14 examine various specific applications of lubricants — internal combustion engines, stationary gas, steam, and hydraulic turbines, nuclear power plants, automotive chassis components and power transmissions, and compressors. Finally, Chapter 15 details the proper handling, storing, and dispensing of lubricants, and the last chapter discusses in-plant conservation methods.

For those who would like to examine any of these subjects further, a list of related Mobil publications is provided at the end of most chapters.

J. George Wills

Acknowledgments

Lubrication Fundamentals, as with all technical publications of this magnitude, is not the work of one person. It is the combined effort of hundreds, even thousands of engineers, designers, chemists, physicists, writers, and artists—the compendium of a broad spectrum of talent over a long period of time. It is the result of the unique cooperation that exists between the machine designer and equipment builders on one side and the lubricant formulators and suppliers on the other, the cooperation that takes place in the many associations such as ASLE, SAE, ASTM, NLGI, AGMA, and API, to name but a few.

Lubrication fundamentals starts with the scientists who study with their microscopes, computers, and other sophisticated equipment the basic nature of the interaction of oil films with bearings, gears, and cams under stresses and loads. It continues with the translation of these data in the research and development laboratories of component and system designers, and lubricating oil and grease formulators and performance evaluators. It culminates in the mating of superior lubricants properly applied to the requirements of the most efficient machines of today's technology.

It would be impossible to list the host of people who have helped to put this book together. We have been fortunate to have had the opportunity to guide it through from draft to publication. It is a compilation of the many technical publications of Mobil Oil Corporation and the cooperative offerings in data and graphics of the foremost international equipment manufacturers. Impossible though it may be to fairly acknowledge the contributions of everyone, the following groups and individuals must be singled out for thanks:

Our supervisors at Mobil — Dalt Schnack, Jack Lane, Max Johnson, and Don Edwards — first for their acceptance of the idea and then for their encouragement to complete it

The marketers, engineers, and researchers at Mobil for their contributions, comments, and constructive criticism which make the book technically adequate and accurate

Norm Danforth, who assisted in the compilation and organization of the manuscript, and Bonnie Higgins, who assisted in its editing

Stan Rup, with whom we have worked for years in the fine arts and graphics fields, for preparing much of the line art and the dust jacket design, which epitomizes the cooperation between engineers, designers, lubricators, and writers — the foundation of successful lubrication.

Contents

Lubrication
Fundamentals

1

Introduction

Petroleum is one of the naturally occurring hydrocarbons that frequently include natural gas, natural bitumen, and natural wax. The name "petroleum" is derived from the Latin *petra* (rock) and *oleum* (oil). According to the most generally accepted theory today, petroleum was formed by the decomposition of organic refuse, aided by high temperatures and pressures, over a vast period of geologic time.

Although petroleum occurs, as its name indicates, among rocks in the earth, it sometimes seeps to the surface through fissures or is exposed by erosion. The existence of petroleum was known to primitive man, since surface seepage, often sticky and thick, was obvious to anyone passing by. Prehistoric animals were sometimes mired in it, but few human bones have been recovered from these tar pits. Early man evidently knew enough about the danger of surface seepage to avoid it.

The first actual use of petroleum seems to have been in Egypt, which imported bitumen, probably from Greece, for use in embalming. The Egyptians believed that the spirit remained immortal if the body was preserved.

About the year 450 B.C., Herodotus, the father of history, described the pits of Kir ab ur Susiana as follows:

At Ardericca is a well which produces three different substances, for asphalt, salt and oil are drawn up from it in the following manner. It is pumped up by means of a swipe; and, instead of a bucket, half a wine skin is attached to it. Having dipped down with this, a man draws it up and then pours the contents into a reservoir, and being poured from this into another, it assumes these different forms: the asphalt and salt immediately become solid, and the liquid oil is collected. The Persians call it Phadinance; it is black and emits a strong odor.

Pliny, the historian, and Dioscorides Pedanius, the Greek botanist, both mention "Sicilian oil," from the island of Sicily, which was burned for illumination as early as the beginning of the Christian Era.

The Scriptures contain many references to petroleum, in addition to the well-known story of Moses, who was set afloat on the river as an infant in a little boat of reeds waterproofed with pitch, and was found by Pharoah's daughter. Some of these references include the following:

> Make thee an ark of gopher wood; rooms shalt thou make in the ark, and shalt pitch it within and without with pitch. (Genesis VI. 14)

> And they had brick for stone, and slime (bitumen) had they for mortar. (Building the Tower of Babel, Genesis XI.3)

> And the Vale of Siddim was full of slime (bitumen) pits; and the kings of Sodom and Gomorrah fled, and fell there . . . . (Genesis XIV.10)

Other references are found in Strabo, Josephus, Diodorus Siculus, and Plutarch, and since then there is much evidence that petroleum was known in almost every part of the world.

Marco Polo, the Venetian traveler and merchant, visited the lands of the Caspian in the thirteenth century. In an account of this visit, he stated:

> To the north lies Zorzania, near the confines of which there is a fountain of oil which discharges so great a quantity as to furnish loading for many camels. The use made of it is not for the purpose of food, but as an unguent for the cure of cutaneous distempers in men and cattle, as well as other complaints; and it is also good for burning. In the neighboring country, no other is used in their lamps, and people come from distant parts to procure it.

Sir Walter Raleigh, while visiting the island of Trinidad off the coast of Venezuela, inspected the great deposit of bitumen there. The following is taken from *The Discoveries of Guiana* (1596):

> At this point called Tierra de Brea, or Piche, there is that abundance of stone pitch that all the ships of the world may be therewith loden from thence, and wee made triall of it in trimming our ships to be most excellent good, and melteth not with the sunne as the pitch of Norway, and therefore for ships trading the south partes very profitable.

On the North American continent, petroleum seepages were undoubtedly known to the aborigines, but the first known record of the substance was made by the Franciscan, Joseph de la Roche D'Allion, who in 1629 crossed the Niagara River from Canada and visited an area later known as Cuba, New York. At this place, petroleum was collected by the Indians who used it medicinally and to bind pigments used in body adornments.

In 1721, Charlevois, the French historian and missionary who descended the Mississippi River to its mouth, quotes a Capitan de Joncaire as follows:

> There is a fountain at the head of a branch of the Ohio River (probably the Allegheny) the waters of which like oil, has a taste of iron and serves to appease all manner of pain.

In the *Massachusetts Magazine*, Volume 1, July 1, 1789, there is an account under the heading "American Natural Curiosities," as follows:

> In the northern parts of Pennsylvania, there is a creek called Oil Creek, which empties into the Allegheny River. It issues from a spring, on the top of which floates an oil similar to that called Barbadoes tar; and from which one man may gather several gallons in a day. The troops sent to guard the western posts halted at this spring, collected some of the oil, and bathed their joints with it. This gave them great relief from the rheumatic complaints with which they were affected. The waters, of which the troops drank freely, operated as a gentle purge.

Although the practice of deriving useful oils by the distillation of bituminous shales and various organic substances was generally known, it was not until the nineteenth century that distillation processes were widely used for a number of useful substances, including tars for waterproofing; gas for illumination; and various chemicals, pharmaceuticals, and oils.

In 1833, Dr. Benjamin Silliman contributed an article to the *American Journal of Science* which contained the following:

> The petroleum, sold in the Eastern states under the name of Seneca Oil, is a dark brown color, between that of tar and molasses, and its degree of consistency is not dissimilar, according to temperature; its odor is strong and too well known to need description. I have frequently distilled it in a glass retort, and the naphtha which collects in the receiver is of a light straw color, and much lighter, more odorous and inflammable than the petroleum; in the first distillation, a little water usually rests in the receiver, at the bottom of the naphtha; from this it is easily decanted, and a second distillation prepares it perfectly for preserving potassium and sodium, the object which led me to distil it, and these metals I have kept under it (as others have done) for years; eventually they acquire some oxygen, from or through the naphtha, and the exterior portion of the metal returns, slowly, to the condition of alkali — more rapidly if the stopper is not tight.

> The petroleum remaining from the distillation is thick like pitch; if the distillation has been pushed far, the residuum will flow only languidly in the retort, and in cold weather it becomes a soft solid, resembling much the maltha or mineral pitch.

Along the banks of the Kanawasha River in West Virginia, petroleum was proving a constant source of annoyance in the brine wells; and one of these wells, in 1814, discharged petroleum at periods of from 1 to 4 days, in quantities ranging from 30 to 60 gallons at each eruption. A Pittsburgh druggist named Samuel M. Kier began bottling the petroleum from these brine wells around 1846 and selling the oil for medicinal purposes. He claimed it was remarkably effective for most ills and advertised this widely. In those days, many people believed that the worse a nostrum tasted the more powerful it was. People died young then, and often did not know what killed them. In the light of today's knowledge, we would certainly not recommend drinking such products. Sales boomed for awhile; but in 1852, there was a falling-off in trade. Therefore, the enterprising Mr. Kier began to distill the substance for its illuminating oil content. His experiment was successful and was a forerunner, in part, of future commercial refining methods.

In 1853, a bottle of petroleum at the office of Professor Crosby of Dartmouth College was noticed by Mr. George Bissel, a good businessman. Bissel soon visited Titusville, Pennsylvania, where the oil had originated, and purchased 100 acres of land in an area known as Watsons Flats and leased a similar tract for the total sum of $5,000. Bissel and an associate, J. D. Eveleth, then organized the first oil company in the United States, the Pennsylvania Rock Oil Company. The incorporation papers were filed in Albany, New York, on December 30, 1854. Bissel had pits dug in his land in the hope of obtaining commercial quantities of petroleum, but was unsuccessful with this method. A new company was formed, which was called the Pennsylvania Rock Oil Company of Connecticut, with New Haven as headquarters. The property of the New York corporation was transferred to the new company, and Bissel began again.

In 1856, Bissel read one of Samuel Kier's advertisements on which was shown a drilling rig for brine wells. Suddenly it occurred to him to have wells drilled, as was being done in some places for brine. A new company, The Seneca Oil Company, succeeded the older company, and an acquaintance of some of its partners, E. L. Drake, was selected to conduct field experiments in Titusville. Drake found that to reach hard rock in which to try the drilling method, some unusual form of shoring was needed to prevent a cavein. It occurred to him to drive a pipe through the loose sand and shale; a plan afterward adopted, in oil well and artesian well drilling.

Drilling then began under the direction of W. A. Smith, a blacksmith and brine well driller, and went down 69½ feet. On Saturday, August 27, 1859, the drill dropped into a crevice about 6 inches deep, and the tools were pulled out and set aside for the work to be resumed on Monday. However, Smith decided to visit the well that Sunday to check on it, and upon peering into the pipe saw petroleum within a few feet of the top. On the following day, the well produced the incredibile quantity of 20 barrels a day.

During the period from 1850 to 1875, many men experimented with the products of petroleum distillation then available, attempting to find uses for them, in addition to providing illumination. Some of the viscous materials were investigated as substitutes for the vegetable and animal oils previously used for lubrication, mainly those derived from olives, rape seed, whale, tallow, lard, and other fixed oils.

As early as 1400 B.C., greases, made of a combination of calcium and fats, were used to lubricate chariot wheels. Traces of this grease were found on chariots excavated from the tombs of Yuaa and Thuiu. During the third quarter of the nineteenth century, greases were made with petroleum oils combined with potassium, calcium, and sodium soaps and placed on the market in limited quantities.

Gradually, as distillation and refining processes were improved, a wider range of petroleum oils was produced to take the place of the fatty oils. These mineral oils could be controlled more accurately in manufacture and were not subject to the rapid deterioration of the fatty oils.

Some of the fatty oils continued to be used in special services as late as the early part of the twentieth century. Tallow was fairly effective in steam cylinders as a lubricant. However, it was not always pleasant to handle, since maggots developed in the tallow particularly in hot weather. Lard oil was used for cutting of metals, and castor oil was used to lubricate the aircraft engines of World War I. Today, some fatty oils are still used as compounding in small percentages with mineral oils, but chemical additives have taken their place for the majority of users.

2
Lubricating Oils

At the end of 1976, the petroleum industry, one of the largest in the world, had a gross investment in property, plant, and equipment that totaled 377 billion dollars — 149 billion dollars in the United States and 228 billion dollars in the free foreign areas of the world. Petroleum companies and their dealer operated service stations employ approximately 1.5 million people in the United States. In 1977, Mobil, one of the largest companies in the industry, averaged approximately 6 percent of the total United States refining capacity in its domestic operations.

Petroleum, or crude oil, is refined to make many essential products used throughout the world in homes and industry. These products include gas burned as fuel, gasoline, kerosine, solvents, fuel oil, diesel fuel, lubricating products, industrial specialty products, waxes, chemicals, asphalt, and coke. Usually, crude oil is refined in two stages — light products refining and refining of lubricating oils and waxes. The refining of light products, which is concerned with all of these except lubricants, specialty products, waxes, asphalts, and coke, is accomplished at, or slightly above, atmospheric pressure. Although all the products discussed are not actually light in weight or color (for example, the heavy fuels, oils, and asphalts), their production is grouped with that of the light products because they are all made in the same or similar equipment.

At approximately 700°F (371°C), the residuum from light products refining has a tendency to decompose. Thus, the refining of lubricating oils and waxes takes place under vacuum conditions and at temperatures under the decomposition point. This type of refining will be discussed in detail in this chapter following brief general discussions of crude oil handling and its initial fractionation into light products and residuum.

CRUDE OIL

Origin and Sources

The petroleum that flows from our wells today was formed many millions of years ago. It is believed to have been formed from the remains of tiny aquatic animals and plants that settled with mud and silt to the bottoms of ancient seas. As successive layers built up, those remains were subjected to high pressures and temperatures and underwent chemical transformations leading to the formation of the hydrocarbons and other constituents of crude oil described herein. In many areas, this crude oil migrated and accumulated in porous rock formations overlaid by impervious rock formations that prevented further travel. Usually a layer of concentrated salt water underlies the oil pool.

The states of Texas, California, and Louisiana, with their offshore areas, are the largest producers of crude oil in the United States, although petroleum was first produced in Pennsylvania. Today, a major portion of this nation's needs is supplied from Canada, South America, and the oil fields in the Middle East.

Production

Crude oil was first found as seepages, and one of the earliest references to it may have been the "fiery furnace of Nebuchadnezzar." This is now thought to have been an oil seepage that caught fire. Currently, holes are drilled into the ground as deep as 5 miles to tap oil bearing strata located by geologists. The crude oil frequently comes to the surface under great pressure and in combination with large volumes of gas. Present practice is to separate the gas from the oil and process the gas to remove from it additional liquids of high volatility to form what is called "natural gasoline," for addition to motor gasoline. The "dry" gas is sold as fuel or recycled back to the underground formations to maintain pressure in the oil pool and, thus, to increase recovery of crude oil. Years ago much of this gas was wasted by burning it in huge flares.

Types and Composition

Crude oils are found in a variety of types ranging from light colored oils, consisting mainly of gasoline, to black, nearly solid asphalts. Crude oils are very complex mixtures containing very many individual hydrocarbons or compounds of hydrogen and carbon. These range from methane, the main constituent of natural gas with one carbon atom, to compounds containing 50 or more carbon atoms; see Fig. 2-1. The boiling ranges of the compounds increase roughly with the number of carbon atoms:

1. Far below 0°F (−18°C) for the light natural gas hydrocarbons with one to three carbon atoms
2. About 80 to 400°F (27−204°C) for gasoline components
3. 400 to 650°F (204−343°C) for diesel and home heating oils
4. Higher ranges for lubricating oils and heavier fuels

The asphaltic materials cannot be vaporized because they decompose when heated and their molecules either 'crack' to form gas, gasoline, and lighter fuels, or unite to form even heavier molecules. The latter form carbonaceous residues called "coke" which, as will be discussed later, can be either a product or a nuisance in refining.

Crude oils also contain varying amounts of compounds of sulfur, nitrogen, oxygen, various metals such as nickel and vanadium, and some entrained water—containing dissolved salts. All of these materials can cause trouble in refining or in subsequent product applications, and their reduction or removal increases refining costs appreciably. Many municipalities and states, for example, have passed laws limiting the sulfur content of fuels.

The carbon atom is much like a four holed Tinker Toy piece, and is even more versatile. Reference to Fig. 2-1 shows that, for molecules containing two or more carbon atoms, a number of configurations can exist for each number of carbon atoms; in fact, the number of such possible shapes increases with the number of carbon atoms. Each configuration has distinct properties. Compounds with the carbon atoms in a straight line (normal paraffins) have low octane ratings when in the gasoline boiling range but

Fig. 2-1 Typical Hydrocarbon Configuration

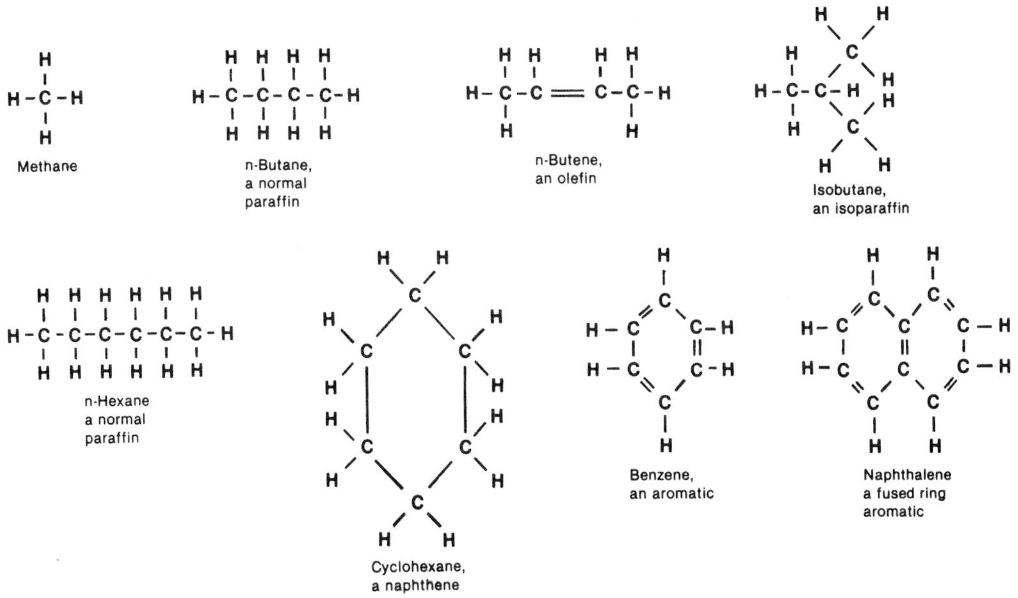

Methane

n-Butane, a normal paraffin

n-Butene, an olefin

Isobutane, an isoparaffin

n-Hexane a normal paraffin

Cyclohexane, a naphthene

Benzene, an aromatic

Naphthalene a fused ring aromatic

make excellent diesel fuels. They consist of waxes when they are in the lubricating oil boiling range. Branched chain and ring compounds with low hydrogen content like benzene burn with a smoky flame in kerosine lamps and may cause knocking in diesel engines.

Crudes range from "paraffinic" types, which are high in paraffin hydrocarbons, through the "intermediate" or "mixed base" types to the "naphthenic" types, which are high in hydrocarbons containing ring structures. Asphalt content varies in different types of crudes.

REFINING

**Crude
Distillation**
Crude oil is sometimes used as such for fuel in power plants; but in most cases, it is separated into many fractions, which in turn require further processing to supply the large number of petroleum products needed. In many cases, the first step is to remove from the crude certain inorganic salts suspended as minute crystals or dissolved in entrained water. These salts break down during processing to form acids that severely corrode refinery equipment, plug heat exchangers and other equipment, and poison catalysts used in subsequent processes. Therefore, the crude is mixed with additional water to dissolve the salts and the resultant brine is removed by settling.

After desalting, the crude is pumped through a tubular furnace (Fig. 2-2) where it is heated and partially vaporized. The refinery furnace usually consists of connected lengths of pipe heated externally by gas or oil burners. The mixture of hot liquid and vapor from the furnace enters a fractionating col-

Fig. 2-2 Crude Distillation Unit

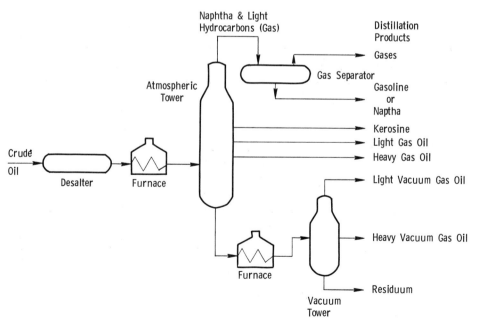

umn. This is a device that operates at slightly above atmospheric pressure and separates groups of hydrocarbons according to their boiling ranges. The fractionating column works because there is a gradation in temperature from bottom to top so that, as the vapors rise toward the cooler upper portion, the higher boiling components condense first. As the vapor stream moves up the column, lower boiling vapors are progressively condensed. Trays are inserted at various levels in the column to collect the liquids that condense at those levels. Naptha, an industry term for raw gasoline that requires further processing, and light hydrocarbons are carried over the top of the column as vapor and are condensed to liquid by cooling. Kerosine, diesel fuel, home heating fuels, and heavy oils (called gas oils) are withdrawn as side cuts from the successively lower and hotter levels of the tower.

A heavy, black, atmospheric residuum is drawn from the bottom of the column. The combination of furnace and atmospheric tower is sometimes called a "pipe still."

Because of the tendency of residuum to decompose at temperatures about 700°F (371°C), heavier (higher boiling) oils such as lubricating oils must be distilled off in a separate vacuum fractionating tower. The greatly reduced pressure in the tower markedly lowers the boiling points of the desired hydrocarbon compounds. Bottom materials from the vacuum tower are used for asphalt, or are further processed to make other products.

The fractions separated by crude distillation are sometimes referred to as "straight run" products. The character of their hydrocarbon constituents is not changed by distillation. If all of the separated fractions were reassembled, we would recover the original crude.

Lubricating Oils

The term lubricating oils is generally used to include all those classes of lubricating materials that are applied as fluids. On a volume basis, nearly all of the world's lubricating oil is obtained by refining distillate or residual fractions obtained directly from crude oil. This section is concerned primarily with these "mineral" oils. Because of their growing importance, synthetic lubricants, both oils and greases, are discussed in Chap. 4.

Lubricating oils are made from the more viscous portion of the crude oil which remains after removal by distillation of the gas oil and lighter fractions. They have been prepared from crude oils obtained from most parts of the world. Although crude oils from various parts of the world differ widely in properties and appearance, there is relatively little difference in their elemental analysis. Thus, crude oil samples will generally show carbon content ranging from 83 to 87 percent, and hydrogen content from 11 to 14 percent. The remainder is composed of elements such as oxygen, nitrogen, sulfur, and various metallic compounds. An elemental analysis, therefore, gives little indication of the extreme range of physical and chemical properties that actually exists, or of the nature of the lubricating oils that can be produced from a particular crude oil.

An idea of the complexity of the lubricating oil refining problem can be obtained from a consideration of the variations that can exist in a single hydrocarbon molecule with a specific number of carbon atoms. For example, the paraffinic molecule containing 25 carbon atoms (a compound falling well within the normal lubricating oil range) has 52 hydrogen atoms. This compound can have about 37,000,000 different molecular arrangements. When it is considered that there are also naphthenic and aromatic hydrocarbon molecules (Fig. 2-1) containing 25 carbon atoms, it will be seen that the possible variations in molecular arrangement for a 25 carbon molecule are immense. This accounts for much of the variation in physical characteristics and performance qualities of lubricating oils prepared from different crude sources. As a result, in order to minimize variations and produce products that will provide consistent performance in specific applications, the refiner follows four main stages in the manufacture of finished lubricating oils from the various available crudes:

1. Selection and segregation of crudes according to the principal types of hydrocarbons present in them
2. Distillation of the crude to separate it into fractions containing hydrocarbons in the same general boiling range
3. Processing to remove undesirable constituents from the various fractions, or to convert some of these materials to more desirable materials
4. Blending to the physical characteristics required in the finished products and incorporating chemical agents which improve performance attributes

Crude Oil Selection

One way to understand the extreme differences that can exist among crude oils is to examine some of the products that are made from different types of crudes.

Table 2-1 shows two base stocks that are similar in viscosity—the most important physical property of a lubricant. The base stock on the left in the table is made from a naphthenic crude. This type of crude is unusual since it contains essentially no wax. In fact, the very low pour point, −50°F (−46°C), of this stock results from the unique composition of the compounds in the crude—no processing has been employed to reduce the pour point. In contrast, the stock on the right in the table required dewaxing to reduce its pour point from about 80°F (27°C) to 0°F (−18°C). One other important difference between these base stocks is shown by the differences in viscosity index. While both oils have similar viscosities at 100°F (38°C), the viscosity of the naphthenic oil will change with temperature much more than the viscosity of the paraffinic stock. This is reflected in the lower viscosity index (VI) of the naphthenic oil. For products that operate over a wide temperature range, such as automotive engine oils, the naphthenic stock would be less desirable. Generally, naphthenic base stocks are used in products that have a limited range of operating temperature and in which the unique composition of

Table 2-1 Lube Base Stocks

Crude Type	Naphthenic	Paraffinic
Viscosity SUS @ 100°F (cSt @ 38°C)	100 (20.53)	100 (20.53)
Pour Point, °F (°C)	−50 (−45.5)	0 (−18)
Viscosity Index	15	100
Flash Point, °F (°C)	340 (171)	390 (199)
Gravity, ° API	24.4	32.7
Color (ASTM)	1$\frac{1}{2}$	$\frac{1}{2}$

naphthenic crudes—with the resultant low pour point—is required. Long-term supply of naphthenic crudes is uncertain, and alternatives are being sought to replace these base stocks as the supply of naphthenic crude diminishes. Recognizing the large differences that can exist between different crudes, the major factors that must now be considered in lube crude selection are supply, refining, product, and marketing.

Supply Factors Supply factors in crude selection are the quantities available, the constancy of composition from shipment to shipment, and the cost and ease of segregating the particular crude from other shipments.

Since about 10 barrels of crude oil are needed to make a barrel of lube base stock, relatively large volume crudes are desirable for processing. Crude oil with variable composition will cause problems in the refinery because of the rather limited ability to adjust processing to compensate for crude changes. Since only some crudes are suitable for lube manufacture, segregation of crude oils is essential. If the cost of segregation is too high, the crude will not be used for lube oil manufacture. For example, if one of the Alaskan North Slope crudes were to be considered for lube processing, the inability to segregate the crude at reasonable cost would clearly eliminate it from consideration, since many of the crudes with which it would be mixed in the Trans-Alaskan Pipeline are extremely poor for lube manufacture.

Refining Factors Refining factors important in selecting a crude oil for lube base stocks are the ratio of distillate to residuum, the processing required to prepare suitable lube base stock, and the final yield of finished lube base stocks. For a crude to be useful for lube manufacture, it must contain a reasonable amount of material in the proper boiling range. For instance, very light crudes (such as condensates) would not be considered as lube crudes since they contain only a few percent of material in the higher boiling range needed for lube base stocks.

Once it has been established that the crude contains a reasonable amount of material in the lube boiling range, the response of the crude to available processes must be examined. If the crude requires very severe refining conditions or exhibits low yields on refining, it will be eliminated. A recent example of this is Gippsland crude, an offshore Australian crude that met the supply criteria, had reasonable distillate yields, and even responded well to furfural extraction. However, in the dewaxing process, Gippsland showed an extremely low yield of dewaxed oil—due to its very high wax content. The dewaxing yield was so far out of line that it was not possible to process this crude economically.

Product Factors Product factors concern the quality aspects of all the products refined from the crude—not only the lube base stock. These product qualities include the base stock quality and its response to presently available additives and, also, the quality of light products and byproducts extracted from the crude.

Since almost 90 percent of the crude will end up in nonlube products, this portion cannot be ignored. In some situations, the quality of a certain byproduct, e.g., asphalt, can be of overriding importance in the evaluation of a crude.

The lube stock produced from a crude must not only be satisfactory in chemical and physical characteristics but must respond to the additives that are readily available on the market. If new or different additives are needed because of a new crude, the economics of using this crude for lubes must support the cost of finding these additives and implementing them within the system.

Marketing Factors Marketing factors to be considered in evaluating a crude oil for lube base stock manufacture include the viscosity range and overall product quality required by the lube oil market, and the operating and investment costs of manufacturing a lube oil based on the market product sales price.

The location (market) in which the crude oil will be used can have a major impact on the economics of its use for lubes. A crude which contains a great deal of low viscosity material would be ideal (in this respect) for use in the United States. Although the U.S. product demand requires a lot of low viscosity product, this requirement might be totally unsuitable for use in a different market which required larger amounts of high viscosity oil. Likewise, the product quality required depends very strongly on the market being served, as do operating and investment costs. Therefore, in addition to the physical and chemical composition of a crude oil, selection finally becomes an economic business decision.

Having discussed the factors that must be considered in selecting a crude oil for lube base stock manufacture, satisfactory crudes are summarized in Table 2-2.

Lube Base Stock Manufacture The manufacture of lube base stocks consists of a series of separation or subtractive processes which remove undesirable components from the feedstock leaving a lube base stock that meets performance requirements.

Almost all the processes used to produce lube base stocks in Mobil re-

Table 2-2 Satisfactory Crudes for Lube Base Stock Manufacture

Amal	Kirkuk
Arabian Extra Light	Kuwait
Arabian Light	Lago Medio
Arabian Medium	Louisiana Light
Arabian Heavy	Luling/Lyton/Karnes
Barco	Mid-Continent Sweet
Basrah	Minagish
Block 17	Mosul
Centro Lago	Murban
Citronelle	Ohanet 105
Dahra	Qatar
Edjele	Raudhatain
Hassi Messaoud	Safaniyah
Hough	Sirtica
Iranian Light	West Texas Bright
Khurais	Zarzaitine

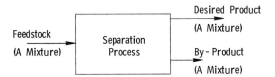

Fig. 2-3 Lube Separating Process

fineries are separation processes; i.e., the processes operate by dividing the feedstock, which is a complex mixture of chemical compounds into products — usually two products. Thus, although the products themselves are still complex mixtures; the compounds in each of the products are similar in either physical or chemical properties.

This concept of a separation process is basic to understanding lube base stock manufacturing. A simple diagram of a separation process is shown in Fig. 2-3.

A simplified block flow diagram (Fig. 2-4) indicates the five processes in lube oil refining:

1. Vacuum distillation
2. Propane deasphalting
3. Furfural extraction
4. Methyl ethyl ketone (MEK) dewaxing
5. Hydrofinishing

The first four items are separation processes. The fifth, hydrofinishing, is a catalytic reaction with hydrogen to decolorize the base oil stocks. The purpose of these processes is to remove materials that are undesirable in the final product.

Fig. 2-4 Lube Processing

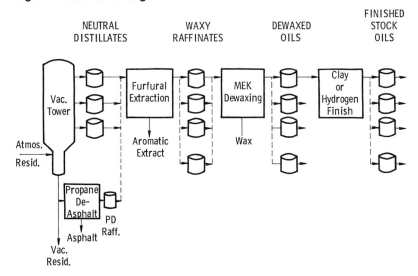

Before discussing these processes in detail, a brief discussion may be useful in clarifying their interrelationship. To help in this discussion, a simplified representation of a crude oil (Fig. 2-5) shows that the crude consists not only of compounds (paraffins, naphthenes, aromatics) that are chemically different but also of compounds that are chemically similar (e.g., paraffins) but which differ in boiling point.

Vacuum Distillation Vacuum distillation is the first step in refining lubricating oils. This is a separation process which segregates crude oil into products that are similar in boiling point range. In terms of the simplified picture of crude oil in Fig. 2-5, distillation can be represented as a vertical cut; see Fig. 2-6. *Note*: In Fig. 2-6, distillation divides the feedstock (crude oil) into products that consist of materials with relatively narrow ranges of boiling points.

Propane Deasphalting Propane deasphalting (PD) operates on the very bottom of the barrel—the residuum. This is the product shown in the simplified distillation on the right of Fig. 2-6, the highest boiling portion of the crude. Note that the residuum in Fig. 2-6 contains some types of compounds not present in the other products from distillation—resins and asphaltenes. PD, which removes these materials, can also be represented as a horizontal cut; see Fig. 2-7. The residuum has now been divided into two products, one containing almost all the resins and asphaltenes, called PD tar, and the other containing compounds that are similar chemically to those in the lube distil-

Fig. 2-5 Crude Oil Composition—Simplified

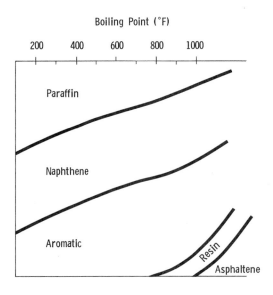

Fig. 2-6 Lube Distillation

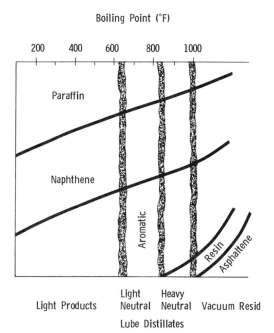

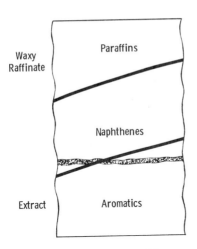

Fig. 2-7   **Propane Deasphalting**

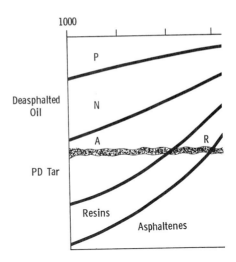

Fig. 2-8   **Furfural Extraction**

lates but of higher boiling point. This material is called deasphalted oil and is refined in the same way as lube distillates.

Furfural Extraction   Furfural extraction, the next process in the sequence, can be represented as a horizontal cut (Fig. 2-8), dividing the distillate or deasphalted oil into a raffinate, which is the desired material, and a byproduct called an extract which is mostly aromatic compounds.

MEK Dewaxing   Methyl ethyl ketone dewaxing (MEK) of the raffinate from the furfural extraction is another horizontal cut (Fig. 2-9), producing a byproduct wax which is almost completely paraffins and a dewaxed oil which contains paraffins, naphthenes, and some aromatics. For many fluid lubricants, this dewaxed oil is the base stock used. For certain premium applications, however, a finishing step is needed.

Hydrofinishing   Hydrofinishing by chemical reaction of the oil with hydrogen changes the polar compounds slightly but retains them in the oil. In this process, the most obvious result is a lighter color oil. From the simplified description of crude oil, the product is indistinguishable from the dewaxed oil feedstock.

The purpose of this oversimplified look at lube processing is to emphasize that it is a series of processes to remove undesirable materials. Thus, in order for a crude oil to be used for lubricating oil manufacture, it must contain some good base oil.

LUBRICATING OIL PROCESSING

Vacuum Distillation   Vacuum distillation is considered the first process in lube base stock manufacturing. This assumes that an acceptable crude has been processed properly in an atmospheric distillation column for recovery of light products.

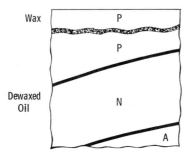

Fig. 2-9 Dewaxing

The residuum from the atmospheric distillation column is the feedstock for the lube vacuum distillation column.

The portion of the residuum from the atmospheric distillation column which boils between 650°F (340°C) and 700°F (371°C), is not used as light product because of its adverse specifications (pour point or cloud point of diesel or heating fuel). An additional upper limitation on the temperature to which an atmospheric column can be operated is imposed by thermal cracking. If a crude is heated to too high a temperature, the molecules will be broken up into smaller, lower boiling molecules. These thermally cracked materials can cause product quality problems as well as operating problems in the atmospheric distillation unit.

Purpose In vacuum distillation, the feedstock is separated into products of similar boiling point to control the physical properties of the lube base stocks that

Fig. 2-10 Lube Distillate Fractionation Viscosities

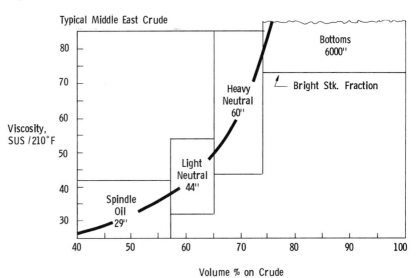

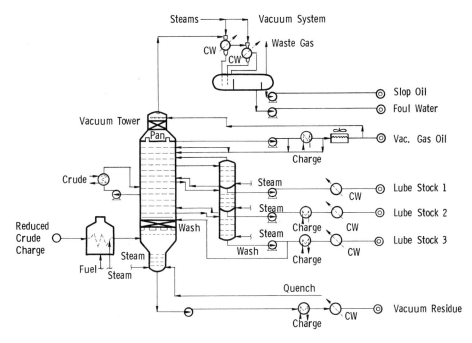

Fig. 2-11   Vacuum Tower Fractionation for Lube Distallates

will be produced from the vacuum tower products. The major properties that are controlled by vacuum distillation are viscosity, flash point, and carbon residue.

The viscosity of the lube oil base stock, the most important physical property for lubrication, is determined by the viscosity of the distillate.

Fig. 2-10 illustrates vacuum distillation in terms of viscosity separation. The dark line represents the distribution of viscosity in a specific crude oil. The figure shows three vacuum tower distillates and a residuum represented by the boxed areas. Notice that each distillate contains material displaying a range of viscosity both higher and lower than the average viscosity. In the example, some overlap is shown, and, in reality, overlap is present in the products of commercial vacuum towers. However, a vacuum tower must be run properly to prevent excessive overlap, otherwise downstream processing problems will result; e.g., a light neutral with too much viscous material will be difficult to dewax. High viscosity material in a heavy neutral — too much overlap with the vacuum residuum — causes not only processing problems but results in inferior product quality.

Fig. 2-11 is a simplified flow diagram of a lube vacuum column. The crude feed (a reduced crude charge consisting of the residuum from the atmospheric distillation) is heated in a furnace and flows into the flash zone of the column where the vapor portion begins to rise and the liquid falls. Temperatures of about 750°F (399°C) are used in the heater — steam is added to assist in the vaporization. A vacuum is maintained in the flash zone by a vacuum system connected to the top of the column. By reducing the pressure to less than one-tenth of atmospheric pressure, materials boiling up to 1000°F (538°C) at atmospheric pressure can be vaporized without thermal cracking. As the hot vapors rise through the column, they are cooled by removing material from the column, cooling it, and returning it to the column.

The liquid in the column wets the packing and starts to flow back down the column where it is revaporized by contacting the rising hot vapors. At various points in the column, special trays, called draw trays, are installed which permit the collection and removal of the liquid from the column. The liquid that is withdrawn contains not only the material that is normally liquid at the temperature and pressure of the draw tray but also a small amount of lower boiling material dissolved in the liquid. To remove this lower boiling material, the distillate, after removal from the vacuum tower, is charged to a stripping column where steam is introduced to strip out these low boiling materials.

In terms of the physical properties of the distillate, the stripping column adjusts the flash point by removing low boiling components. The low boiling materials and the steam are returned to the vacuum column. Other lube distillates are stripped similarly. The vacuum residuum is also steam stripped; however, this is generally done internally in the vacuum tower in a stripping section below the flash zone.

Propane Deasphalting

The highest boiling portions of most crude oils contain resins and asphaltenes. These materials must be removed in order to provide an oil with acceptable performance. Traditionally, deasphalting of vacuum residuum has been used to remove these materials. Fig. 2-12 is a simplified illustration of deasphalting. As shown here, a solvent (generally propane) is mixed with the vacuum residuum. The paraffins, naphthenes, and aromatics have a high solubility in the solvent compared to the resins and asphaltenes. After mixing, the system is allowed to settle, and two liquid phases form (an analogy to a system of oil and water may be helpful in visualizing the two liquid layers). The top layer is primarily propane containing the soluble

Fig. 2-12 Propane Deasphalting

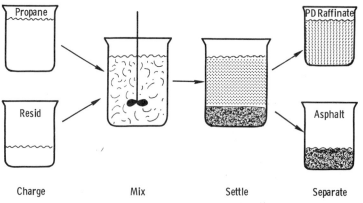

Charge   Mix   Settle   Separate

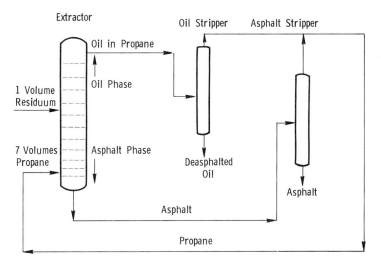

Fig. 2-13   Propane Deasphalting

components of the residuum. The bottom layer is primarily the asphaltic material with some dissolved propane. The two phases can be separated by decanting, and after the solvent is removed from each, the separation is accomplished.

The separation is carried out in the refinery in one continuous operation (see Fig. 2-13). The residuum (usually diluted with a small amount of propane) is pumped to the middle of the extraction column. Propane (usually about 6–8 volumes per volume of residuum) is charged to the bottom of the column. Since the residuum is more dense than the propane, the residuum will flow down the column, the propane rising up in a counterflow. The mixing is provided within the column, by either perforated plates or a rotor with discs attached. The rising propane dissolves the more soluble components which are carried out the top of the column with the propane. The insoluble, asphaltic material is removed from the bottom of the column. Temperatures used in the column range from approximately 120°F (50°C) to 180°F (80°C). The column must be operated under pressure (about 500 psig–35 atm) in order to maintain the propane as a liquid at the temperatures used. Propane is vaporized from the products and is then recovered and liquefied for recycling by compressing and cooling it.

Fig. 2-14 shows the types of chemical compounds separated in deasphalting. In this figure, the residuum composition is defined in terms of component groups measured by a solid-liquid chromatographic method. The components are saturates (paraffins and naphthenes); mononuclear, dinuclear, and polynuclear aromatics; resins; and asphaltenes. The tar is composed mostly of resins and asphaltenes.

The propane deasphalted (PD) oil is composed primarily of saturates and mononuclear, dinuclear, and polynuclear aromatics. However, since the separation is not ideal, a small amount of these components is lost to the tar while the PD oil is left with small amounts of resins and asphaltenes. After

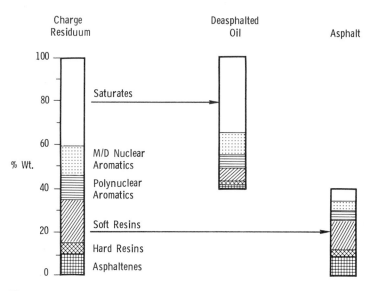

Fig. 2-14 Propane Deasphalting of Mid-Continent Sweet Residuum

deasphalting, the PD oil is processed in the same manner as the lube distillates. The next process in the sequence is furfural extraction.

Furfural Extraction

Before discussing furfural extraction and the subsequent processing, it should be pointed out that, from this stage on, a lube refinery differs from most fuel refineries in another important aspect. Lube units process very different feedstocks at various times. For example, a furfural extraction unit will process not only the deasphalted oil from the PD unit but the distillate feeds from the vacuum tower as well. These different feedstocks are processed in "blocked out" operation. This means that while one of the feeds is being processed, the others are collected in intermediate tankage and are processed later. Runs of various lengths, from days to weeks, are scheduled so that demands are met and intermediate tankage requirements are not excessive. During the switching from one stock to another, some transition oil is produced and must be disposed of. One other consequence of this blocked out type operation is that each section of the process units must be designed to handle the most demanding service. As a result, different sections of a unit may limit its capacity for various stocks.

Furfural extraction separates aromatic compounds from nonaromatic compounds. In its simplest form, the process consists of mixing furfural with the feedstock, allowing the mixture to settle into two liquid phases, decanting, and removing the solvent from each phase. This can be demonstrated in a glass graduated cylinder where the more dense furfural dissolves the dark colored aromatic materials from a distillate leaving a lighter colored raffinate product. The resultant product from the furfural extraction shows an increase in thermal and oxidative stability as well as an improvement in viscosity and temperature characteristics, as measured by a higher viscosity index (VI).

Fig. 2-15 is a simplified flow diagram of a commercial furfural extraction unit. The feedstock is charged to the middle of the extraction column, the furfural near the top. The density difference causes a counterflow in the column; the downward flowing furfural dissolves the aromatic compounds. The furfural raffinate rises and is removed from the top of the column. The furfural extract is removed from the bottom of the column. Each of the products is passed to its solvent recovery system, the furfural being recycled as feed to the extractor. While the solvent recovery system is not discussed in any detail herein, it should be pointed out that this portion of the unit is much more complicated than shown here.

The major effect of furfural extraction on the physical properties of a base stock is an increase in viscosity index—an improvement in the viscosity-termperature relationship of the oil. However, equally important, but less obvious, changes in the base stock result. Although oxidation and thermal stability are improved, there is no physical property that can be related to these characteristics. Thus, while viscosity index is sometimes used to monitor the day to day operations of a furfural extraction unit, it is only an indication of the continuity of the operation and not an absolute criterion of quality. The quality of base stocks for a given product (and the refining conditions needed to produce the base stocks) is arrived at by extensive testing, ranging from bench scale to full fleet testing programs. Careful control of the operation of the refining units is essential to assure the continuous production of base stocks meeting all the quality criteria needed in the final products.

MEK Dewaxing The next process in lube base stock manufacture is the removal of wax to reduce the pour point of the base stock. A simple representation of the process Fig. 2-16 shows the waxy oil being mixed with methyl ethyl ketone-toluene solvent. The mixture is then cooled to a temperature between 10°F (-12°C) and 20°F (-6°C) below the desired pour point. The wax crystals that form are kept in suspension by stirring during the cooling. The wax is then

Fig. 2-15 Furfural Extraction

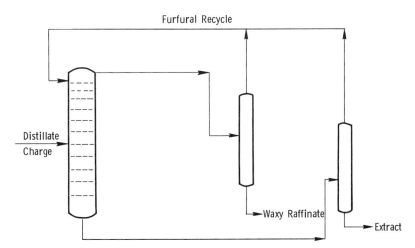

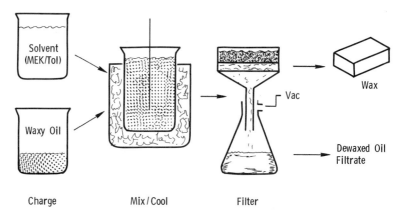

Charge   Mix / Cool   Filter

Fig. 2-16   MEK/Toluene Dewaxing

removed from the oil by filtration. Solvent is removed from both the oil and the wax and recycled for reuse.

A simplified flow diagram for a commercial dewaxing unit is shown in Fig. 2-17. In this unit, the waxy oil is mixed with solvent and heated to a temperature high enough to dissolve all the oil and wax. The purpose of this step is to destroy all the crystals that are in the oil so that the crystals that will be separated at the filter are formed under carefully controlled conditions. The solution is then cooled, first with cooling water, then by heat exchange with cold product, and finally by a refrigerant. In some cases, additional solvent is added at various points in the heat exchange train.

Fig. 2-17   Dewaxing for Lubes — MEK Process

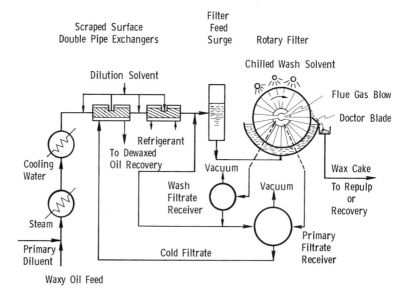

One other distinctive feature of this cooling train is the use of scraped wall, double pipe heat exchangers. These exchangers consist of a pipe inside a pipe. The inner pipe carries the solvent-oil-wax mixture; the outer pipe the cooling medium, either cold product or refrigerant. The inner pipe is equipped with a set of scraper blades that rotate and scrape away any wax that plates out on the walls of the inner pipe. This action is necessary to maintain a reasonable rate of heat transfer. Although this method is efficient, the scraped wall double pipe heat exchangers are expensive and costly to maintain.

The cooled slurry is passed to a filter feed surge drum and then to the filter itself where the actual separation is accomplished. Rotary vacuum filters used in dewaxing plants are large drums covered with a filter cloth which prevents the wax crystals from passing through to the inside of the drum as the drum rotates in a vat containing the slurry of wax, oil, and solvent. A vacuum applied inside the drum pulls the solvent-oil mixture (filtrate) through the cloth, thus separating the oil from the wax.

The next series of figures illustrates the operation of this type of filter. In these figures we will be looking at the end of the drum and will follow one segment as the drum rotates through one revolution. In Fig. 2-18, the segment is submerged in the slurry vat and is building up a wax cake on the filter cloth. The filtrate is being pulled into the interior of the drum. As the drum rotates through the slurry, the wax cake will increase in thickness and, as the segment leaves the vat, the vacuum state is maintained for a short period to dry the cake and remove as much oil and solvent as possible from the wax cake. Fig. 2-19 shows the cold wash solvent being applied to the cake. A vacuum inside the drum pulls the wash solvent through the cake, displacing more of the oil in the cake. The wash portion of the cycle is followed by another short period of drying. The wax cake (Fig. 2-20) is lifted from the filter by means of flue gas. This is accomplished by applying a positive pressure to the inside of the drum. As the drum rotates, the wax cake is guided from the drum by means of a blade (see Fig. 2-21), which

Fig. 2-18 Dewaxing Cycle Filtration

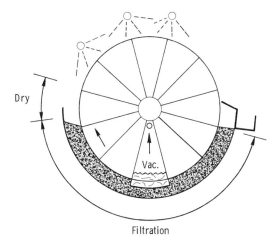

Dry

Vac.

Filtration

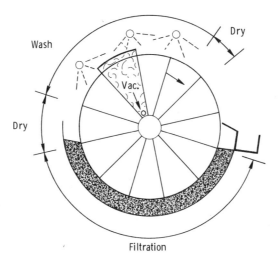

Fig. 2-19 Dewaxing Cycle Wash

directs it to a conveyor and then to the solvent recovery system. This segment of the drum is then ready to reenter the vat and continue with another cycle of pickup, dry, wash, dry, cake lift, and wax removal.

The oil and solvent are removed continuously from the inside of the drum through a complicated valving system on one end of the drum, and then are pumped to a solvent recovery system. After removal of solvent for recycle, the base stock is ready for use in many applications. The wax from the dewaxing filter, after its solvent is removed, is the starting material for wax manufacture.

Fig. 2-20 Dewaxing Cycle Flue Gas Blow

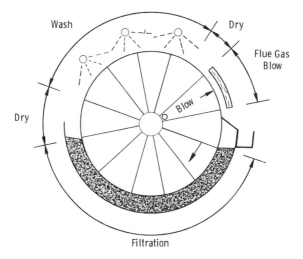

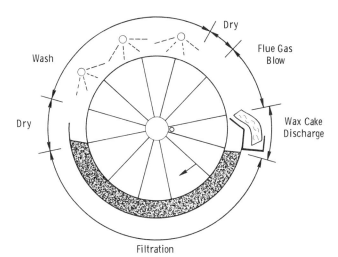

Fig. 2-21 Dewaxing Cycle Wax Cake Discharge

Hydrofinishing For many base stocks, dewaxing is the final process. The stocks are then shipped to a blending plant where the final products are made by blending the base stocks with additives. Some stocks — particularly premium stocks — require a finishing process to improve the color, oxidation, or thermal stability of the base stock.

As shown in Fig. 2-22, the hydrofinishing process consists of a bed of catalyst through which hot oil and hydrogen are passed. The catalyst slightly changes the molecular structure of the color bodies and unstable components in the oil — resulting in a lighter colored oil which is improved in certain performance qualities. The process operates similarly to processes that are used to disulfurize kerosines and diesel fuels.

Fig. 2-22 Lube Hydrofinishing Process

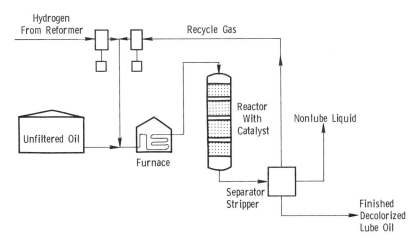

ALTERNATIVE PROCESSES FOR LUBRICATING OIL MANUFACTURE

Although the process sequence described herein is typical of modern refineries, many alternative processes are available. Those processes that are currently of significant commercial interest are briefly discussed in the following paragraphs.

Deasphalting While no widely accepted substitute for propane deasphalting has been available, there is a combination process called Duo-Sol which combines in one process both propane deasphalting and extraction. The solvent extraction portion of the Duo-Sol process employs either phenol or cresylic acid as the extraction solvent.

Solvent Extraction Another solvent extraction process that is widely used employs phenol as the solvent. The capacity and the number of phenol extraction units in operation (if Duo-Sol units are included with phenol) exceeds furfural extraction. However, recent trends indicate that the use of phenol extraction is likely to decline. Mobil's recent solvent extraction units are based on furfural extraction. The action of these solvents on the oil charge is quite similar, although different operating conditions are used with each solvent.

Dewaxing In addition to MEK-toluene, several other dewaxing solvents are used. Propane, the same solvent used for deasphalting, may be used for dewaxing. The solubility characteristics of propane are unique, since at the temperatures needed for dewaxing, wax is insoluble and may be filtered from the propane-oil mixture. Methyl isobutyl ketone (MIBK) also has been used as a dewaxing solvent. The operation of these solvents is generally similar to that of MEK-toluene.

Finishing The alternative processes to hydrofinishing use adsorbents, generally clays of various types. In these processes, the dark colored and unstable molecules are removed from the oil and remain on the solid clay. In the "contact" process, very fine clay is mixed with the oil at a high temperature. The clay is then filtered from the oil and the used clay discarded. In the percolation process, the oil percolates down through a bed packed with coarse clay. At the start of percolation, the dark colored charge exits the clay bed almost water white. Progressively, the discharged oil colors as the filter becomes fouled with the dark colored unstable molecules. When the collected product reaches specification limits, the process is stopped and the clay regenerated.

The clay is then washed free of oil and is removed from the percolation drum and regenerated by burning off the adsorbed material. Some loss of effectiveness results from the regeneration. This is compensated for by the regular replacement of spent clay with some fresh clay. Because of the lower yield relative to hydrofinishing, along with the environmental problems, clay processes are being replaced by hydrofinishing.

**Other
Hydrogen
Processes**

Several hydrogen processes have been used in addition to hydrofinishing. Hydrotreating, a more severe process, is sometimes used prior to solvent extraction. The objective is to improve the yield from the extraction process by converting some of the aromatic molecules, which would end up in the extract, into nonaromatic molecules that will remain in the raffinate. In this process, a high degree of desulfurization usually is realized, as well as a moderate degree of nitrogen removal. When this process is employed, hydrofinishing of the product oil is not needed.

Catalytic hydrodewaxing, in which the wax molecules are catalytically cracked to lower boiling products, has been proposed. While few commercial units have been built, it is likely that this process will be utilized, especially to produce very low pour point base stocks that are now produced from increasingly scarce naphthenic crudes.

A variation on this process involves hydrotreating the raffinate from solvent extraction. With this technique, it may be possible to reduce or eliminate the need for wax hydrofinishing in addition to oil hydrofinishing.

A totally different process approach to lube oil manufacture involves an even more severe hydrogen process—hydrocracking. In this process many of the molecules in the feedstock are changed. In addition to converting aromatics to naphthenes, many of the naphthene rings are broken open and many of the paraffinic molecules are rearranged and/or broken up. This massive "reforming" of the charge stock produces molecules that have improved viscosity and temperature behavior (higher VI) as well as converting a significant portion of the feedstock to lower boiling materials. The VI improvement by hydrocracking enables it to compete with solvent extraction. The other processes—deasphalting, dewaxing—are not replaced and hydrofinishing is not needed. However, since the product contains a wide boiling range of materials, an additional distillation is required. Several commercial lube hydrocracking processes are in operation or under construction.

ADDITIVES

Additives are chemical compounds added to lubricating oils to impart specific properties to the finished oils. Some additives impart new and useful properties to the lubricant, some enhance properties already present, while some act to reduce the rate at which undesirable changes take place in the product during its service life.

Additives, in improving the performance characteristics of lubricating oils, have aided significantly in the development of improved prime movers and industrial machinery. Modern passenger car engines, automatic transmissions, hypoid gears, railroad and marine diesel engines, high speed steam turbines, and industrial processing machinery, as well as many other types of equipment would have been greatly retarded in their development were it not for additives and the performance benefits they provide.

Additives for lubricating oils were used first during the 1920s and their use has since increased tremendously. Today, practically all types of lubricating oils contain at least one additive, and some oils contain several different types of additives. The amount of additive used varies from a few hundredths of a percent to 30 percent or more.

In addition to their primary beneficial effects, additives can have detrimental side effects, especially if the dosage is excessive or if interactions with other additives occur. It is the responsibility of the oil formulator to achieve a balance of additives for optimum performance, and to ensure by testing that this combination does not exhibit undesirable side effects. When this is achieved, it is usually unnecessary and undesirable for the oil user to add oil additive supplements.

The more commonly used additives are discussed in the following sections. Although some are multifunctional, as in the case of certain VI improvers which also function as pour point depressants or dispersants, they are discussed only in terms of their primary function.

Pour Point Depressants These are high molecular weight polymers that function by inhibiting the formation of a wax crystal structure that would prevent oil flow at low temperatures. Two general types are used:

1. Alkylaromatic polymers adsorb on the wax crystals as they form, preventing them from growing and adhering to each other.
2. Polymethacrylates cocrystallize with wax prevent crystal growth.

The additives do not entirely prevent wax crystal growth, but rather lower the temperature at which a rigid structure is formed. Depending on the type of oil, pour point depression of up to 50 Fahrenheit degrees (28 Celsius degrees) can be achieved by these additives, although a lowering of the pour point by about 20 to 30 Fahrenheit degrees (11 – 17 Celsius degrees) is more common.

Viscosity Index Improvers These are long chain, high molecular weight polymers that function by increasing the relative viscosity of an oil more at high temperatures than they do at low temperatures. Generally this results from the polymer changing its physical configuration with increasing temperature of the mixture. It is postulated that in cold oil the molecules of the polymer adopt a coiled form so that their effect on viscosity is minimized. In hot oil, the molecules tend to straighten out, and the interaction between these long molecules and the oil produces a proportionally greater thickening effect. *Note:* although the oil-polymer mixture still decreases in viscosity as the temperature increases, it does not decrease as much as would the oil alone.

Among the principal VI improvers are methacrylate polymers and copolymers, acrylate polymers, olefin polymers and copolymers, and styrene-butadiene copolymers. The degree of VI improvement from these materials is a function of the molecular weight distribution of the polymer.

The long molecules in VI improvers are subject to degradation due to mechanical shearing in service. Shear breakdown occurs by two mechanisms. Temporary shear breakdown occurs under certain conditions of moderate shear stress and results in a temporary loss of viscosity. Apparently, under these conditionsn the long molecules of the VI improver align themselves in the direction of the stress so there is less resistance to flow. When the stress is removed, the molecules return to their usual random arrangement and the temporary viscosity loss is recovered. This effect can be beneficial in that it can temporarily reduce oil friction to permit easier starting, as when cranking a cold engine. Permanent shear breakdown occurs when the shear stresses actually rupture the long molecules, converting them into lower molecular weight materials that are less effective VI improvers. This results in a permanent viscosity loss, which can be significant. It is generally the limiting factor controlling the maximum amount of VI improver that can be used in a particular oil blend.

VI improvers are used in engine oils, automatic transmission fluids, multipurpose tractor fluids, and hydraulic fluids. They are also used in gear lubricants to some extent. Their use permits the formulation of products that provide satisfactory lubrication over a much wider temperature range than is possible with straight mineral oils alone.

Defoamants The ability of oils to resist foaming varies considerably depending on the type of crude oil, type and degree of refining, and viscosity. In many applications, there may be considerable tendency to agitate the oil and cause foaming, while in other cases even small amounts of foam can be extremely troublesome. In these cases, a defoamant may be added to the oil.

Silicone polymers used at a few parts per million are the most widely used defoamants. These materials are marginally soluble in oil, and the correct choice of polymer size is critical if settling during long term storage is to be avoided. Also, they may increase air entrainment in the oil. Organic polymers are sometimes used to overcome these difficulties with the silicones, although much higher concentrations are required.

It is thought that the defoamant molecules attach themselves to the air bubbles in the foam, producing points of weakness. The bubbles, therefore, coalesce into larger bubbles which rise more readily to the surface of the foam layer and there collapse, releasing the air.

Oxidation Inhibitors When oil is heated in the presence of air, oxidation occurs. As a result of this oxidation, the oil viscosity and concentration of organic acids in the oil both increase, and varnish and lacquer deposits may form on hot metal surfaces exposed to the oil. In extreme cases, these deposits may be further oxidized to hard, carbonaceous materials.

The rate at which oxidation proceeds is affected by several factors. As the temperature increases, the rate of oxidation increases exponentially. Greater exposure to air (and the oxygen it contains), or more intimate mixing with it, will also increase the rate of oxidation. Many materials, such as metals, particularly copper, and organic and mineral acids, may act as catalysts or oxidation promoters.

Although the complete mechanism of oil oxidation is not too well defined, it is generally recognized as proceeding by free radical chain reaction. Reaction chain initiators are formed first from unstable oil molecules, and these react with oxygen to form peroxy radicals which in turn attack the unoxidized oil to form new initiators and hydroperoxides. The hydroperoxides are unstable and divide, forming new initiators to expand the reaction. Any materials that will interrupt this chain reaction will inhibit oxidation. Two general types of oxidation inhibitors are used: those that react with the initiators, peroxy radicals and hydroperoxides, to form inactive compounds, and those that decompose these materials to form less reactive compounds.

At temperatures below 200°F (93°C), oxidation proceeds slowly and inhibitors of the first type are effective. Examples of this type are hindered (alkylated) phenols such as 2,6-ditertiary-butyl-4-methylphenol (also called 2,6-ditertiary-butylparacresol, DBPC), and aromatic amines such as N-phenyl-α-naphthylamine. These are used in products such as turbine, circulation, and hydraulic oils which are intended for extended service at moderate temperatures.

When the operating temperature exceeds about 200°F (93°C), the catalytic effects of metals become important factors in promoting oil oxidation. Under these conditions, inhibitors that reduce the catalytic effect of the metals must be used. These materials usually react with the surfaces of the metals to form protective coatings and for that reason are sometimes called metal deactivators. Typical of this type of additives are the dithiophosphates, primarily zinc dithiophosphate. The dithiophosphates also act to decompose hydroperoxides at temperatures above 200°F (93°C), so they inhibit oxidation by this mechanism as well.

Oxidation inhibitors may not entirely prevent oil oxidation when conditions of exposure are severe, and some types of oils are inhibited to a much greater degree than others. Oxidation inhibitors are not, therefore, cure-alls, and the formulation of a satisfactorily stable oil requires proper refining of a suitable base stock combined with careful selection of the type and concentration of oxidation inhibitor.

Corrosion Inhibitors

A number of kinds of corrosion can occur in systems served by lubricating oils. Probably the two most important types are corrosion by organic acids that develop in the oil itself, and corrosion by contaminants that are picked up and carried by the oil.

One of the areas where corrosion by organic acids can occur is the high strength bearing inserts used in internal combustion engines. Some of the metals used in these inserts, such as the lead in copper-lead or lead-bronze, are readily attacked by organic acids in oil, as illustrated in Fig. 2-23. The corrosion inhibitors form a protective film on the bearing surfaces that prevents the corrosive materials from reaching or attacking the metal (Fig. 2-24). The film may be either adsorbed on the metal or chemically bonded to it. The additive used for this purpose is primarily zinc dithiophosphate, but other sulfur and phosphorus containing materials are also used.

During combustion in gasoline or diesel engines, certain materials in the fuel, such as sulfur and antiknock scavengers, can burn to form strong acids. These acids can then condense on the cylinder walls and be carried to other parts of the engine by the oil. Corrosive wear of rings and cylinder walls, and corrosion of crankshafts, rocker arms, and other engine components can then occur.

It has been found that the inclusion of highly alkaline materials in the oil will help to neutralize these strong acids as they are formed, greatly reducing this corrosion and corrosive wear. These alkaline materials are also used to provide detergency. See the detailed discussion in the section on Detergents and Dispersants.

Rust Inhibitors

Rust inhibitors are usually compounds having a high polar attraction toward metal surfaces. By physical or chemical interaction at the metal surface,

Fig. 2-23   Heavily Corroded Copper-Lead Bearing   Fig. 2-24   Satisfactorily Protected Copper-Lead Bearing

they form a tenacious, continuous film which prevents water from reaching the metal surface. Typical materials used for this purpose are amine succinates and alkaline earth sulfonates. The effectiveness of properly selected rust inhibitors is illustrated in Fig. 2-25.

Rust inhibitors can be used in most types of lubricating oils, but the selection must be made carefully in order to avoid problems such as corrosion of nonferrous metals or the formation of troublesome emulsions with water. Because rust inhibitors are adsorbed on metal surfaces, an oil can be depleted of rust inhibitor in time. In certain cases, it is possible to correct this by adding more inhibitor.

Detergents and Dispersants

In internal combustion engine service, a variety of effects tends to cause oil deterioration and the formation of harmful deposits. These deposits can interfere with oil circulation, build up behind piston rings to cause ring sticking and rapid ring wear, and affect clearances and proper functioning of critical components, such as hydraulic valve lifters. Once formed, such deposits are generally hard to remove except by mechanical cleaning. The use of detergents and dispersants in the oil can delay the formation of deposits and reduce the rate at which they deposit on metal surfaces. An essential factor with this approach is regular draining and replacement of the oil so that the contaminants in it are removed from the engine before the oil's capacity to hold them is exceeded.

Detergents are generally considered to be those chemical compounds which chemically neutralize deposit precursors that form under high temperature conditions or as the result of burning fuels with high sulfur content or other materials that form acidic combustion products. Dispersants, on the

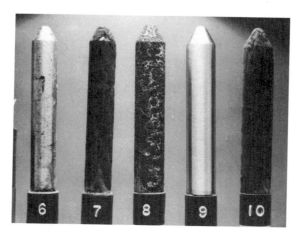

Fig. 2-25   ASTM Rust Test Specimens

other hand, are chemical compounds that disperse or suspend in the oil potential sludge forming materials, particularly those formed during low temperature operation when condensation and partially burned fuel find their way into the oil. These contaminants are removed from the system when the oil is drained. There is no sharp line of demarcation between detergents and dispersants. Detergents have some ability to disperse and suspend contaminants, while dispersants have some ability to prevent the formation of high temperature deposits.

The principal detergents used today are organic soaps and salts of alkaline earth metals such as barium, calcium, and magnesium. These materials are often referred to as metallo-organic compounds. Barium, calcium, and magnesium sulfonates, and barium and calcium phenates (or phenol sulfides) are widely used, and barium phosphonates are still used in some applications. The sulfonates and phenates may be neutral or overbased, that is, contain more of the alkaline metal than is required to neutralize the acidic components used in making the additive. Overbased materials are widely used in diesel engine oils to neutralize the strong acids formed from combustion of the sulfur in the fuel. This neutalization reduces corrosion and corrosive wear and minimizes the tendency of these acids to cause oil degradation.

Overbased materials are generally not used at as high concentration in gasoline engine oils where the fuel sulfur is much lower, but are usually included to help reduce corrosion in low temperature operation. Both neutral and overbased materials also act to disperse and suspend potential varnish forming materials resulting from oil oxidation, preventing these materials from depositing on engine surfaces.

Metallo-organic detergents, on combustion, leave an ashy residue (see Sulfated Ash). In some cases, this may be detrimental in that the ash can contribute to combustion chamber deposits. In other cases, it may be beneficial in that the ash provides wear resistant coatings for surfaces such as valve faces and seats.

Typical dispersants (also called polymeric dispersants and ashless dispersants) in use today are described as polymeric succinimides, olefin/P_2S_5 reaction products, polyesters, and benzylamides. These are based on long chain hydrocarbons which are acidified and then neutralized with a compound containing basic nitrogen. The hydrocarbon portion provides oil solubility, while the nitrogen portion provides an active site that attracts and holds potential deposit forming materials to keep them suspended in the oil.

While the primary use of detergents and dispersants is in engine oils, increasing quantities of them are being used in products such as automatic transmission fluids and circulation oils for high temperature service. In these applications, the detergents and dispersants help to prevent the deposition of lacquer and varnish resulting from oil oxidation, and in this way supplement the effects of the oxidation inhibitors.

Antiwear Additives   Antiwear additives are used in many lubricating oils to reduce friction, wear, and scuffing and scoring under boundary lubrication conditions, that is, when full lubricating films cannot be maintained. As the oil film becomes progressively thinner due to increasing loads or temperatures, contact through the oil film is first made by minute surface irregularities or asperities. As these opposing asperities make contact, friction increases and welding can occur. The welds break immediately as sliding continues, but this can form new roughnesses through metal transfer, and also form wear particles which can cause scuffing and scoring. Two general classes of materials are used to prevent metallic contact, depending on the severity of the requirements.

Mild Wear and Friction Reducing Additives   These compounds, sometimes called boundary lubrication additives, are polar materials such as fatty oils, acids, and esters. They are long chain materials that form an adsorbed film on the metal surfaces with the polar ends of the molecules attached to the metal and the molecules projecting more or less normal to the surface. Contact is then between the projecting ends of the layers of molecules on the opposing surfaces. Friction is reduced, and the surfaces move more freely relative to each other. Wear is reduced under mild sliding conditions, but under severe sliding conditions the layers of molecules can be rubbed off so that their wear reducing effect is lost.

Extreme Pressure Additives   At high temperatures or under heavy loads where more severe sliding conditions exist, compounds called extreme pressure (EP) additives are required to reduce friction, control wear, and prevent severe surface damage. These materials function by chemically reacting with the sliding metal surfaces to form relatively oil insoluble surface films. The kinetics of the reaction are a function of the surface temperatures generated by the localized high temperatures that result from rubbing between opposing surface asperities, and breaking of junctions between these asperities.

Even with extreme pressure additives in the lubricant, wear of new surfaces may be high initially. In addition to the normal break-in wear, nascent metal (freshly formed, chemically reactive surfaces), time, and temperature are required to form the protective surface films. After the films are formed, relative motion is between the layers of surface films rather than the metals. The sliding process can lead to some film removal, but replacement by further chemical reaction is rapid so that the loss of metal is extremely low. This process gradually depletes the amount of EP additive available in the oil although the rate of depletion is usually slow. Thus, there will be sufficient additive left to provide adequate protection for the metal surfaces.

The severity of the sliding conditions dictates the reactivity of the EP additives required for maximum effectiveness. The optimum reactivity occurs when the additives minimize the adhesive or metallic wear without leading to appreciable corrosive or chemical wear. Additives that are too reactive lead to the formation of excessively thick surface films which have less resistance to attrition, so some metal is lost by the sliding action. Since a particular EP additive may have different reactivity with different metals, it is important to match additive metal reactivity to the additives not only with the severity of the sliding system but also with the specific metals involved. For example, some additives that are excellent for steel-on-steel systems may not be satisfactory for bronze-on-steel systems operating at similar sliding severity because they are too reactive with the bronze.

Another important function of EP additives is that, because the chemical reaction is greatest on the asperities where contact is made and localized temperatures are highest, they lead to polishing of the surfaces. The load is then distributed more uniformly over a greater contact area which allows for a reduction in sliding severity, more effective lubrication, and a reduction in wear.

Extreme pressure agents are usually compounds containing sulfur, chlorine, or phosphorus, either alone or in combination. The compounds used depend on the end use of the lubricant and the chemical activity required in it. Sulfur compounds, sometimes with chlorine or phosphorus compounds, are used in many metal cutting fluids. Sulfurphosphorus combinations are used in some industrial gear lubricants, and in special cases sulfur-chlorine-lead compounds are used in some automotive gear lubricants. The major amounts of both automotive and industrial gear lubricants are now formulated with sulfur-phosphorus compounds. These materials provide excellent protection against gear tooth scuffing and have the advantages of better oxidation stability, lower corrosivity, and often lower friction than other combinations that have been used in the past.

PHYSICAL AND CHEMICAL CHARACTERISTICS

There are a multitude of physical and chemical tests which yield useful information on the characteristics of lubricating oils. However, the quality or the performance features of a lubricant cannot be adequately described on the basis of physical and chemical tests alone. Thus, major consumers of lubricating oils, such as military purchasing agencies and many commercial consumers, include performance tests as well as physical and chemical tests in their purchase specifications. Physical and chemical tests are of considerable value in maintaining uniformity of products during manufacture. Also, they may be applied to used oils to determine changes that have occurred in service, and to indicate possible causes for those changes.

Some of the most commonly used tests for physical or chemical properties of lubricating oils are outlined in the following sections, with brief explanations of the significance of the tests from the standpoint of the refiner and consumer. For detailed information on methods of test, the reader is referred to the American Society for Testing and Materials handbooks of "Annual Standards for Petroleum Products and Lubricants," the British Institute of Petroleum handbook "Standard Methods for Testing Petroleum and Its Products," The United States Federal Test Method Standard No. 791, and similar types of publications used in a number of other countries.

Carbon Residue

The carbon residue of a lubricating oil is the amount of deposit, in percentage by weight, left after evaporation and pyrolysis of the oil under prescribed conditions. In the test, oils from any given type of crude oil show increasing residues with increasing viscosities. Distillate oils show lower values than those of similar viscosity containing residual stocks. Oils of naphthenic type usually show lower residues than those of similar viscosity made from paraffinic crudes. The more severe the refining treatment an oil is given—whether by solvents, hydrotreating, filtration, or acid treatment—the lower the carbon residue value will be. Although many finished lubricating oils now contain additives that may contribute significantly to the amount of residue in the test, their effect on performance is distinctly beneficial.

Originally the carbon residue test was developed to determine the carbon forming tendency of steam cylinder oils. In the years that followed, unsuccessful attempts were made to relate carbon residue values to the amount of carbon formed in the combustion chambers and on the pistons of internal combustion engines. Since such factors as fuel composition and engine operating and mechanical conditions, as well as other lubricating oil properties are of equal or greater importance, carbon residue values alone have only limited significance. The carbon residue determination is now made mainly on base oils for engine oil manufacture; straight mineral engine oils, such as aircraft engine oils; and some of the heavy cylinder oil type products. In these cases, the determination is an indication of the degree of refining to which the product has been subjected.

Color

The color of lubricating oils as observed by light transmitted through them varies from practically clear or transparent to opaque or black. Usually, the various methods of measuring color are based on a visual comparison of the amount of light transmitted through a specified depth of oil with the amount of light transmitted through one of a series of colored glasses. The color is then given as a number corresponding to the number of the colored glass.

Color variations in lubricating oils result from differences in crude oils, viscosity, method and degree of treatment during refining, and in the amount and nature of additives included in them. During processing, color is a useful guide to the refiner to indicate if processes are operating properly. In finished lubricants, color has little significance except in the case of medicinal and industrial white oils, which are often compounded into, or applied to, products where staining or discoloration would be undesirable.

Density and Gravity

The density of a substance is the mass of a unit volume of it at a standard temperature. The specific gravity (relative density) is the ratio of the mass of a given volume of a material at a standard temperature to the mass of an equal volume of water at the same temperature. API gravity is a special function of specific gravity that is related to it by the following equation:

$$\text{Gravity }°\text{API} = \frac{14.5}{\text{sp. gr. }60/60°\text{F}} - 131.5$$

The API gravity value, therefore, increases as the specific gravity decreases. Since both density and gravity change with temperature, determinations are made at a controlled temperature and then corrected to a standard temperature by use of special tables.

Density and gravity can be determined by means of hydrometers; see Fig. 2-26. The hydrometer can be calibrated to read any of the three properties—density, gravity, or API gravity.

Gravity determinations are quickly and easily made. Because products of a given crude oil—having definite boiling ranges and viscosities—will fall into definite ranges, this property is widely used for control in refinery operations. It is also useful for identifying oils, provided that the distillation range or viscosity of the oils is known. Its primary use, however, is to convert weighed quantities to volume and measured volumes to weight.

In testing used oils, particularly used engine oils, a decrease in specific gravity (increase in API gravity) may indicate fuel dilution, whereas an increase in specific gravity might indicate the presence of contaminants such as fuel soot or oxidized materials. Additional test information is necessary to fully explain changes in gravity since some effects tend to cancel others.

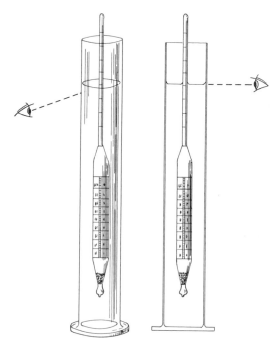

Fig. 2-26 Use of Hydrometer to Determine Density and Gravity

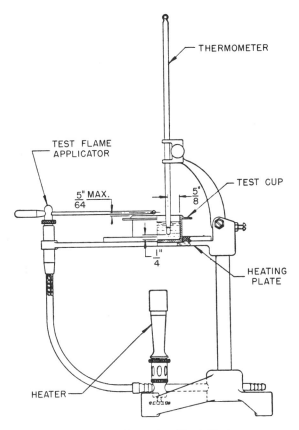

THERMOMETER

TEST FLAME
APPLICATOR

TEST CUP

$\frac{5"}{64}$ MAX.

$\frac{5}{8}$"

$\frac{1}{4}$"

HEATING
PLATE

HEATER

Fig. 2-27 Cleveland Open Cup Flash Tester

Flash and Fire Points

The flash point of an oil is the temperature at which the oil releases enough vapor at its surface to ignite when an open flame is applied. For example, if a lubricating oil is heated in an open container, ignitible vapors are released in increasing quantities as the temperature rises. When the concentration of vapors at the surface becomes great enough, exposure to an open flame will result in a brief flash as the vapors ignite. When a test of this type is conducted under certain specified conditions, as in the Cleveland Open Cup method (Fig. 2-27), the bulk oil temperature at which this happens is reported as the flash point. The release of vapors at this temperature is not sufficiently rapid to sustain combustion, so the flame immediately dies out. However, if heating is continued, a temperature will be reached at which vapors are released rapidly enough to support combustion. This temperature is called the fire point. For any specific product, both flash and fire points will vary depending on the apparatus and the heating rate.

The flash point of new oils varies with viscosity—higher viscosity oils have higher flash points. Flash points are also affected by the type of crude, with naphthenic oils generally having lower flash points than paraffinic oils of similar viscosity.

Flash and fire tests are of value to refiners for control purposes and are significant to consumers under certain circumstances for safety considerations. Also, in certain high temperature applications, use of an oil with a low flash point, indicating higher volatility, may result in higher oil consumption rates. Flash and fire points are of little value in determining whether fire resistant fluids are safe near possible points of ignition.

In the inspection of a used oil, a significant reduction in the flash point usually indicates contamination with lower flash material. An exception to this occurs when certain products are used for long periods at high temperatures and undergo thermal cracking with the formation of lighter hydrocarbons and a reduction in the original flash point.

Neutralization Number

In the acid treating processes, it is essential that all the sulfuric acid for treating be neutralized and the residues washed from the oil. As a result, tests were developed to provide a quick determination of the amount of acid in an oil by neutralizing it with a base. The amount of acid in the oil was expressed in terms of the amount of a standard base required to neutralize a specified volume of oil. This quantity of base came to be called the neutralization number of the oil.

Some of the products of oxidation of petroleum hydrocarbons are organic acids. An oil oxidizes in service, caused by exposure to elevated temperatures, and there is a tendency for it to become more acidic. Thus, measuring the acidity of an oil in service was one way of following the progress of oxidation. This technique is still used with a number of products, primarily those intended for extended service, such as steam turbine oils and electrical insulating oils.

Many of the additives now used to improve performance properties may make an oil acidic or basic, depending on their composition. In other cases, the additives may contain weak acids and weak bases which do not react with each other in the oil solution, but react with both the strong acid and the strong base used in the neutralization number tests to give both acidic and basic neutralization numbers. There are also other additives which undergo exchange reactions with the base used for neutralization to give false acid values in the test. These effects of the additives tend to obscure any changes in acidity occurring in the oil itself, so neutralization number determinations may have little significance for some oils containing additives.

Many of the additives now used in engine oils contain alkaline materials intended to neutralize acidic products of combustion. The rate of consumption of these alkaline materials is an indication of the projected serviceable life of the oil. This measure of alkalinity is called the total base number and is now reported for most diesel engine oils.

Pour Point

The pour point of a lubricating oil is the lowest temperature at which it will pour or flow when it is chilled without disturbance under prescribed conditions. Most oils contain some dissolved wax and, as an oil is chilled, this wax begins to separate as crystals that interlock to form a rigid structure which traps the oil in small pockets in the structure. When this wax crystal structure becomes sufficiently complete, the oil will no longer flow under the conditions of the test. Mechanical agitation can break up the wax structure so that it is possible to have an oil flow at temperatures considerably below its pour point. Cooling rates also affect wax crystallization; it is possible to cool an oil rapidly to a temperature below its pour point and still have it flow.

While the pour point of most oils is related to the crystallization of wax, certain oils, which are essentially wax free, have viscosity limited pour points. In these oils the viscosity becomes progressively higher as the temperature is lowered until at some temperature no flow can be observed. The pour points of such oils cannot be lowered with pour point depressants, since these agents act by interfering with the growth and interlocking of the wax crystal structure.

Untreated lubricating oils show wide variations in pour points. Distillates from waxy paraffinic or mixed base crudes typically have pour points in the range of 80°F to 120°F (27–49°C), while raw distillates from naphthenic crudes may have pour points on the order of 0°F (−18°C) or lower. After solvent dewaxing, the paraffinic distillates will have pour points on the order of 20°F to 0°F (−7 to −18°C); where lower pour points oils are required, pour point depressants are normally used.

From the consumer's viewpoint, the importance of the pour point of an oil is almost entirely dependent on its intended use. For example, the pour point of a winter grade engine oil must be low enough so that the oil can be dispensed readily, and will flow to the pump suction in the engine at the lowest anticipated ambient temperatures. On the other hand, there is no particular need for low pour points for oils to be used inside heated plants or in continuous service such as steam turbines, or for many other applications.

Sulfated Ash The sulfated ash of a lubricating oil is the residue, in percent by weight, remaining after burning the oil, treating the initial residue with sulfuric acid, and burning the treated residue. It is a measure of the noncombustible constituents (usually metallic materials) contained in the oil.

New, straight mineral lubricating oils contain essentially no ash forming materials. Many of the additives used in lubricating oils contain metallo-organic components, which will form a residue in the sulfated ash test, so the concentration of such materials in an oil is roughly indicated by the test. Thus, during manufacture, the test gives a simple method of checking to ensure that the additives have been incorporated in approximately the correct amounts. However, since the test combines all metallic elements into a single residue, additional testing may be necessary to determine if the various metallic elements are in the oil in the correct proportions.

Several manufacturers now include a maximum limit on sulfated ash content in their specifications for certain types of internal combustion engine oils. This is done in the belief that, while the sulfated ash content results from the incorporation of materials intended to improve overall oil performance, excessive quantities of some of these materials may contribute to such problems as combustion chamber deposits and top ring wear.

With used oils, an increase in ash content usually indicates a buildup of contaminants such as dust and dirt, wear debris, and possibly other contamination such as lead salts, which are derived from the combustion of leaded gasolines in internal combustion engines.

Viscosity Probably the most important single property of a lubricating oil is its viscosity. It is a factor in the formation of lubricating films under both thick and

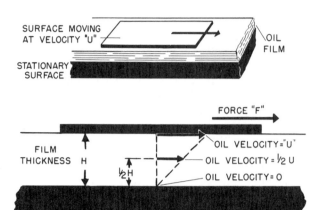

Fig. 2-28   Concept of Dynamic Viscosity

thin film conditions; it affects heat generation in bearings, cylinders, and gears; it governs the sealing effect of the oil and the rate of consumption or loss; and it determines the ease with which machines may be started under cold conditions. For any piece of equipment, the first essential for satisfactory results is to use an oil of proper viscosity to meet the operating conditions.

The basic concept of viscosity is shown in Fig. 2-28, where a plate is shown being drawn at a uniform speed over a film of oil. The oil adheres to both the moving surface and the stationary surface. Oil in contact with the moving surface travels with the same velocity (U) as that surface, while oil in contact with the stationary surface is at zero velocity. In between, the oil film may be visualized as being made up of many layers, each being drawn by the layer above it at a fraction of velocity U that is proportional to its distance above the stationary plate (see Fig. 2-28, lower view). A force (F) must be applied to the moving plate to overcome the friction between the fluid layers. Since this friction is the result of viscosity, the force is proportional to viscosity. Viscosity can be determined by measuring the force required to overcome fluid friction in a film of known dimensions. Viscosity determined in this way is called *dynamic* or *absolute* viscosity.

Dynamic viscosities are usually reported in poise (P) or centipoise (cP; 1 cP = 0.01 P), or in SI* units in Pascal seconds (Pa s; 1 Pa s = 10 P). Dynamic viscosity, which is a function only of the internal friction of a fluid, is the quantity used most frequently in bearing design and oil flow calculations. Because it is more convenient to measure viscosity in a manner such that the measurement is affected by the density of the oil, *kinematic* viscosities normally are used to characterize lubricants.

The kinematic viscosity of a fluid is the quotient of its dynamic viscosity divided by its density, both measured at the same temperature and in consistent units. The most common units for reporting kinematic viscosities now are the Stokes (St) or centistokes (cSt; 1 cSt = 0.01 St), or in SI, square

*SI — Le Système International d'Unités, the modern metric system, abbreviated to SI in all languages.

millimetres per second (mm²/s; 1 mm²/s = 1 cSt). Dynamic viscosities, in centipoise, can be converted to kinematic viscosities, in centistokes, by dividing by the density in grams per cubic centimetre (g/cm³) at the same temperature. Kinematic viscosities, in centistokes, can be converted to dynamic viscosities, in centipoise, by multiplying by the density in grams per cubic centimetre. Kinematic viscosities, in square millimetres per second (mm²/s), can be converted to dynamic viscosities, in Pascal seconds, by multiplying by the density, in grams per cubic centimetre, and dividing the result by 1000.

Other viscosity systems, including the Saybolt, Redwood, and Engler, are in wide use and will probably continue in use for many years because of their familiarity to many people. However, the instruments developed to measure viscosities in these systems are rarely used. Most actual viscosity determinations are made in centistokes and converted to values in the other systems by means of published SI conversion tables.

The viscosity of any fluid changes with temperature — increasing as the temperature is decreased, and decreasing as the temperature is increased. Thus, it is necessary to have some method of determining the viscosities of lubricating oils at temperatures other than those at which they are measured. This is usually accomplished by measuring the viscosity at two temperatures, then plotting these points on special viscosity-temperature charts developed by ASTM. A straight line can then be drawn through the points and viscosities at other temperatures read from it with reasonable accuracy; see Fig. 2-29. The line should not be extended below the pour point or above approximately 300°F (for most lubricating oils) since it may no longer be straight in these regions.

The two temperatures most used in the United States for reporting viscosities are 100 and 210°F. In the interests of international standardization, reporting temperatures of 40°C (104°F) and 100°C (212°F) will be adopted worldwide. Conversion to these temperatures is now in progress.

In selecting the proper oil for a given application, viscosity is a primary consideration. It must be high enough to provide proper lubricating films but not so high that friction losses in the oil will be excessive. Since viscosity varies with temperature, it is necessary to consider the actual operating temperature of the oil in the machine. Other considerations, such as whether a machine must be started at low ambient temperatures, must also be taken into account.

Three viscosity numbering systems are in use to identify oils according to viscosity ranges. Two of these are for automotive lubricants and one for industrial oils.

Engine Oil Viscosity Classification   SAE (Society of Automotive Engineers) recommended Practice J300d classifies oils for use in automotive engines by viscosities determined at either 212°F (100°C) or 0°F (−18°C). The ranges for the grades* in this classification are shown in Table 2-3.

In these classifications, grades with the suffix letter W are intended primarily for use where low ambient temperatures will be encountered, while

---

*The words grade and viscosity are used interchangeably.

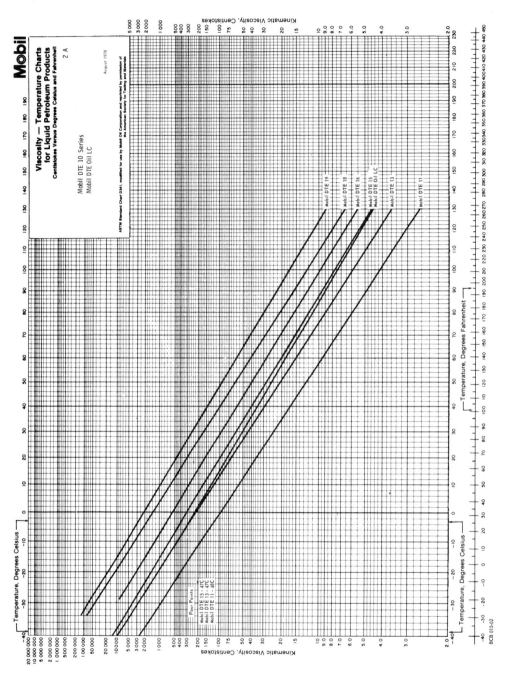

Fig. 2-29 ASTM Viscosity-Temperature Chart

Table 2-3   Engine Oil Viscosity Classification—SAE J300d

| SAE Viscosity Grade | Viscosity Range[a] | | |
| | Centipoise (cP) @ −18°C (ASTM D2602) Max | Centistokes (cSt) @ 100°C (ASTM D445) | |
		Min	Max
5W	1250	3.8	—
10W	2500	4.1	—
20W[b]	10,000	5.6	—
20	—	5.6	less than 9.3
30	—	9.3	less than 12.5
40	—	12.5	less than 16.3
50	—	16.3	less than 21.9

[a]*Note:* 1 cP = 1 mPa s; 1 cSt = 1 mm²/s.
[b]SAE 15W may be used to identify SAE 20W oils which have a maximum viscosity at −18°C of 5000 cP.

grades without the suffix letter are intended for use where low ambient temperatures will not be encountered.

Oils can be formulated that will meet the −18°C or 0°F limits of one of the W grades and the 100°C or 212°F limits of one of the non-W grades. For example, an oil can be formulated to meet the −18°C limits for the 10W grade and the 100°C limits for the 40 grade. It can then be designated an SAE 10W-40 grade and is referred to as a multigrade or multiviscosity oil. Oils of this type generally require the use of VI improvers (see Additives) in conjunction with a petroleum or synthetic lubricating oil base.

This SAE system is widely used by engine manufacturers to determine suitable oil viscosities for their engines, and by oil marketers to indicate the viscosities of internal combustion engine oils.

Axle and Manual Transmission Lubricant Viscosity Classification   SAE Recommended Practice J306c classifies lubricants for use in automotive manual transmissions and drive axles by viscosity measured at 100°C (212°F), and by the maximum temperature at which they reach a viscosity of 150,000 cP (150 Pa s) when cooled and measured in accordance with ASTM D 2983, Method of Test for Apparent Viscosity at Low Temperature Using the Brookfield Viscometer.

The limits for lubricant oil viscosity are given in Table 2-4. Multigrade oils such as 80W-90 or 85W-140 can be formulated under this system. This limiting viscosity of 150,000 cP was selected on the basis of test data which indicated that lubrication failures of pinion bearings of a specific axle design could be experienced when the lubricant viscosity exceeded this value. Since other axle designs, as well as transmissions, may have higher or lower limiting viscosities, it is the responsibility of the gear manufacturer to specify the actual grades that will provide satisfactory service under different ambient conditions.

Viscosity System for Industrial Fluid Lubricants   This system was developed jointly by ASTM and ASLE (American Society of Lubrication Engineers) to establish a series of definite viscosity levels that could be used as a common basis for specifying or selecting the viscosity of industrial fluid lubricants, and to eliminate unjustified intermediate grades. The system was originally based on viscosities measured at 100°F but was converted to viscosities measured at 40°C in the interests of international standardization. In this form, the system now appears as ASTM D 2422, American National Standard Z11.232, BSI (British Standards Institute) BS 4231, DIN (Deutsches Institut fur Normung) No. 51519, and ISO (International Standards Organization) Std. 3448. The viscosity ranges and identifying grade numbers are shown in Table 2-5.

This viscosity system is now coming into wide use in the industrial field.

Viscosity Index　　Different oils have different rates of change of viscosity with temperature. For example, a distillate oil from a naphthenic base crude would show a greater rate of change of viscosity with temperature than would a distillate oil from a paraffinic crude. The viscosity index (VI) is a method of applying a numerical value to this rate of change, based on comparison with the relative rates of change of two arbitrarily selected types of oils that differ widely in this characteristic. A high VI indicates a relatively low rate of change of viscosity with temperature; a low VI indicates a relatively high rate of change of viscosity with temperature. In other words, if a high VI oil and a low VI oil had the same viscosity at, say, room temperature, as the temperature increased the high VI oil would thin out less and, therefore, would have a higher viscosity than the low VI oil at higher temperatures.

The VI of an oil is calculated from viscosities determined at two temperatures by means of tables published by ASTM. Tables based on viscosities determined at both 100°F and 212°F, and 40°C and 100°C are available.

Finished lubricating oils made by conventional methods range in VI from somewhat below 0 to slightly above 100. Some synthetic lubricating oils have VIs both below and above this range, and additives called VI improvers can be blended into oils to produce VIs well above 100.

In many types of service, where the operating temperature remains more or less constant, VI is of little concern. However, in services where the operating temperature may vary over a wide range, such as in passenger car engines, it is probably safe to say that the VI of the oil used should be as high as practicable, consistent with other performance properties.

Table 2-4　Axle and Manual Transmission Lubricant Viscosity Classification — SAE J306c*

SAE Viscosity Grade	Maximum Temperature for Viscosity of 150,000 cP, °C	Viscosity @ 100°C	
		Minimum cSt	Maximum cSt
75W	−40	4.1	—
80W	−26	7.0	—
85W	−12	11.0	—
90	—	13.5	< 24.0
140	—	24.0	< 41.0
250	—	41.0	—

Note: 1 cP = 1 mPa s; 1 cSt = 1 mm²/s.

Table 2-5 Viscosity System for Industrial Fluid Lubricants

Viscosity System Grade Identification	Mid-Point Viscosity cSt (mm²/s) @ 40.0°C	Kinematic Viscosity Limits cSt (mm²/s) @ 40.0°C	
		Min	Max
ISO VG 2	2.2	1.98	2.42
ISO VG 3	3.2	2.88	3.52
ISO VG 5	4.6	4.14	5.06
ISO VG 7	6.8	6.12	7.48
ISO VG 10	10	9.00	11.0
ISO VG 15	15	13.5	16.5
ISO VG 22	22	19.8	24.2
ISO VG 32	32	28.8	35.2
ISO VG 46	46	41.4	50.6
ISO VG 68	68	61.2	74.8
ISO VG 100	100	90.0	110
ISO VG 150	150	135	165
ISO VG 220	220	198	242
ISO VG 320	320	288	352
ISO VG 460	460	414	506
ISO VG 680	680	612	748
ISO VG 1000	1000	900	1100
ISO VG 1500	1500	1350	1650

EVALUATION AND PERFORMANCE TESTS

Testing under actual service conditions is the best way to evaluate the performance of a lubricant and is the final evaluation used before a new product or new formulation is placed on the market. Service testing is not practical, however, during the early phases of lubricant development because of the time and cost involved. Therefore, shorter and less costly tests are used in the laboratory during development of both commercial and experimental lubricants. The correlation of results from these laboratory tests with field performance has been a major effort of petroleum laboratories for many years. Many tests have been proposed, but a large proportion of them have been discarded for various reasons. Most of the tests that have survived have done so because they have shown the ability to predict with a fair degree of reliability some of the aspects of how a lubricant will perform in service.

Since the process of test development and refinement is continuing, the following discussion treats the tests in a fairly general manner.

Oxidation Tests Oxidation of lubricating oils depends on the temperature, amount of oxygen contacting the oil, and the catalytic effects of metals. If the oil's service conditions are known, these three variables can be adjusted to provide a test that closely represents actual service. However, oxidation in service is often an extremely slow process, so the test may be time consuming. In order to shorten the test time, the test temperature is usually raised so that more rapid oxidation will result. Unfortunately, this then tends to make the test a less reliable indication of expected field performance. As a result, very few oxidation tests have received wide acceptance, although a considerable number are used by specific laboratories which have developed satisfactory correlations for them.

One oxidation test that is widely used is ASTM D943, "Oxidation Characteristics of Inhibited Steam-Turbine Oils." This test is commonly known as TOST, Turbine Oil Stability Test. While it is intended mainly for use on inhibited steam turbine oils, it is has been used for hydraulic and circulation oils, and for base oils for use in the manufacture of turbine, hydraulic, and circulation oils. The test is operated at a moderate temperature (95°C, 203°F). Iron and copper catalyst wires are immersed in the oil sample, to which water is added. Oxygen is bubbled through the sample at a prescribed rate. The test is either run for a prescribed number of hours, after which the neutralization number of the oil is determined, or until the neutralization number reaches a value of 2.0. The result in the latter case is than reported as the hours to a neutralization number of 2.0.

Objections to ASTM D 943 are that extremely long test times, often on the order of several thousand hours, are required for stable oils, and that the only criterion for acceptability is the neutralization number. Severe sludging and deposits on the catalyst wires can occur with some oils without excessive increase in the neutralization number. A modification of the procedure, called Procedure B, which requires a determination of the sludge content, is now being introduced to overcome some of these latter objections.

Two oxidation tests that are receiving increasing attention in Europe are the IP 280 procedure for turbine oils and the PNEUROP procedure of the Comite European des Constructeurs de Compresseurs et d'Outillage for compressor oils. Because of ISO activities in international standardization, these tests are also receiving attention in the United States.

In the IP 280 procedure, often referred to as the CIGRE (Conference Internationale des Grandes Reseaux Electriques) test, oxygen is passed through a sample of oil containing soluble iron and copper catalysts. The sample is held at 120°C (248°F) and test time is 164 h. During the test, volatile acids formed are absorbed in an absorption tube. At the end of the test, the acid numbers of the oil and the absorbent are determined and combined to give the total acidity; the sludge is determined as a weight percent. These may then be further combined to give the total oxidation products. Compared to ASTM D943, the CIGRE test requires a short, fixed test time, and the amount of sludge formed during the test is an important criterion of the evaluation. Where the limits for satisfactory performance in the test are properly set, correlation with performance in modern turbines has been reported to be good.

In the PNEUROP test, a sample of oil containing iron oxide as a catalyst is aged by being held at 200°C (392°F) for 24 h while air is bubbled through it. At the end of the test, the evaporation loss and the Conradson Carbon Residue of the remaining sample are determined. The evaporation loss is only significant in that it must not exceed 20 percent, so the main criterion is the CCR value. The test is believed to correlate to some extent with the tendency of oils to form carbonaceous deposits on compressor valves.

Thermal Stability

Thermal stability, as opposed to oxidation stability, is the ability of an oil or additive to resist decomposition under prolonged exposure to high temperatures. Decomposition may result in thickening, increasing acidity, the formation of sludge, or any combination of these.

Thermal stability tests usually involve static heating, or circulation over hot metal surfaces. Exposure to air is usually minimized, but catalyst coupons of various metals may be immersed in the oil sample. While no tests have received wide acceptance, a number of proprietary tests are used to evaluate the thermal stability of products such as hydraulic fluids and system oils for large diesel engines.

Rust Protection Tests

The rust protective properties of lubricating oils are difficult to evaluate. Rusting of ferrous metals is a chemical reaction which is initiated almost immediately when a specimen is exposed to air and moisture. Once initiated, the reaction is difficult to stop. Thus, when specimens are prepared for rust tests, extreme care must be taken to minimize exposure to air and moisture so that rusting will not start before the rust protective is applied and the test begun. Even with proper precautions, rust tests do not generally show good repeatability or reproducibility.

Most laboratory rust tests involve polishing or sandblasting a test specimen, coating it with the oil to be tested, then subjecting it to rusting conditions. Testing may be in a humidity cabinet, by atmospheric exposure, or by some form of dynamic test. In the latter category is ASTM D665, "Rust Preventive Characteristics of Steam Turbine Oil in the Presence of Water." In this test, a steel specimen is immersed in a mixture of distilled or synthetic sea water and the oil under test. The oil and water mixture is stirred continuously during the test, which usually lasts for 24 h. The specimen is then examined for rusting.

Other dynamic tests may involve testing of actual mechanisms lubricated with a test oil, as in the L-33 test described in Automotive Gear Lubricant Tests.

Foam Tests

The most widely used foam test is ASTM D892, "Foaming Characteristics of Lubricating Oils." Air is blown through an oil sample held at a specified temperature for a specified period of time. Immediately after blowing is stopped, the amount of foam is measured and reported as the foaming tendency. After allowing 10 min for the foam to collapse, the volume of foam is again measured and reported as the foam stability.

This test gives a fairly good indication of the foaming characteristics of new, uncontaminated oils, but service results may not correlate well if contaminants such as moisture or finely divided rust, which can aggravate foaming problems, are present in a system. For a number of applications, such as automatic transmission fluids, special foam tests involving severe agitation of the oil have been developed. These are generally proprietary tests used for specification purposes.

EP and Antiwear Tests

One of the main functions of a lubricant is to reduce mechanical wear. Closely related to wear reduction is the ability of extreme pressure (EP) type lubricants to prevent scuffing, scoring, and seizure as applied loads are increased. As a result, a considerable number of machines and procedures have been developed to try to evaluate EP and antiwear properties. In a number of cases, the same machines are used for both purposes, although different operating conditions may be used.

Wear can be divided into four classifications based on the cause:

1. Abrasive wear
2. Corrosive (chemical) wear
3. Adhesive wear
4. Fatigue wear

Abrasive Wear Abrasive wear is caused by abrasive particles, either contaminants carried in from outside or wear particles formed due to adhesive wear. In either case, oil properties do not have much direct influence on the amount that occurs.

Corrosive or Chemical Wear Corrosive or chemical wear results from chemical action on the metal surfaces combined with rubbing action that scuffs off the corroded metal. A typical example is the wear that may occur on cylinder walls and piston rings of diesel engines burning high sulfur fuels. The strong acids formed by combustion of the sulfur can attack the metal surfaces, forming compounds that can be fairly readily removed as the rings rub against the cylinder walls. Direct measurement of this wear requires many hours of test unit operation, which is often done as the final stage of testing new formulations. However, useful indications have been obtained in relatively short periods of time by means of sophisticated electronic techniques.

Adhesive wear in lubricated systems occurs when, due to load, speed, or temperature conditions, the lubricating film becomes so thin that opposing surface asperities can make contact through it, provided that adequate extreme pressure additives are not present. If loads are increased further, scuffing and scoring can result, and eventually seizure may occur. Adhesive wear can also occur with extreme pressure lubricants when the reaction kinetics of the additives with the surfaces are such that metal-to-metal contact is not fully controlled.

A number of machines, such as the Almen, Alpha LFW-1, Falex, Four-Ball, SAE, and Timken, are used to determine the loading conditions under which a lubricant will permit seizure, welding, or drastic surface damage to test specimens. Of these, the Alpha LFW-1, Four-Ball, Falex, and Timken are also used to measure wear at loads below the failure load. The results obtained with these machines do not necessarily correlate with field performance, but in a number of cases, the results from certain machines have been found to provide useful information for specific applications. As shown in Fig. 2-30, as additive reactivity increases, adhesive wear decreases and chemical wear increases. To some extent, the machines used to judge additive-surface reactivity also can be used in selecting additives for optimum effectiveness in balancing adhesive and chemical wear for certain metal combinations.

Either actual machines or scale model machines are used in testing the antiwear properties of lubricants. Antiwear type hydraulic fluids are now

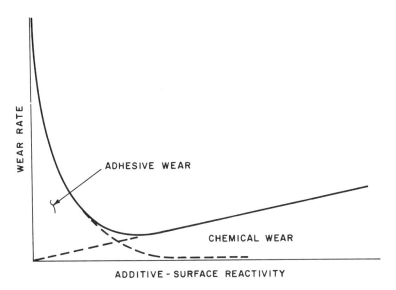

Fig. 2-30 Balance of Adhesive-Chemical Wear

routinely tested for antiwear properties in pump test rigs using commercial vane type hydraulic pumps. Shortcomings of this approach are being reviewed in order to develop a more reliable test to correlate with lubricant performance in service. Frequently, industrial gear lubricants are tested in the FZG Spur Gear Tester, a scale model machine that closely approximates commercial practice as to gear tooth loading and sliding speeds. Again, correlation with field performance has been quite good.

Fatigue Wear Fatigue wear occurs under certain conditions where the lubricating oil film is intact and metallic contact of opposing asperities is negligible. Due to cyclic stressing of the surfaces, fatigue cracks form in the metal and lead to fatigue spalling or pitting. Fatigue wear occurs in rolling element bearings and gears where there is a high degree of rolling, and adhesive wear, associated with sliding, is negligible.

Studies of the influence of lubricant composition and characteristics on fatigue wear are comparatively new, and are still at the research level rather than at the product application level. Two tests have been introduced by Institute of Petroleum (IP) for studies of this type. In IP 300/75, "Pitting Failure Tests for Oils in a Modified Four-Ball Machine," oils are tested under conditions that cause cyclic stressing of steel bearing balls. In order to shorten the test time, the load and stress are set higher than is usual for ball bearings. In IP 305/74 "Tentative, The Assessment of Lubricants by Measurement of Their Effect on the Rolling Fatigue Resistance of Bearing Steel Using the Unisteel Machine," a ball thrust bearing with a flat thrust ring is used. The flat thrust ring is the test specimen. With a flat thrust ring, stresses are higher than would be normally encountered if a raceway providing better conformity was machined in the ring. Both tests are run in replicate. Results to date indicate that both tests provide useful information on the effect of lubricant physical properties and chemical composition on fatigue life under cyclic stressing conditions.

Emulsion and Demulsibility Tests

In some cases, lubricating oils are expected to mix with water to form emulsions. Products such as marine steam engine oils, rock drill oils, soluble cutting oils, and certain types of metal rolling lubricants are examples of products that should form stable emulsions with water. Moisture is usually present in the lubricated areas served by marine steam engine oils and rock drill oils, and emulsification with this water is necessary to form adherent lubricating films that protect the metal surfaces against wear and corrosion. Soluble cutting oils and some types of rolling oils are designed to be mixed with water to provide the high cooling capability of water combined with some lubricating properties.

Various proprietary tests for emulsion stability are in use. In addition, ASTM D1479, "Emulsion Stability of Soluble Cutting Oils," and ASTM D3342, "Measuring the Emulsion Stability of New (Unused) Rolling Oil Dispersions in Water" are used to some extent. Both tests require agitation of the oil sample with water under prescribed conditions. In ASTM D1479, the emulsion is then allowed to settle for 24 h, a bottom portion of it is broken with acid, centrifuged, and the oil content compared with the oil content of a freshly prepared emulsion. The difference is reported as the oil depletion of the stored sample. In ASTM D3324, bottom samples are withdrawn periodically, broken with acid, centrifuged, and the oil content determined. A logarithmic plot of the oil content versus time is then prepared to determine an average rate of separation of the emulsion.

Lubricating oils used in circulating systems should separate readily from water that may enter the system as a result of condensation, leakage, or splash of process water. If the water separates easily, it will settle to the bottom of the reservoir, where it can be periodically drawn off. Steam turbine oils, hydraulic fluids, and industrial gear oils are examples of products where it is particularly important to have good water separation properties. These properties are usually described as the emulsion, demulsibility, or demulsification characteristics.

To determine the emulsion and demulsibility characteristics of lubricating oils, three tests are widely used: ASTM D1401, D2711; and IP 19.

ASTM D1401, "Emulsion Characteristics of Petroleum Oils and Synthetic Fluids," was developed specifically for steam turbine oils with viscosities of 150 to 450 SUS at 100°F (32–97 cSt at 38°C), but can be used for testing lower and higher viscosity oils and synthetic fluids. The normal test temperature is 130°F (54°C), but for oils more viscous than 450 SUS at 100°F, it is recommended that the test temperature be increased to 180°F (82°C). In the test, equal parts of oil (or fluid) and water are stirred together for 5 min. The time for the separation of the emulsion thus formed is recorded. If complete separation does not occur after standing for 1 h, the volumes of oil (or fluid), water, and emulsion remaining are reported.

ASTM D2711, "Demulsibility Characteristics of Lubricating Oils," is intended for use in testing medium and high viscosity oils that are prone to water contamination and that may encounter the turbulence of pumping and circulation capable of producing water-in-oil emulsions. A modification of the test is suitable for evaluation of oils containing extreme pressure additives. In the test, water and the oil under test are stirred together for 5 min at 180°F (82°C). After 5 h settling, the percentage of water in the oil and the percentages of water and emulsion separated from the oil are measured and recorded.

Institute of Petroleum (IP) 19, "Demulsification Number," is intended for testing new inhibited and straight mineral turbine oils. A sample of the oil held at about 90°C is emulsified with dry steam. The emulsion is then held in a bath at 94°C and the time in seconds for the oil to separate is recorded and reported as the demulsification number. If the complete separation has not occurred in 20 min, the demulsification number is reported as 1200+.

Properly used, these three tests have value in describing the emulsion or demulsibility characteristics of lubricating oils. The IP 19 procedure is sometimes used in describing products that emulsify readily as well as products that easily separate from water. A demulsification number of 1200+ is sometimes specified for such products as rock drill oils where the formation of lubricating emulsions is desired.

Engine Tests A large number of engine tests have been and are used for evaluation of the performance properties of engine oils. Most of the standardized tests are designed to evaluate oils intended for automotive type engines, that is, relatively small, high speed diesel engines and four stroke cycle gasoline engines. Considerable effort has been devoted to standardizing the tests, both in the United States and in Europe. In the United States, ASTM publishes the "Multicylinder Engine Test Sequences for Evaluating Automotive Engine Oils" (Sequences IID, IIID and VD)* in STP (Special Technical Publication) 315G, and the "Single Cylinder Engine Tests" (L-38, 1-D- 1-G, and 1-H) in STP 509, "Testing for Evaluating Crankcase Oils." In Europe, the Coordinating European Council (CEC) publishes a number of tests under CEC designations. In most cases, these tests are better known by the names applied to them by their original sponsors. In addition to these tests, the Mack engine tests have come into fairly wide use in the United States, although they are not yet standardized by a national organization. All of these tests can be conveniently grouped into six categories:

1. Tests for oxidation stability and bearing corrosion protection
2. Single cylinder high temperature tests
3. Multicylinder high temperatures tests
4. Low operating temperature tests
5. Rust and corrosion protection tests
6. Preignition tests

Oxidation Stability and Bearing Corrosion Protection Two tests are used to evaluate the ability of oils to resist oxidation under high temperature operating conditions and to protect sensitive bearing metals from corrosion. The tests are operated under conditions that promote oil oxidation and the formation of oil oxyacids which cause bearing corrosion. Copper-lead inserts are used

*Sequences IID and IIID have recently been approved, updating IIC and IIIC and are in the process of publication. Sequence VD has been updated as a tentative test awaiting final approval.

in the connecting rod bearing. After the test, these inserts are examined for surface condition and weight loss to determine the protection afforded by the oil. The engine is also rated for varnish and sludge deposits. The operating conditions for the two tests are summarized in Table 2-6.

Single Cylinder High Temperature Tests These tests are used primarily for the evaluation of detergency, that is, the ability to control piston deposits and ring sticking under operating conditions or with fuels that tend to promote the formation of piston deposits. All of the tests are run in diesel engines. After completion of a test, the engine is disassembled and rated for piston deposits, top ring groove filling, and for evidence of wear. The operating conditions for these tests are summarized in Table 2-7.

Multicylinder High Temperature Tests The two categories of tests just discussed are run in single cylinder engines, either special test engines or commercial engines that were specially adapted for oil testing. While these tests provide useful data and have been extremely valuable in the development of improved oil formulations, there is a trend toward the use of full scale commercial engines for oil testing. Four tests of this type are now used to evaluate high temperature deposit control. The tests are ASTM Sequence IIID, Ford Cortina, and Mack T-1 and T-5. Sequence IIID is also used to evaluate the ability of an oil to resist oxidation and thickening during extended periods of high temperature operation. Operating conditions of these tests are summarized in Table 2-8.

Low Operating Temperature Tests Since dispersancy is more or less specific to deposit forming materials resulting from low temperature operation, two tests (ASTM Sequence VD and Fiat 600D) are used primarily for this property. The tests are operated on cycles designed to promote condensation and the formation of low temperature sludge. After a test, the engine is rated for sludge on various engine surfaces. Sequence VD is also used to evaluate the ability of an oil to keep a closed PCV system clean and functioning properly. Test conditions are summarized in Table 2-9.

Table 2-6 Oxidation Stability and Bearing Corrosion Protection Tests

Test Identification	CRC L-38	Petter W-1 CEC L-02-A-69
Main Area of Use	USA	Europe
Engine Make	CLR	Petter
Test Duration, h	40	36
Speed, rpm	3150	1500
Fuel Input, lb/h (kg/h)	4.75 (2.15)	—[a]
Air Temperature, °F (°C)	80 (27) min	Room
Coolant Outlet Temperature, °F (°C)	200 (93)	302 (150)
Oil Temperature, °F (°C)	290 (143)[b]	280 (137.5)[c]
Hot Spot Temperature, °F (°C)	—	392 (200)
Air:Fuel Ratio	14.0:1	11.9:1

[a]113 s for 50 cm^3.
[b]275°F (135°C) for SAE 10W oils.
[c]266°F (130°C) for SAE 10W oils.

Table 2-7 Single Cylinder High Temperature Tests

Test Identification	L-1	1-H	1-D	1-G	AV-1 CEC L-01-A-69	AV-B CEC L-13-T-74	MWM KD 12E CEC L-05-A-70
Main Area of Use	USA	USA	USA	USA	Europe	Europe	Europe
Engine Make	Caterpillar	Caterpillar	Caterpillar	Caterpillar	Petter	Petter	MWM
Test Duration, h	480	480	480	480	120	50	50
Oil Drain Interval, h	120	120	120	120	None	None	None
Speed, rpm	1000	1800	1200	1800	1500	2250	1850
Fuel Input							
Btu/min (kcal/min)	2950 (744)	4950 (1250)	5600 (1410)	5850 (1470)	—	—	—
lb/h (kg/h)	—	—	—	—	2.4 (1.09)	8.16 (3.7)	4.7 (2.14)
Air Temperature, °F (°C)	100 (38) max	170 (77)	200 (93)	255 (124)	Room	167 (75)	—
Coolant Outlet Temperature							
°F	175-180	160	200	190	185	212	194
°C	79-82	71	93	88	85	100	90
Oil Temperature							
°F	145-150	180	175	205	131	194	194
°C	63-66	82	79	96	55	90	90
Fuel Sulfur, wt %	0.35 min or 1.0*	0.35 min	1.0	0.35 min	0.4 or 1.0	1.0	1.0
Manifold Pressure							
in. Hg abs	Atmos	40	44-45	52.7-53.3	Atmos	67.3	Atmos
mm Hg abs	Atmos	1016	1118-1143	1339-1353	Atmos	1710	Atmos

*Usually referred to as the L-1, S-1.

Table 2-8 Multicylinder High Temperature Tests

Test Identification	Sequence IIID[a]	Ford Cortina CEC L-03-A-70	Mack T-1	Mack T-5
Main Area of Use	USA	Europe	USA	USA
Test Engine	V8 Oldsmobile 5.7L (350 in.³) 2V	Ford Cortina 120E	Mack ENDT675 Maxidyne	Mack ETAZ673
Test Stage	—	—	1; 2; 3	1; 2; 3
Stage Duration, h	—	—	4; 4; 4	4; 4; 4
Stage Repetition	—	—	17; 17; 16	50; 50; 50
Test Duration, h	64[b]	100	200 (total)	600 (total)
Speed, rpm	3000	3500	1400; 1800; 2100	1400; 1800; 2100
Torque, lb-ft (N-m)	—	—	821 (113); 668 (906); 567 (769)	—
Load, bhp (kW)	100 (74.6)	Full 46 (34.3) approx.	219 (163); 229 (171); 227 (169)	—
Coolant Outlet Temperature, °F (°C) — Engine	245 (118)	180 (82)	165–170 (74–77)	185–195 (85–91)
Rocker Covers / Breather Tube	240 (116) / 100 (38)	—	—	—
Oil Temperature, °F (°C)	300 (149)	194 (90)	235 (113)	235 (113)
Intake Air Temperature, °F (°C)	80 (27)	Room	235 (113)	75–85 (24–29)
Fuel Inlet Temperature, °F (°C)	—	—	—	95 (35)
Air Humidity, gr/lb (g/kg)	80 (11.4)	—	95 (35)	95 (35)
Blowby, efm (mm³/s) at 100°F (38°C) 29.7 in. Hg	2.0 (0.94)	—	—	—
Fuel Consumption, litres/h	—	16.2–16.5	—	—
Air:Fuel Ratio	16.5:1	—	—	—
Diesel Fuel Sulfur, % max	—	—	0.3	0.3

[a]Sequence IIID has been approved recently updating Sequence IIIC and now in process of publication.
[b]Following four stages of 1 h duration each gradually increasing speed, load, and media temperatures to test parameters.

Table 2-9 Low Operating Temperature Tests

Test Identification	Sequence VD[a]			Fiat 600D CEC L-04-A-70				
Main Area of Use	USA			Europe				
Test Engine	Ford 4 Cyl, 2.3 L (140 in.³)			Fiat 600D				
Test Duration, h	← 192 →			← 378 →				
Test Stage	I	II	III	I	II	III	IV	V
Stage Duration, min	120	75	45	5	55	120	120	60
Stage Repetition	← 48 →			← 63 →				
Speed, rpm	2500	2500	750	1200	800	2500	4000	0
Load, bhp (kw)	33.5 (25)	33.5 (25)	1 (0.7)	none	none	12.5 (9.3)	20 (15)	—
Oil Temperature								
°F	175	187	120	77	77	77	176	77
°C	79	86	49	25	25	25	80	25
Coolant Outlet Temperature								
°F	135	155	120	77	77	77	194	77
°C	57	68	49	25	25	25	90	25
Air:Fuel Ratio	—[b]	—[b]	—[b]	full choke	12.5:1	14.5:1	14.5:1	—
Intake Air Temperature, °F (°C)	80 (27)	80 (27)	85 (29)					
Intake Air Humidity, gr/lb (g/kg)	← 80 (11.4) →							
Blowby, cfm (m³/h) 100°F, 29.7 in. Hg	← 1.8 (3.0) avg. →							

[a]Sequence VC has been updated and is a tentative test awaiting final approval and publication.
[b]Air:fuel ratio controlled to maintain exhaust gas analysis within prescribed limits. Stages I and II are lean conditions, Stage III is a rich condition.

Rust and Corrosion Protection Tests   This property has received more attention in the United States than in many other countries, probably because of the high proportion of stop-and-go driving in combination with severe winter weather. These conditions tend to promote condensation and accumulation of partially burned fuel in the crankcase, both of which promote rust and corrosion. Only one test, Sequence IID, is now used to evaluate the ability of oils to control rust and corrosion. At the end of the test, valve lifters, pushrods, and the oil pump relief valve are rated for rust, and any sticking of lifters or the oil pump relief valve is noted. Operating conditions are summarized in Table 2-10.

Preignition Test   In gasoline engines, preignition can be a problem leading to power loss, increased fuel consumption, and in severe cases, engine damage. The nature and quantity of combustion chamber deposits derived from the lubricating oil can affect preignition tendencies. This problem has received more attention in Europe than it has in the United States. In Europe, the Fiat 124AC Preignition Test has been standardized by Coordinating European Council.

In order to determine the tendency of an oil to cause preignition, one cylinder of the four cylinder engine is provided with a separate fuel induction system. During the test, this cylinder is fed with a mixture of gasoline and 2 percent of the test oil. The engine is run until preignition in this cylinder is

Table 2-10   Rust and Corrosion Protection Test

Test Identification	Sequence IID[a]		
Main Area of Use	USA		
Test Engine	V8 Oldsmobile 5.7 L (350 in.³) 2V		
Test Duration, h	⟵——— 32 ———⟶		
Test Stage	I	II[b]	III
Stage Duration, h	28	2	2
Speed, rpm	1500	—	3600
Load, bhp (kw)	25 (18.6)	—	100 (75)
Oil Temperature, °F (°C)	120 (49)	—	260 (127)
Intake Air Temperature, °F (°C)	80 (27)	—	80 (27)
Intake Air Humidity, gr/lb (g/kg)	80 (11.4)	—	80 (11.4)
Coolant Outlet Temperature, °F (°C)			
Jacket	110 (43.3)	120 (49)	190 (88)
Crossover	109 (42.8)	119 (48)	197 (92)
Rocker Covers	60 (15.6)	—	198 (92)
Breather Tube	60 (15.6)	—	199 (93)
Blowby, cfm (dm³/h) at 100°F (38°C)			
29.7 in. Hg	0.8 (0.4)	—	1.4 (0.7)
Air:Fuel Ratio	13.0:1	—	16.5:1

[a]Sequence IID has been approved recently updating Sequence IIC and is now in process of publication.
[b]All values same as Stage I except those shown.

Table 2-11 Preignition Test

Test Identification	Fiat 124 AC Preignition CEC L-09-T71
Main Area of Use	Europe
Test Engine	Fiat 124 AC
Test Duration	200 h or until preignition is detected
Speed, rpm	3500
Fuel Consumption, litres/h	
Cylinders 1, 2, 3	9
Cylinder 4	3
Air:Fuel Ratio	13.8:1
Fuel Sulfur, % wt. max	0.20
Oil Temperature, °F (°C)	176–185 (80–85)
Coolant Temperature, °F (°C)	194 (90)
Crankcase Oil Charge	SAE 30 straight mineral oil
Oil Change Interval, h	24

detected, or for 200 h if no preignition is detected. Other operating conditions are shown in Table 2-11.

Automotive Gear Lubricant Tests

The most widely used tests for automotive gear lubricants are those required for approval against U.S. Military Specification MIL-L-2105C and its predecessor MIL-L-2105B. All of the tests are run in commercial type drive axles, operated under conditions designed to evaluate specific lubricant properties. A brief summary of these tests is shown in Table 2-12.

In addition to the automotive gear lubricant tests, various car manufacturers have gear tests, many of which are conducted in cars either on a chassis dynamometer or on the road. These tests frequently represent special requirements such as the ability of lubricants to provide satisfactory per-

Table 2-12 MIL-L-2105C Gear Lubricant Tests

Test Designation Federal Test Method Standard No. 79lb	CRC No.	Characteristics Being Evaluated
6506.1	L-37	Load carrying, wear and extreme pressure characteristics of gear lubricants in axles operating under conditions of high speed, low torque operation and low speed, high torque operation
6507.1	L-42	Antiscoring properties of gear lubricants under high speed and shock conditions
2504	—	Deterioration of lubricants under severe oxidation conditions
5326.1	L-33	Corrosion preventive properties of gear lubricants

formance in limited slip axles. Generally, all of the tests have shown good ability to predict field performance.

Automatic Transmission Fluid Tests

Automatic transmission fluids are among the most complex lubricants now available. In the converter section, these fluids are the power transmission and heat transfer medium; in the gear box, they lubricate the gears and bearings and control the frictional characteristics of the clutches and bands; and in the control circuits, they act as hydraulic fluids. All of these functions must be performed satisfactorily over temperatures ranging from the lowest expected ambient temperatures to operating temperatures on the order of 300°F (149°C) or higher, and for extended periods of service. Obviously, very careful evaluation is required before a fluid can be considered acceptable for such service.

Most automatic transmission fluids are manufactured to meet specifications issued by either Ford Motor Company or General Motors Corporation. Qualification against these specifications involves extensive testing in automatic transmissions operated under high temperature and cycling conditions, as well as various bench tests for such properties as oxidation stability, antiwear properties, viscosity retention after service. Products qualified against these specifications generally have given excellent performance in service.

Bibliography

Mobil Technical Bulletins
Refining of Lubricating Oils and Waxes
Light Products Refining
Engine Oil Specifications and Tests — Significance and Limitations
Additives for Petroleum Oils
Extreme Pressure Lubricant Test Machines

3

Lubricating Greases

The American Society for Testing and Materials (ASTM) D288, "Standard Definitions of Terms Relating to Petroleum," defines a lubricating grease as: "A solid to semifluid product of dispersion of a thickening agent in liquid lubricant. Other ingredients imparting special properties may be included." This definition indicates that a grease is a liquid lubricant thickened to some extent in order to provide properties not available in the liquid lubricant alone.

WHY GREASES ARE USED

The reasons for the use of greases in preference to fluid lubricants are well stated in the Society of Automotive Engineers (SAE) Information Report J310a on Automotive Lubricating Grease. This report states:

> Greases are most often used instead of fluids where a lubricant is required to maintain its original position in a mechanism, especially where opportunities for frequent relubrication may be limited or economically unjustifiable. This requirement may be due to the physical configuration of the mechanism, the type of mo-

tion, the type of sealing, or to the need for the lubricant to perform all or part of any sealing function in the prevention of lubricant loss or the entrance of contaminants. Because of their essentially solid nature, greases do not perform the cooling and cleaning functions associated with the use of a fluid lubricant. With these exceptions, greases are expected to accomplish all other functions of fluid lubricants.

A satisfactory grease for a given application is expected to:

1. Provide adequate lubrication to reduce friction and to prevent harmful wear of bearing components
2. Protect against corrosion
3. Act as a seal to prevent entry of dirt and water
4. Resist leakage, dripping, or undesirable throwoff from the lubricated surfaces
5. Resist objectionable change in structure or consistency with mechanical working (in the bearing) during prolonged service
6. Not stiffen excessively to cause undue resistance to motion in cold weather
7. Have suitable physical characteristics for the method of application
8. Be compatible with elastomer seals and other materials of construction in the lubricated portion of the mechanism
9. Tolerate some degree of contamination, such as moisture, without loss of significant characteristics

While this statement is concerned primarily with the use of lubricating greases in automotive equipment, the same considerations apply to the use of greases in other applications.

COMPOSITION OF GREASE

In the definition of a lubricating grease given here, the liquid portion of the grease may be a mineral oil or any fluid that has lubricating properties. The thickener may be any material that, in combination with the selected fluid, will produce the solid or semifluid structure. The other ingredients are additives or modifiers that are used to impart special properties or modify existing ones.

Fluid Components

Most of the greases produced today have mineral oils as their fluid components. These oils may range in viscosity from as light as kerosine up to the heaviest cylinder stocks. In the case of some specialty greases, products such as waxes, petrolatums, or asphalts may be used. Although these latter materials may not be too accurately described as "liquid lubricants," they perform the same function as the fluid components in conventional greases.

Greases made with mineral oils provide satisfactory performance in most automotive and industrial applications. In very low or high temperature applications or in applications where temperature may vary over a wide range, greases made with synthetic fluids generally are now used. For a detailed discussion, see Chap. 4.

Thickeners

The principal thickeners used in greases are metallic soaps. The earliest greases were made with calcium soaps, then greases made with sodium soaps were introduced. Later, soaps such as aluminum, lithium, and barium came into use. Some greases which are made with mixtures of soaps, such as sodium and calcium, are usually referred to as mixed base greases. Soaps made with other metals have been proposed for use in greases but have not received commercial acceptance, either because of cost or performance problems. Some lead soaps are used as thickeners, but most lead and zinc soaps are used as modifying agents in greases.

Modifications of metallic soap greases, called "complex" greases, are becoming increasingly popular. These complex greases are made by using a combination of a conventional metallic soap forming material with complexing agent. The complexing agent may be either organic or inorganic and may or may not involve another metallic constituent. Among the most successful of the complex greases are the calcium complex greases. These are made with a combination of conventional calcium soap forming materials and a low molecular weight organic acid as the complexing agent. Greases of this type are characterized by very high dropping points, usually above 500°F (250°C), and may also have excellent load carrying properties. Other complex greases—aluminum, barium, and lithium—are also manufactured for certain applications.

A number of nonsoap thickeners are in use, primarily for special applications. Modified bentonite (clay) and silica aerogel are used to manufacture nonmelting greases for high temperature applications. Since oxidation can still cause the oil component of these greases to deteriorate, regular relubrication is required. Thickeners such as polyurea, pigments, dyes, and various other synthetic materials are used to some extent. However, since they are generally more costly, their use is somewhat restricted to applications where performance requirements are critical. Table 3-1 outlines lubricating grease characteristics as determined by thickener type for various major grease soaps.

Additives and Modifiers

Additives and modifiers commonly used in lubricating greases are oxidation or rust inhibitors, pour point depressants, extreme pressure antiwear agents, lubricity or friction reducing agents, and dyes or pigments. Most of these materials have much the same function as similar materials added to lubricating oils.

Molybdenum disulfide is used in many greases for applications where loads are heavy, surface speeds are low, and restricted or oscillating motion is involved. In these applications, the use of "molysulfied," (or "moly" as it is sometimes called) reduces friction and wear. Polyethylene and modified tetrafluorethane (Teflon) may also be used for applications of this type.

MANUFACTURE OF GREASE

The manufacture of a grease, usually a batch process, involves the dispersion of the thickener in the fluid and the incorporation of any additives or modifiers. This is accomplished in a number of ways. In some cases, the thickener is purchased by the grease manufacturer in a finished state and then mixed with oil until the desired grease structure is obtained. In most cases with metallic soap thickeners, the thickener is produced, through reaction, during the manufacture of the grease.

Table 3-1 Lubricating Grease Characteristics By Thickener Type

Characteristic	Calcium (Lime) Soap	Sodium Soap	Lithium Soap	Calcium Lead Complex	Lithium Complex	Inorganic (Nonsoap)
Type of Service	Limited Service	Limited Service	Multiservice	Multiservice	Multiservice	Special Service at High Temperatures
Thickener Temperature Limitation	Loss of Water at 160°F (71°C) (Soap Dehydrates)	Phase Change at 200–250°F (93–121°C)	Phase Change at 300–350°F (149–177°C)	None	None	None
Texture-Structure as Manufactured	Smooth	Fibrous, Long Fibre, Short Fibre, Sometimes Smooth	Smooth	Smooth	Smooth	Smooth
Stability in Service	Fair	Fair to Good	Good	Good	Good	Good
Oxidation Resistance[a]	Fair	Good	Fair	Good	Fair	Good
Protection against Wear[a]	Good	Good	Good	Good	Good	Good
Effects of Water[a]	Water Resistant	No Water Resistance Good Antirust	Water Resistant	Water Resistant Good Antirust	Water Resistant	Water Resistant
Maximum Operating Temperature[b]	160°F (71°C)	250°F (121°C)	250°F (121°C)	300°F (149°C)	300°F (149°C)	300°F (149°C)
Principal Uses: Automobive vehicle, Aircraft, Small Marine	Chassis, Water-Pump, Cup, Special Aircraft	Wheel and other Antifriction Bearings, Universal Joints, High-Temperature Applications, Special Aircraft	Multipurpose, Special Aircraft	Multipurpose	Multipurpose, Wheel Bearing and High Temperature Applications	High Temperature, Special Aircraft

[a]Oil-Soap combination only: characteristics may be different for a specific grease as a result of use of additives or the manufacturing process employed.
[b]Normal commercial greases; higher maximum, ranging from 225 to 500°F (107–280°C), are possible for limited periods and in aviation and other special purpose greases for longer periods.

In the manufacture of a lithium soap grease, for example, hydrogenated castor oil, fatty acids, and/or glycerides are dissolved in a portion of the oil and then saponified with an aqueous solution of lithium hydroxide. This produces a wet lithium soap that is partially dispersed in the mineral oil and is then dehydrated by heating. After drying, the mixture is cut back with additional oil to produce the consistency desired in the finished grease. In this case, the dehydrated soap-oil mixture would be a plastic mass with a grainy structure. During or following the cutback operation, the grease might be further processed by kettle milling or homogenization to modify this structure. Once the proper structure and consistency are obtained, the grease is ready for finishing and packaging.

Finishing may or may not involve deaeration to remove entrained air. Immediately before the filling operation, greases are filtered to remove any contaminants that may have been picked up from the raw materials or processing reaction that might affect the performance of the grease.

As noted in the preceding discussion, manufacture of one of the basic greases involves all or some of the following six steps:

1. Saponification
2. Dehydration
3. Cutback
4. Milling
5. Deaeration
6. Filtering

These basic processing steps are used in the manufacture of most greases. In certain manufacturing processes some of these processing operations may be accomplished simultaneously, while in others they are distinct and separate steps. The main equipment required for the manufacture of grease is a vessel suitable for the saponification step. This vessel often is a heated kettle equipped with some form of agitation. Heating may be by direct firing or by means of a steam or hot oil jacket. Agitation is usually of the double acting type, that is, two sets of paddles rotating in opposite directions. One set of paddles is equipped with scraper blades to remove soap masses from the sides of the kettle. Usually, grease kettles are of the open type. However, to make some types of greases, closed or pressure kettles are used in order to speed up the saponification step or to obtain the required reaction.

After saponification, the grease is cooled in the same manner as it was heated—namely, by a cooling medium in the kettle jacket. The rate of cooling after the soap is formed is very important in the development of the proper structure for many greases. Therefore, close temperature control is required.

As mentioned earlier, the structure may be modified by milling. This milling may be continuous in the kettle during the cooling period or it may be accomplished in a separate operation. If milling is done in a separate operation, a high shear rate pump, homogenizer, or colloid mill may be used. Usually, the purpose of milling is to break a fibrous structure or to improve the dispersion of the soap in the lubricating fluid. Kettle milling will break a fibrous structure, but milling in a homogenizer or other milling equipment is required to improve dispersion.

During processing, grease may become aerated. Generally, aeration does not detract from the performance of a grease as a lubricant, but it does affect the appearance. To improve customer appeal, some modern greases are deaerated. Various types of equipment are used for this purpose, but basically they are similar in that they expose a thin film of grease to a vacuum. The vacuum draws off the entrained air, giving a much brighter appearance to the grease.

The final processing step is filtration. This is done with commercial edge type or screen type filters. The screen size used varies according to the end use of the grease. Some cheaper commodity greases may be filtered through openings corresponding to about 40 mesh; however, greases for most applications, including rolling element bearing lubrication, require the use of filters with openings of 100 mesh or finer.

This discussion has considered the manufacture of one type of grease in the simplest type of equipment. Many greases are still made in this manner. However, grease plants are now using newer types of equipment and more complex formulations in the manufacture of many greases. Saponification is accomplished in pressure vessels that have a much higher rate of heat transfer and improved agitation compared to the older open kettles. The Stratco Contactor is typical of the new equipment. In this equipment, a mixture to be saponified and/or reacted to form the thickener is subjected to agitation and circulation as a thin film while it is being heated. Many of the newer greases require processing at much higher temperatures than can be obtained economically with steam. This has led to increasing use of heating systems using oil or synthetic fluid as the heat transfer medium.

Considerable work has been done on the development of continuous processes for grease manufacture. Originally, during start up, the adjustment and stabilization of the grease flow resulted in a significant amount of off specification product. This waste grease had an adverse effect on the overall economics. However, improved sensory equipment, which has minimized greatly the start up time, and improved process control are overcoming this problem.

GREASE CHARACTERISTICS

The general description of a grease is in terms of the materials used in its formulation and physical properties, some of which are visual observations. The type and amount of thickener and the viscosity of the fluid lubricant are formulation properties. Color and texture, or structure, are observed visually. There is some correlation between these descriptive items and performance, for example:

1. Certain types of soaps usually impart certain properties to a finished grease.
2. The viscosity of the fluid lubricant is important in selecting greases for some applications.
3. Light colored or white greases may be desirable in certain applications.

This description normally is supplemented by tests for the consistency and dropping point of the grease, and sometimes by data on the apparent viscosity of the grease. Most of the other tests that are used to describe greases come under the category of evaluation and performance tests.

Consistency Consistency is defined as the degree to which a plastic material resists deformation under the application of a force. In the case of lubricating greases, it is a measure of the relative hardness or softness, and may indicate something of flow and dispensing properties. Consistency is reported in terms of ASTM D217, "Cone Penetration of Lubricating Grease" or National Lubricating Grease Institute (NLGI) Grade, Consistency, similar to viscosity, varies with temperature and, therefore, must be reported at a specific temperature.

Cone Penetration The cone penetration of greases is determined with the ASTM Penetrometer, see Fig. 3-1. After the sample is prepared in accordance with ASTM D217, the cone is released and allowed to sink into the grease, under its own weight, for 5 s. The depth the cone has penetrated is then read, in tenths of a millimetre, and reported as the penetration of the grease. Since the cone will sink farther into softer greases, higher penetrations indicate softer greases. ASTM penetrations are normally measured at 77°F (25°C).

In addition to the standard equipment (ASTM D217) shown in Fig. 3-1, ¼ and ½ scale cone equipment (ASTM D1403) is available for determining the penetrations of small samples. An equation is used to convert the penetrations obtained by ASTM D1403 to equivalent penetrations for the full scale test.

Penetrations are reported as *undisturbed penetrations, unworked penetrations, worked penetrations,* or *prolonged worked penetrations. Undisturbed penetrations* are measured in the original container, without disturbance, to determine hardening or softening in storage. *Unworked penetrations* are measured on samples transferred to the grease cup with minimum disturbance. This value may have some significance with regard to transferring greases from the original containers to application equipment. The value normally reported is the worked penetration, measured after the sample has been worked 60 double strokes in the ASTM Grease Worker; see Fig 3-2. It is considered to be the most reliable test since the amount of disturbance of the sample is controlled and repeatable. *Prolonged worked penetrations* are discussed under Mechanical or Structural Stability Tests.

NLGI Grease Grade Numbers On the basis of ASTM worked penetrations, the NLGI has standardized a numerical scale for classifying the consistency of greases. The NLGI grade and corresponding penetration ranges, in order of increasing hardness, are shown in Table 3-2. This system has been well accepted by both manufacturers and consumers. It has proved adequate for specifying the preferred consistency of greases for most applications.

Apparent Viscosity Newtonian fluids, such as normal lubricating oils, are defined as those materials for which the shear rate (or flow rate) is proportional to the applied

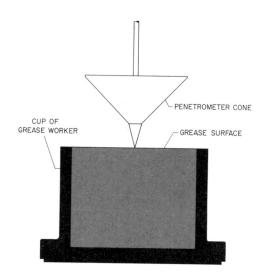

CUP OF
GREASE WORKER

PENETROMETER CONE

GREASE SURFACE

Fig. 3-1 Grease Consistency by Penetrometer
In the drawing the cone is shown in its initial position just touching the surface of the grease in the cup. In the photograph the cone has penetrated into the grease, and the amount of penetration is recorded on the dial.

shear stress (or pressure) at any given temperature. That is, the viscosity, which is defined as the ratio of shear stress to shear rate, is constant at a given temperature. Grease is a non-Newtonian material that does not begin to flow until a shear stress exceeding a yield point is applied. If the shear stress is then increased further, the flow rate increases more proportionally, and the viscosity, as measured by the ratio of shear stress to shear rate, decreases. The observed viscosity of a non-Newtonian material such as grease is called its *apparent viscosity*. Apparent viscosity varies with both temperature and shear rate; thus, it must always be reported at a specific temperature and flow rate.

Apparent viscosities of greases are determined in accordance with ASTM D1092. In this test, samples of a grease are forced through a set of capillary tubes, at predetermined flow rates. From the dimensions of the capillaries, the known flow rates, and the pressure required to force the grease through the capillaries at those flow rates, the apparent viscosity of the grease, in poise, can be calculated. Results usually are reported graphically as apparent viscosity versus shear rate at a constant temperature, or apparent viscosity versus temperature at a constant shear rate.

Apparent viscosity is used to predict the handling and dispensing properties of a grease. In addition, it can be related to starting and running torque, in grease lubricated mechanisms, and is useful in predicting leakage tendencies.

Fig. 3-2   Worked Penetration Equipment

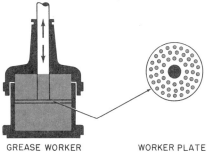  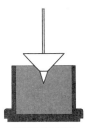

GREASE WORKER   WORKER PLATE   PENETRATION AFTER 60 STROKES   PENETRATION AFTER 10,000 OR MORE STROKES

Table 3-2 NLGI Grease Classification

NLGI Grade	ASTM Worked Penetration*
000	445–475
00	400–430
0	355–385
1	310–340
2	265–295
3	220–250
4	175–205
5	130–160
6	85–115

*Ranges are the penetration in tenths of a millimeter after 5 s at 77°F (25°C).

Dropping Point

The dropping point of a grease is the temperature at which a drop of material falls from the orifice of a test cup under prescribed test conditions; see Fig. 3-3. Two procedures are used (ASTM D566 and ASTM D2265) which differ in the type of heating units and, therefore, the upper temperature limits. An oil bath is used for ASTM D566 with a measureable dropping point limit of 500°F (260°C); ASTM D2265 uses an aluminum block oven with a dropping point limit of 625°F (330°C). Conventional soap thickened greases do not have a true melting point but have a melting range during which they become progressively softer. Some other types of greases may, without change in state, separate oil. In either case, only an arbitrary, controlled test procedure can provide a temperature that can be established as a characteristic of the grease.

The dropping point of a grease is not considered to have any bearing on service performance other than that, at temperatures above the dropping point, the grease may approach the operating limits of satisfactory performance. It does not establish the maximum usable temperature for a grease since so many other factors must be taken into account in high temperature lubrication with grease. It is useful for characterization, and also as a quality control during grease manufacture.

EVALUATION AND PERFORMANCE TESTS

The tests described in the previous section characterize greases. Most of the other tests for lubricating greases are designed to be useful in predicting performance under certain conditions.

Mechanical or Structural Stability Tests

The ability of a grease to resist changes in consistency during mechanical working is termed its mechanical or structural stability. This is important in most applications since if a grease softens excessively as a result of the mechanical shearing it is subjected to in service, it can lead to leakage. This loss of lubricant may cause equipment failure. Hardening as a result of shearing can be equally harmful in that it can prevent the grease from feeding oil properly to the equipment and can also result in its failure.

Generally, two methods are used for determining the structural stability of greases. Determinations of prolonged worked penetration (see Consistency) are made after a grease has been worked 10,000, 50,000 or 100,000 double strokes in the ASTM Grease Worker. In the Roll Stability Test (ASTM D1831), a small sample of grease is milled in a cylindrical chamber by a heavy roller for 2 h at room temperature; see Fig. 3-4. The penetration after milling is then determined with the $1/2$ or $1/4$ scale cone equipment.

In both of these tests, the change in consistency with mechanical working is reported as either the absolute change in penetration or the percent change in penetration. The significance of the tests is somewhat limited because of the differences in test shear rates and the actual rates of shearing in a bearing. The shear rates in the tests range between 100 and 1000 reciprocal seconds (s^{-1}), while the shear rates in bearings may be as high as, or higher than, 1,000,000 s^{-1}.

Oxidation Tests

Resistance to oxidation is an important characteristic of greases intended for use in rolling element bearings. Improvement in this property through the use of oxidation inhibitors has enabled the development of the so-called "packed for life" bearings.

Fig. 3-3 Dropping Point Test Photo shows complete apparatus with viewing window. Assembled grease cup and thermometer at lower right with assembly rig. Enlarged view of cup and thermometer shown in sketch at left.

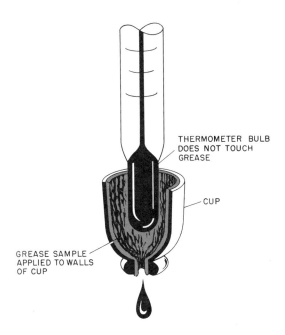

THERMOMETER BULB
DOES NOT TOUCH
GREASE

CUP

GREASE SAMPLE
APPLIED TO WALLS
OF CUP

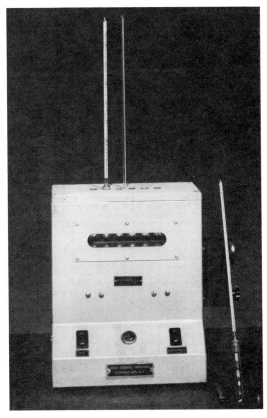

Fig. 3-4   Roll Stability Test   The heavy cylindrical roller shown standing at the left rolls freely inside the tubular chamber which is driven by the motor and gears at the right. This machine can run two rollers simultaneously.

Both the oil and the fatty constituents in a grease oxidize; the higher the temperature the faster the rate of oxidation. When grease oxidizes it generally acquires a rancid or oxidized odor and darkens in color. Simultaneously, organic acids usually develop and the lubricant becomes acid in reaction. These acids are not necessarily corrosive but may affect the grease structures causing hardening or softening.

Laboratory tests have been developed to evaluate oxidation stability under both static and dynamic conditions. In the static test, ASTM D942, "Oxidation Stability of Lubricating Greases by the Oxygen Bomb Method," the grease is placed in a set of five dishes. The dishes are then placed in a pressure vessel, or bomb, which is pressurized with oxygen to 110 psi (758 kPa), and placed in a bath held at 210°F (99°C) where it is allowed to remain for a period of time, usually 100, 200, or 500 h. At the end of this time, the pressure is recorded and the amount of pressure drop reported. For specification purposes, pressure drops of 5 to 25 psi (34−172 kPa) are usually allowed, depending on the test time and the intended use of the grease.

The results of this test are probably most indicative of the stability of thin films of a grease in extended storage, as on prelubricated bearings. It is not intended for the prediction of the stability of a grease under dynamic conditions or in bulk storage in the original containers.

A number of tests are used to evaluate oxidation stability under dynamic conditions. Two tests that were formerly used, and are still used to some extent, are ASTM D1741, "Functional Life of Ball Bearing Greases," and Method 333 of Federal Test Method (FTM) Standard No. 79lb. A new test, ASTM D3336, "Performance Characteristics of Lubricating Greases in Ball Bearings at Elevated Temperatures," is designed to replace both of these earlier tests. All of the tests are run in ball bearing test rigs with the bearings loaded and heated. The tests differ principally in the maximum operating temperature—ASTM D1741 provides for operation at temperatures up to 125°C (247°F), Method 333 FTM 79lb to 450°F (232°C), and ASTM D3336 to 371°C (700°F). The tests are run until the bearing fails or for a specific number of hours if failure has not occurred. All of the tests are considered to be useful screening methods for determining projected service life of ball bearing greases operating at elevated temperatures.

Oil Separation Tests

The resistance of a grease to separation of oil from the thickener involves certain compromises. When greases are used to lubricate rolling element bearings, a certain amount of bleeding of the oil is necessary in order to perform the lubrication function. On the other hand, if the oil separates too readily from a grease in application devices, a hard, concentrated soap residue may build up which will clog the devices and prevent or retard the flow of grease to the bearings. In bearings, excessive oil separation may lead to the buildup of a hard soap in bearing recesses which in time could be troublesome. Further, leakage of separated oil from bearings can damage materials in production or equipment components such as electric motor windings.

In application devices, such as central lubrication systems and spring loaded cups where pressure is applied to the grease on a more or less continuous basis, oil can be separated from greases by a form of pressure filtration. The pressure forces the oil through the clearance spaces around plungers, pistons, or spool valves; but since the soap cannot pass through the small clearances, it is left behind. This may result in blockage of the devices and lubricant application failure.

Some oil release resulting in free oil on the surface of the grease in containers in storage is normal. However, excessive separation, while not usually harmful, may result in loss of the user's confidence in the product.

Generally, there is no accepted method for evaluating the oil separation properties of a grease in service. Trials in typical dispensing equipment may be conducted. Some useful information may also be obtained from tests such as ASTM D1741 and D3336, and Method 333 of FTM 79lb; see Oxidation Stability. The tendency of a grease to separate oil during storage can be evaluated by ASTM D1742, "Oil Separation from Lubricating Grease During Storage." In this test, air pressurized to 0.25 psi (1.72 kPa) is applied to a sample of grease held on a 75 μm (No. 200) mesh screen. After 24 h at 77°F (25°C) the amount of oil separated is determined and reported. The test correlates directly with oil separation in other sizes of containers.

The tendency of a grease to separate oil at elevated temperatures under static conditions can be evaluated by Method 321.2 of FTM 79lb. In this test, a sample of grease is held in a wire mesh cone suspended in a beaker. The beaker is placed in an oven, approximately at 212°F (100°C), for the desired time, usually 30 h. After the test, the oil collected in the beaker is weighed and calculated at a percentage of the original sample. Sometimes the test is used for specification purposes.

Water Resistance Tests

The ability of a grease to resist washout under conditions where water may splash or impinge directly on a bearing is an important property in such applications as automobile front wheel bearings. Comparative results between different greases can be obtained with ASTM D1264, "Water Washout Characteristics of Lubricating Greases."

In this test, a ball bearing with increased clearance shields is rotated with a jet of water impinging on it; see Fig. 3-5. Resistance to washout is measured by the amount of grease lost from the bearing during the test. This test is considered to be a useful screening test for greases that are to be used wherever water washing may occur.

In many cases, direct impingement of water may not be a problem, but a moist atmosphere or water leakage may expose a grease to water contamination. One method of evaluating a grease for use under such conditions is to homogenize water into it. The grease may then be reported on the basis of the amount of water it will absorb without loss of grease structure, or in the amount of hardening or softening resulting from the admixture of a specific proportion of water.

Rust Protection Tests

In many applications, greases are not only expected to provide lubrication, but are also expected to provide protection against rust and corrosion. Some types of greases have inherent rust protective properties, while others do

Fig. 3-5   Water Washout Test

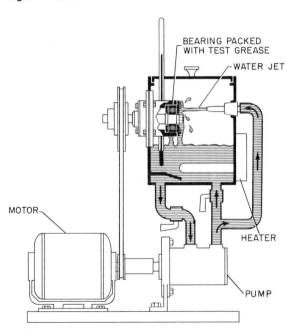

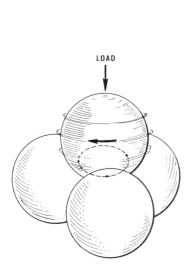

LOAD

Fig. 3-6   Four Ball EP Test

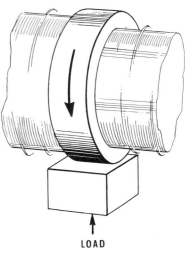

LOAD

Fig. 3-7   Timken Load Tester

not. Rust inhibitors can be incorporated in greases to improve rust protective properties.

Both static and dynamic tests are used to evaluate the rust protective properties of greases. Often, the test specimen is a rolling element bearing lubricated with the grease under test and then exposed to conditions designed to promote rusting. One typical static test is ASTM D1743, "Rust Preventive Properties of Lubricating Greases." In this test, tapered roller bearings are packed with the test grease which is distributed by rotating the bearings for 60 s under light load. The bearings are then dipped in distilled water and stored for 48 h at 125°F (52°C) and 100 percent relative humidity. After storage, the bearings are cleaned and examined for rusting or corrosion. A bearing that shows no corrosion is rated 1. Incipient corrosion (no more than three spots of visible size) is rated 2, and anything more is rated 3.

This test was developed as a cooperative project to correlate with difficulties experienced in aircraft wheel bearings some years ago. The correlation with service performance, particularly under static conditions and without water washing, is considered quite good.

EP and Wear Prevention Tests

While the results of laboratory extreme pressure and wear prevention tests do not necessarily correlate with service performance, the tests presently provide the only means to evaluate these properties at a reasonable cost. ASTM has standarized the test procedures to determine EP properties of greases using the "Four-Ball Extreme Pressure" (ASTM D2596) and the Timken (ASTM D2509) Machines; see Fig. 3-6 and Fig. 3-7, respectively. Also, ASTM has standarized tests for wear prevention properties using the "Four-Ball Wear Tester Machine" (ASTM D2266).

The two extreme pressure tests (ASTM D2509 and D2596) are considered to be capable of differentiating between greases having low, medium, or high levels of extreme pressure properties. The wear prevention test (ASTM D2266) is intended to compare only the relative wear preventive characteristics of greases in sliding steel-on-steel applications. It is not intended to predict wear characteristics with other metals.

Bibliography

Mobil Technical Bulletins
Extreme Pressure Lubricant Test Machines
Lubricating Grease Tests — Significance and Applicability

4
Synthetic Lubricants

Considerable attention has been focused on synthetic lubricants since the recent introduction into the retail market of synthetic or synthesized base automotive engine oils. Although these products are relatively new, the use of synthetic base lubricants in aviation and industrial applications extends back over many years.

The terms "synthetic" and "synthesized" are both used to describe the base fluids used in these lubricants. A synthesized material is one that is produced by combining or building individual units into a unified entity. The production of synthetic lubricants starts with synthetic base stocks that are often manufactured from petroleum. The base fluids are made by chemically combining (synthesizing) low molecular weight compounds that have adequate viscosity for use as lubricants. Unlike mineral oils, which are a complex mixture of naturally occurring hydrocarbons, synthetic base fluids are man-made and tailored to have a controlled molecular structure with predictable properties.

As developed in Chap. 2, the properties of a mineral lubricating oil result from the selection of those crude oil compounds that have the best properties for the intended application. This is accomplished through fractionation, solvent refining, hydrogen processing, solvent dewaxing, and filtration. However, even with extensive treatment, the finished product is a mixture of many compounds. There is no way to select from this mixture only those materials with the best properties, and if there were, the yield would be so low that the process would be uneconomical. Thus, the mineral oils produced have properties that are the average of the mixture, including both the most and the least suited components.

With synthetic lubricant base stocks, on the other hand, the process of combining individual units can be controlled so that a large proportion of the finished base fluid is either one or only a few compounds. Depending on the starting materials and the combining process, the compound (or compounds) can have the properties of the best compounds in a mineral base oil. It can also have unique properties, such as miscibility with water or complete nonflammability, that are not found in any mineral oil.

In either case, the special properties of the finished synthetic lubricants prepared from the compound may justify the additional cost in applications where mineral oil lubricants do not provide adequate performance. Some of the primary applications for synthetic lubricants are listed in Table 4-1.

The primary performance advantage of synthetic lubricants is the extended range of service temperatures. Their outstanding flow characteristics at extremely low temperatures and their stability at extremely high temperatures mark the preferred use of these lubricants. The comparative operating temperature limits of mineral oil and synthetic lubricants are shown in Fig. 4-1. Other advantages, as well as limiting properties, are outlined in Table 4-2.

Various ways have been used to classify synthetic base fluids. Some of these classification approaches neglect similarities between certain types of materials, while others may lead to confusion by grouping materials that

Table 4-1 Primary Applications for Synthetic Lubricants

Field of Service	Synthetic Fluids Used
Industrial	
Circulating Oils	Polyglycol, SHF, Organic Ester
Gear Lubricants	Polyglycol, SHF
Hydraulic Fluid (Fire Resistant)	Phosphate Ester, Polyglycol
Compressor Oils	Polyglycol, Organic Ester, SHF
Gas Turbine Oils	SHF, Organic Ester
Greases	SHF
Automotive	
Passenger Car Engine Oil	SHF, Organic Ester
Commercial Engine Oil	SHF, Organic Ester
Gear Lubricant	SHF
Brake Fluids	Polyglycols
Aviation	
Gas Turbines	Organic Ester
Hydraulic Fluids	Phosphate Ester, Silicones, SHF
Greases	Silicones, Organic Ester, SHF

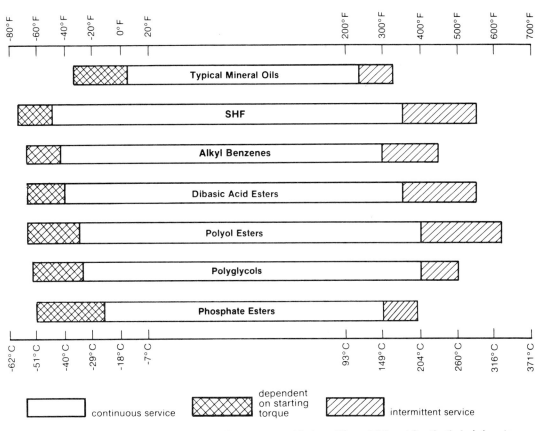

Fig. 4-1 Comparative Temperature Limits – Mineral Oil and Synthetic Lubricants

have similarities in chemical structure but are totally dissimilar from a performance or application standpoint. For the purposes of this discussion, the materials are classified as follows:

1. Synthesized hydrocarbons
2. Organic esters
3. Polyglycols
4. Phosphate esters
5. Other synthetic lubricating fluids

The first four classes account for over 90 percent of the volume of synthetic lubricant bases now used. The fifth class includes a number of materials, some of which are quite high in cost, that are generally used in somewhat lower volumes.

SYNTHESIZED HYDROCARBONS

These materials are probably the fastest growing type of synthetic lubricant base stocks. They are pure hydrocarbons, manufactured from raw materials derived from crude oil. Three types are used in considerable volume: olefin

Table 4-2 Advantages and Limiting Properties of Synthetic Base Stocks

Synthetic	Advantages vs. Mineral Oil	Limiting Properties
SHF	High Temperature Stability Long Life Low Temperature Fluidity High Viscosity Index Improved Wear Protection Low Volatility, Oil Economy Compatibility with Mineral Oils and Paints No Wax	Solvency/Detergency* Seal Compatibility*
Organic Esters	High Temperature Stability Long Life Low Temperature Fluidity Solvency/Detergency	Seal Compatibility* Mineral Oil Compatibility* Antirust* Antiwear and Extreme Pressure* Hydrolytic Stability Paint Compatibility
Phosphate Esters	Fire Resistant Lubricating Ability	Seal Compatibility Low Viscosity Index Paint Compatibility Metal Corrosion* Hydrolytic Stability
Polyglycols	Water Versatility High Viscosity Index Low Temperature Fluidity Antirust No Wax	Mineral Oil Compatibility Paint Compatibility Oxidation Stability*

*Limiting properties of synthetic base fluids which can be overcome by formulation chemistry.

oligomers, alkylated aromatics, and polybutenes. A fourth type, cycloaliphatics, is used in small volumes in specialized applications.

Olefin Oligomers These materials, also called poly-α olefins, are formed by combining a low molecular weight material, usually ethylene, into a specific olefin. This olefin is oligomerized into a lubricating oil type material, and then hydrogen stabilized. In this sense, oligomerization is a polymerization process in which a few, usually three to ten, of the basic building block molecules are combined to form the finished material. Therefore, the product may be formed with varying molecular weights and attendant viscosities to meet a broad range of requirements. Polymerization in the usual sense involves condensation of the same basic building blocks (monomers). The structure of a typical olefin oligomer base oil molecule is shown in Fig. 4-2.

Olefin oligomers can be considered to be a special type of paraffinic mineral oil, comparable in properties to the best components found in petroleum derived base oils. They have high viscosity indexes, usually above 135, excellent low temperature fluidity, and very low pour points. Their shear stability is excellent, as is their hydrolytic stability. Due to the saturated nature of the hydrocarbons, both their oxidation and thermal stability are good. Volatility is lower than with comparably viscous mineral oils; thus, evaporation loss at elevated temperatures is lower. Olefin oligomers have low solubility for ammonia and Refrigerant 22 (R22).* In many applications, it is important that olefin oligomers are similar in composition to mineral oils and that they are compatible with mineral oils, additive systems developed for use in mineral oils, and machines designed to operate on mineral oils. The olefin oligomers do not cause any softening or swelling of typical seal materials, which may be a disadvantage in systems where slight swelling of the seals is desirable to keep them tight and prevent leakage. However, proper formulation of finished lubricants can overcome this problem.

Application   Olefin oligomers are used widely as automotive lubricants. They are often combined with one of the organic esters as the base fluid in engine oils, gear oils, and hydraulic fluids; for equipment operating in extremely cold climates; and for premium oils for the service station market in temperate climates.

In industrial applications, olefin oligomers may be combined with an organic ester as the base fluid in high temperature gear and bearing oils and as lubricants for small land based gas turbines. They are also formulated as wide temperature range hydraulic fluids, refrigeration compressor oils for severe service (particularly low temperature ammonia systems and screw compressors operating on R22), power transmission fluids, and heat transfer fluids. Wide temperature range greases made from an olefin oligomer combined with an inorganic thickener are finding increasing acceptance as long life rolling element bearing greases for severe duty applications.

*This designation, R22, is part of a nomenclature system for refrigerants standardized by the American Society for Refrigerating Engineers.

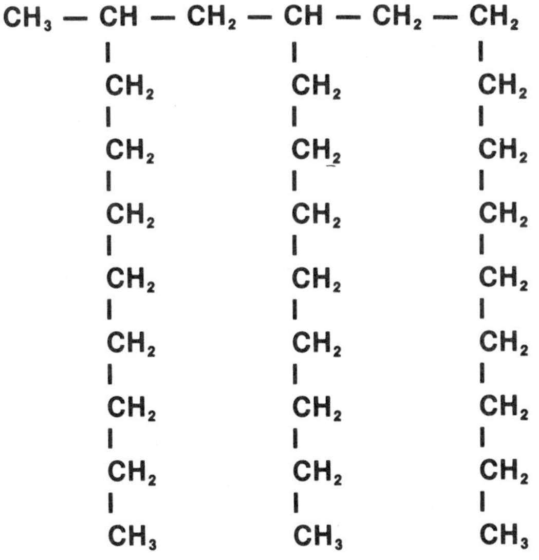

$$CH_3 - CH - CH_2 - CH - CH_2 - CH_2$$

Fig. 4-2 **Olefin Oligomer** This compound, an oligomer of 1-decene, is a low viscosity oil suitable for blending low viscosity engine oils, as well as a variety of other products.

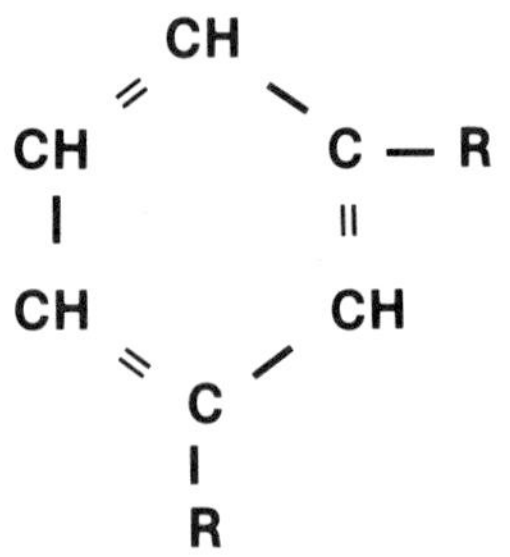

CH
CH   C — R
CH   CH
C
R

Fig. 4-3 Alkylated Aromatic This structure represents dialkylated benzene. The R is an alkyl group which usually contains from 10 to 14 carbon atoms for products in the lubricating oil range.

In commercial aviation applications, greases formulated from olefin oligomers and inorganic thickener are widely used as general purpose aircraft greases. These greases are also used for military aircraft applications, as are less flammable hydraulic fluids formulated with olefin oligomer base fluids.

Alkylated Aromatics These synthesized hydrocarbons are formed by the alkylation of an aromatic compound, usually benzene. The alkylation process involves the joining to the aromatic molecule of substituent alkyl groups. Generally, the alkyl groups used contain from 10 to 14 carbon atoms and are of normal paraffinic configuration. The properties of the final product can be altered by changing the structure and position of the alkyl groups. Dialkylated benzene, a typical alkylated aromatic in the lubricating oil range, is shown in Figure 4-3.

Alkylated aromatics have excellent low temperature fluidity and low pour points. Their viscosity indexes (VI) are about the same as, or slightly higher than, high VI mineral oils. They are less volatile than comparably viscous mineral oils, and more stable to oxidation, high temperatures, and hydrolysis. However, it is more difficult to incorporate inhibitors, and the lubrication properties of specific structures may be poor. As with the olefin oligomers, the alkylated aromatics are compatible with mineral oils and systems designed for mineral oils.

Application Alkylated aromatics are used as the base fluid in engine oils, gear oils, hydraulic fluids, and greases in subzero applications. They also are used as the base fluid in power transmission fluids and gas turbine, air compressor, and refrigeration compressor lubricants.

Polybutenes Polybutenes are not always considered in discussions of synthesized lubricant bases, although there are a number of lubricant related applications for the materials. Polybutenes are produced by controlled polymerization of

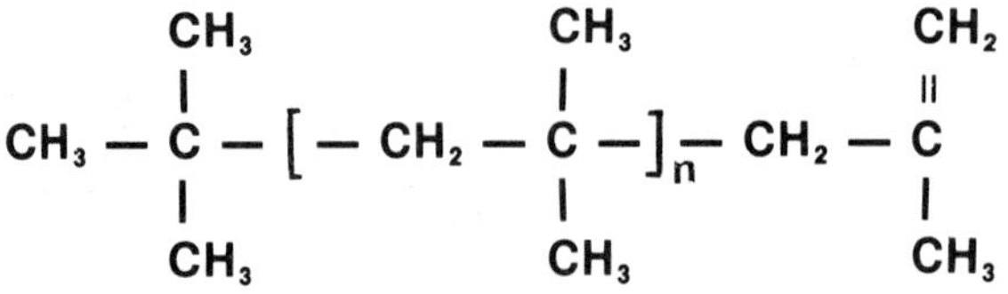

$$CH_3 - \underset{\underset{CH_3}{|}}{\overset{\overset{CH_3}{|}}{C}} - [\, - CH_2 - \underset{\underset{CH_3}{|}}{\overset{\overset{CH_3}{|}}{C}} -]_n - CH_2 - \underset{\underset{CH_3}{|}}{\overset{\overset{CH_2}{\|}}{C}}$$

Fig. 4-4 Polybutene This is a typical molecular structure found in polybutenes. Viscosity is a function of the factor n, higher values giving higher viscosities.

butenes and isobutene (isobutylene). A typical structure is shown in Fig. 4-4. The lower molecular weight materials produced by this process have lubricating properties, while higher molecular weight materials, usually referred to as polyisobutylenes, are used as VI improvers and thickeners.

Polybutenes in the lubricating oil range have viscosity indexes between 70 and 110 and fair lubricating properties and can be manufactured to have excellent dielectric properties. An important characteristic in a number of applications is that, above their decomposition temperature, which is approximately 550°F (288°C), the products decompose completely to gaseous materials.

Application   The major use of polybutenes in the lubricant related field is as electrical insulating oils. They are used as cable oils in high voltage underground cables, as impregnants for insulating paper for cables, as liquid dielectrics, and as impregnants for capacitors. Significant volumes are also used as lubricants for rolling, drawing, and extrusion of aluminum when the aluminum is to be annealed afterward. Other applications include gas compressor lubrication and as carriers for solid lubricants.

Cycloaliphatics   These comparatively new materials are now used in small quantities because of certain special properties they possess. One typical structure is shown in Fig. 4-5. Cycloaliphatics are sometimes referred to as traction fluids. Under high stresses they develop a glasslike structure and can transmit shear forces; that is, they have high traction coefficients. At the same time, they perform somewhat of a lubricating function in that they prevent welding and metal transfer from one surface to the other. Their stability is excellent under these conditions.

Application   The main application for the cycloaliphatics at present is in stepless, variable speed drives in which the torque is transmitted from the driving member to the driven member by the resistance to shear of the lubricating fluid. The high traction coefficients of the cycloaliphatics permit higher power ratings than do conventional lubricants.

Also, cycloaliphatics have also found some application in rolling element bearings where, due to speed and load conditions, skidding of the rolling elements may occur with conventional lubricants.

Fig. 4-5   Typical Cycloaliphatic Structure

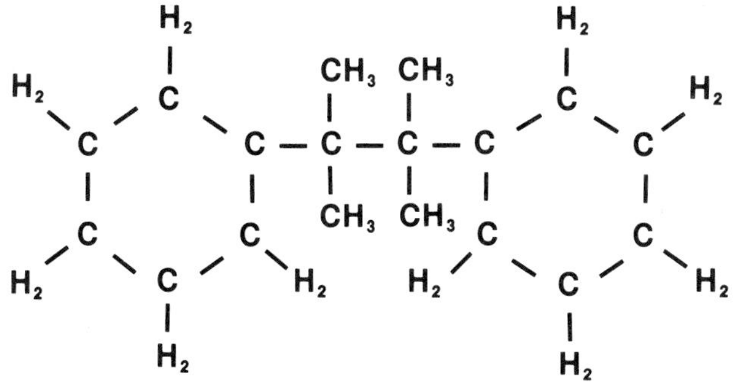

ORGANIC ESTERS

Organic esters have been an important class of synthesized base fluids longer than any of the other materials now in use. Their use dates from World War II in Germany when they were used in mineral oil blends to improve low temperature properties, and to supplement scarce supplies of mineral oils. They were first used as aircraft jet engine lubricants in the 1950s, and are now used as the base for essentially all aircraft jet engine lubricants. They are also used as the base fluid in many wide temperature range aircraft greases. Two types of organic esters are used — dibasic acid esters and polyol esters.

Dibasic Acid Esters These esters, sometimes called diesters, are prepared by reacting a dibasic acid with an alcohol containing one reactive hydroxyl group (a monohydric alcohol). As shown in Fig. 4-6, the backbone of the structure is formed by the

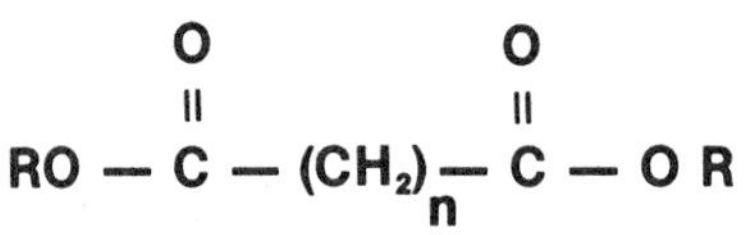

$$RO - \underset{\displaystyle \overset{\displaystyle O}{\|}}{C} - (CH_2)_n - \underset{\displaystyle \overset{\displaystyle O}{\|}}{C} - O\,R$$

Fig. 4-6 Diabasic Acid Ester Commonly used acids are adipic ($n = 4$), azelaic ($n = 7$) and sebacic ($n = 8$). Alcohols used (R) include 2-ethylhexyl (C_8), trimethylhexyl (C_9), isodecyl (C_{10}), and tridecyl (C_{13}).

acid, with the alcohol radicals joined to its ends. The physical properties of the final product can be varied by using different alcohols or acids. Acids that are usually used are adipic, azelaic, and sebacic. Among the commonly used alcohols are 2-ethylhexyl, trimethylhexyl, isodecyl, and tridecyl.

Dibasic acid esters have excellent low temperature fluidity and very low pour points. Viscosity indexes are usually high, some above 140, and the products are shear stable. Generally, their hydrolytic stability is not as good as mineral oils. Diesters have good lubricating properties, good thermal and oxidation stability, and lower volatility than comparably viscous mineral oils. They also have the ability to suspend potential deposit forming materials so that hot metal surfaces in contact with them will remain relatively clean. On the negative side, diesters cause somewhat more seal swelling than mineral oils. Also, they may not have good solubility for additives developed for use in mineral oils, although good results are obtained with additives developed especially for these fluids. They may also affect paints and finishes more than mineral oils.

Application Dibasic acid esters are used as the base for Type I jet engine oils. Generally, the use of these oils is now mainly restricted to older military jet engines, and some jet engines in industrial service. Diesters are used as the base oil, or as a component of the base oil, for automotive engine oils and air compressor lubricants. They are also used as the base fluid in low temperature aircraft greases, and some relatively wide temperature range aircraft greases.

Polyol Esters These esters are formed by reacting an alcohol with two or more hydroxyl groups, a polyhydric alcohol, with a monobasic acid. In contrast to the dies-

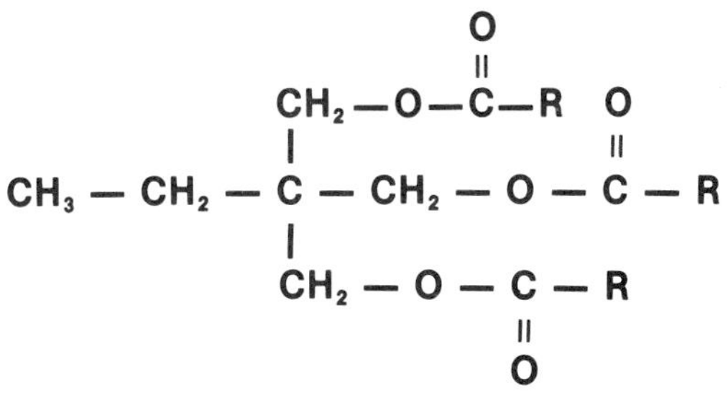

$$
\begin{array}{c}
\overset{\textstyle O}{\overset{\|}{}} \\
CH_2\!-\!O\!-\!C\!-\!R \quad O \\
| \qquad\qquad \| \\
CH_3 \!-\! CH_2 \!-\! C \!-\! CH_2 \!-\! O \!-\! C \!-\! R \\
| \\
CH_2 \!-\! O \!-\! C \!-\! R \\
\underset{\textstyle O}{\underset{\|}{}}
\end{array}
$$

Trimethylol Propane Ester

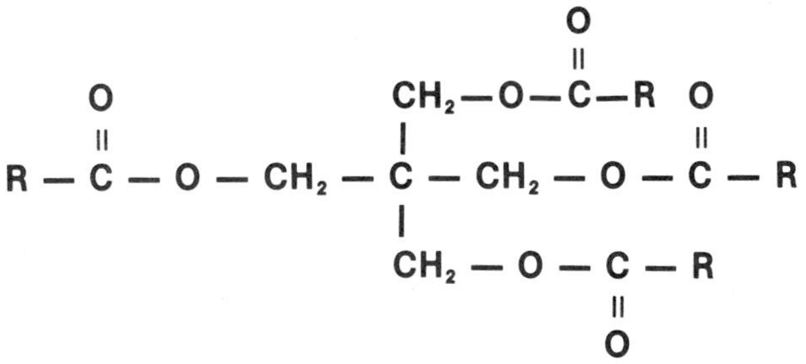

$$
\begin{array}{c}
\overset{\textstyle O}{\overset{\|}{}} \\
O CH_2\!-\!O\!-\!C\!-\!R \quad O \\
\| | \qquad\qquad \| \\
R \!-\! C \!-\! O \!-\! CH_2 \!-\! C \!-\! CH_2 \!-\! O \!-\! C \!-\! R \\
| \\
CH_2 \!-\! O \!-\! C \!-\! R \\
\underset{\textstyle O}{\underset{\|}{}}
\end{array}
$$

Pentaerithritol Ester

Fig. 4-7 Typical Polyol Esters Typical acid radicals (R) used contain from six to ten carbons, with those containing an odd number of carbon atoms being generally favored.

ters, as shown in Fig. 4-7, in the polyol esters the polyol forms the backbone of the structure with the acid radicals attached to it.

As with diesters, the physical properties of polyol esters may be varied by using different polyols or acids. Trimethylol propane and pentaerithritol are two of the polyols that are commonly used. Usually, the acids used are obtained from animal or vegetable oils and are those containing from five to ten carbon atoms.

Polyol esters have better high temperature stability than the diesters. Their low temperature properties and hydrolytic stability are about the same, but their viscosity indexes may be lower. Their volatility is equal or lower. The polyol esters also have somewhat more effect on paints and finishes, and cause somewhat more seal swelling.

Application Type II, or second generation, jet engine oils that are now used almost universally in commercial jet aircraft and the majority of later model military aircraft are formulated with polyol ester base oils. These same lubricants are used in many industrial and marine gas turbines that use aircraft jet engines as gas generators.

Polyol esters are also used as air compressor oil bases, and as components in ester synthesized hydrocarbon blends for automotive engine oils. Additional quantities are used in low temperature gear oils, instrument oils, and as the base fluid in wide temperature range greases.

POLYGLYCOLS

Polyglycols cover a wide range of products and properties. Presently, they are the largest single class of synthetic lubricant bases.

Polyglycols are variously described as polyalkylene glycols, polyethers, polyglycol ethers, and polyalkylene glycol ethers. The latter term is the most complete and accurate for the bulk of the materials used in lubricants. Small quantities of simple glycols, such as ethylene and polyethylene glycol, are also used as hydraulic brake fluids. Typical structures for the two types are shown in Fig. 4-8.

Polyglycols are available in water soluble and water insoluble types. Both types are used, depending on the application.

One of the major advantages of polyglycols is that they decompose completely to volatile compounds under high temperature oxidizing conditions. This results in low sludge buildup under moderate to high operating temperatures, or complete decomposition without leaving deposits in certain extremely hot applications.

Polyglycols have good viscosity-temperature characteristics, although at low temperatures they tend to become somewhat more viscous than some of the other synthesized bases. Pour points are relatively low. High temperature stability ranges from fair to good and may be improved with additives. Thermal conductivity is high. Polyglycols are not compatible with mineral oils or additives developed for use in mineral oils, and may have considerable effect on paints and finishes. They have low solubility for hydrocarbon gases and some refrigerants. Seal swelling is low, but with the water soluble types some care must be exercised in seal selection to be sure that the seals are compatible with water. Even if the glycol fluid does not initially contain any water, it has a tendency to pick up moisture from the atmosphere.

Application The applications for the polyglycols are divided into those for the water soluble types and those for the water insoluble types.

The largest volume application of the water soluble polyglycols is in hydraulic brake fluids. Other major applications are in metal working lubricants, where they can be removed by water flushing or being burned off, and in fire resistant hydraulic fluids. In the latter application, the polyglycol is mixed with water, which provides the fire resistance. Water soluble polyglycols are also used in the preparation of water diluted lubricants for rubber bearings and joints.

Water insoluble polyglycols are used as heat transfer fluids and as the base fluid in certain types of industrial hydraulic fluids and high temperature and bearing oils. They are also finding application as lubricants for screw

Fig. 4-8 Glycols Ethylene glycol is the simplest glycol. When it is polymerized the oxygen bond is formed and it becomes a polyglycol ether.

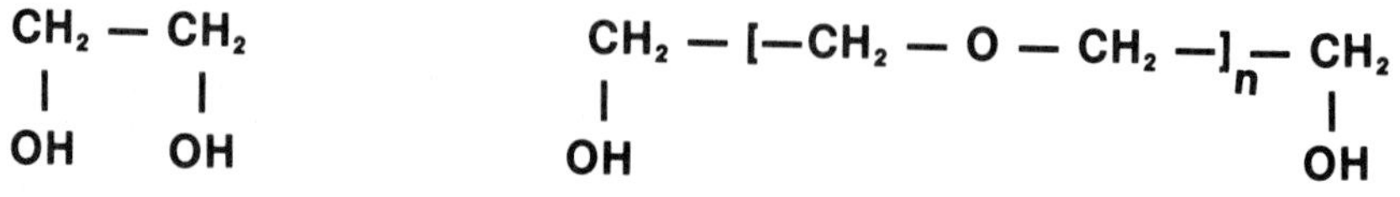

$$CH_2 - CH_2$$
$$|\qquad |$$
$$OH\quad OH$$

Ethylene Glycol

$$CH_2 - [-CH_2 - O - CH_2 -]_n - CH_2$$
$$|\qquad\qquad\qquad\qquad\qquad\qquad |$$
$$OH\qquad\qquad\qquad\qquad\qquad\qquad OH$$

Polyethylene Glycol Ether

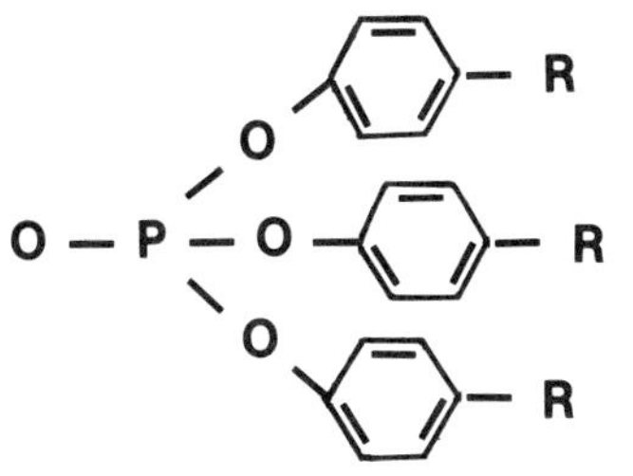

Fig. 4-9 Phosphate Ester The R group can be either an aryl or an alkyl type. If, for example, the methyl group (CH_3^-) is used, the ester is tricresyl phosphate.

type refrigeration compressors operating on R12 and hydrocarbon gases, and screw type compressors handling hydrocarbon gases. In these latter applications, the low solubility of the gases in the polyglycol minimizes the dilution effect, contributing to better high temperature lubrication.

PHOSPHATE ESTERS

Phosphate esters are one of the larger volume classes of synthetic base fluids. A typical phosphate ester structure is shown in Fig. 4-9.

One of the major features of the phosphate esters is their fire resistance, which is superior to mineral oils. Their lubricating properties are also generally good. The high temperature stability of phosphate esters is only fair, and the decomposition products can be corrosive. Generally, they have poor viscosity-temperature characteristics, although pour points are reasonably low and their volatility is quite low. Phosphate esters have considerable effect on paints and finishes, and may cause swelling of many seal materials. Their compatibility with mineral oils ranges from poor to good, depending on the ester. Their hydrolytic stability is only fair. They have specific gravities greater than 1, which means that water contamination tends to float rather than settle to the bottom, and pumping losses are high.

Application The major application of phosphate esters is in fire resistant fluids of various types. Hydraulic fluids for commercial aircraft are phosphate ester based, as are many industrial fire resistant hydraulic fluids. These latter fluids are used in applications such as the electrohydraulic control systems of steam turbines and industrial hydraulic systems, where hydraulic fluid leakage might contact a source of ignition. In some cases they may also be used in the turbine bearing lubrication system.

Considerable quantities of phosphate esters are used as lubricants for compressors where discharge temperatures are high, to prevent receiver fires that might occur with conventional lubricants.

Some quantities of phosphate esters are also used in greases and mineral oil blends as wear and friction reducing additives.

OTHER SYNTHETIC LUBRICATING FLUIDS

Brief descriptions and principle applications of some of the other synthetic base fluids follow.

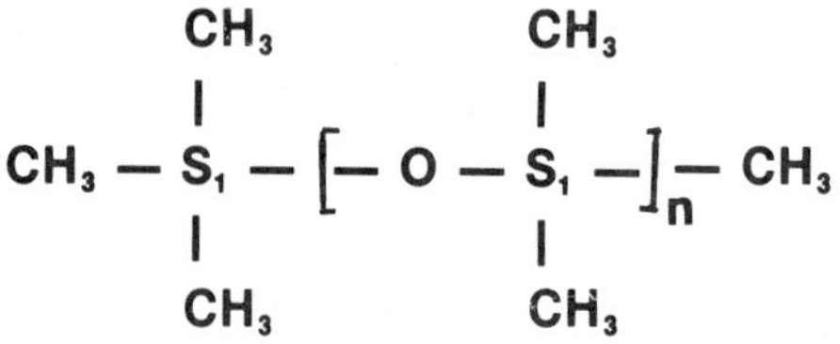

$$CH_3 - S_1 - \left[- O - S_1 - \right]_n - CH_3$$

with CH_3 groups above and below each S_1.

Fig. 4-10   Silicone Polymer   This structure represents dimethyl polysiloxane, one of the more widely used silicone fluids.

Silicones   These are one of the older types of synthetic fluids. As shown in Fig. 4-10, their structure is a polymer type with the carbons in the backbone replaced by silicon.

Silicones have high viscosity indexes, some on the order of 300 or more. Their pour points are low and their low temperature fluidity is good. They are chemically inert, nontoxic, fire resistant, and water repellant, and have low volatility. Seal swelling is low. Compressibility is considerably higher compared to mineral oils. Thermal and oxidation stability of silicones are good up to quite high temperatures. If oxidation does occur, their oxidation products include silicon oxides, which can be abrasive. A major disadvantage of the common silicones is that they have low surface tension that permits extensive spreading on metal surfaces, especially steel. As a result, effective adherent lubricating films are not formed. Unfortunately, the silicones that exhibit this characteristic also show poor response to wear and friction reducing additives. Some newer silicones show promise of overcoming these deficiencies.

Application   Silicones are used as the base fluid in both wide temperature range and high temperature greases. They are also used in specialty greases designed to lubricate elastomeric materials that would be adversely affected by other types of lubricants. Silicones are also used in specialty hydraulic fluids for such applications as liquid springs and torsion dampers where their high compressibility and minimal change in viscosity with temperature are beneficial. They are also being developed for use as hydraulic brake fluids. Some newer silicones are also offered as compressor lubricants.

Silicate Esters   Silicate esters have excellent thermal stability, and with proper inhibitors, show good oxidation stability; see Fig. 4-11. They have excellent viscosity-

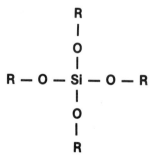

Fig. 4-11   Silicate Ester   The R groups can be either alkyl or aryl groups. The physical properties are dependent on the nature of these groups.

temperature characteristics, and their pour points are low. Their volatility is low and they have fair lubricating properties. A major factor is their resistance to hydrolysis, which is poor.

Application Small quantities of silicate esters are used as heat transfer fluids and dielectric coolants. Some specialty hydraulic fluids are formulated with silicate esters.

Polyphenyl Ethers

These organic materials have excellent high temperature properties and outstanding radiation resistance. They are thermally stable to above 800°F (450°C) and have excellent resistance to oxidation at elevated temperatures. However, they have high viscosities at normal ambient temperatures, which tends to restrict their use.

Application Small quantities of polyphenyl ethers are used as heat transfer fluids, as lubricants for high vacuum pumps, and as the fluid component of radiation resistant greases.

Halogenated Fluids

Chlorine or fluorine or combinations of the two are used to replace part (or all) of the hydrogen in hydrocarbon or other organic structures to form lubricating fluids. Generally, these fluids are chemically inert, essentially nonflammable, and often show excellent resistance to solvents. Some have outstanding thermal and oxidation stability, being completely unreactive even in liquid oxygen. Also, their volatility may be extremely low.

Application Some of the lower cost halogenated hydrocarbons are used alone or in blends with phosphate esters as fire resistant hydraulic fluids. Other halogenated fluids are used for such applications as oxygen compressor lubricants, lubricants for vacuum pumps handling corrosive materials, solvent resistant lubricants, and other lubricant applications where highly corrosive or reactive materials are being handled.

5
Machine Elements

The elements of machines that require lubrication are bearings — plain, rolling element, slides, guides, and ways; gears; cylinders; flexible couplings; chains; cams and cam followers; and wire ropes. These elements have fitted or formed surfaces that move with respect to each other by sliding, rolling, approaching and receding, or by combinations of these motions. If actual contact between surfaces occurs, high frictional forces leading to high temperatures and wear will result. Therefore, the elements are lubricated in order to prevent or reduce the actual contact between surfaces.

Without lubrication, most machines would run for only a short time. With inadequate lubrication, excessive wear is usually the most serious consequence, since a point will be reached, usually after a short period of operation, when the machine elements cannot function and the machine must be taken out of service and repaired. Repair costs — material and labor — may be high, but the cost of the lost production or lost availability of the machine may be by far the greatest cost. With inadequate lubrication, even before failure of elements occurs, frictional forces between surfaces may be so great that drive motors will be overloaded or frictional power losses excessive. Finally, with inadequate lubrication machines will not run smoothly and quietly.

Machine elements are lubricated by interposing and maintaining between moving surfaces, films that minimize actual contact between the surfaces and that shear easily so that the frictional force opposing motion of the surfaces is low.

TYPES OF LUBRICATING FILMS

Lubrication films may be classified as follows:

1. *Fluid films* are thick enough that during normal operation they completely separate surfaces moving relative to each other.
2. *Thin films* are not thick enough to maintain complete separation of the surfaces all the time.
3. *Solid films* are more or less permanently bonded onto the moving surfaces.

Fluid Films

Fluid film lubrication is the most desirable form of lubrication since, during normal operation, the films are thick enough to completely separate the load carrying surfaces. Thus friction is at a practical minimum, being due only to shearing of the lubricant films, and wear does not occur since there is essentially no mechanical contact. Fluid films are formed in three ways:

1. Hydrodynamic film is formed by motion of lubricated surfaces through a convergent zone such that sufficient pressure is developed in the film to maintain separation of the surfaces.
2. Hydrostatic film is formed by pumping fluid under pressure between surfaces that may or may not be moving with respect to each other.
3. Squeeze film is formed by movement of lubricated surfaces toward each other.

Two types of hydrodynamic film lubrication are now recognized. In plain journal bearings and tilting pad or tapered land thrust bearings, what can be best described as thick hydrodynamic films are formed. These films are usually in excess of 0.001 in. (25 μm) thick. In applications, the loads are low enough, and the areas over which the loads are distributed are large enough, that the amount of deformation of the load carrying area is not great enough to significantly alter that area. Load carrying surfaces of this type are often referred to as "conforming," although it is obvious that in the case of tapered land thrust bearings, for example, the surfaces do not conform in the normal concept of the word. However, the term is a convenient opposite for the term "nonconforming," which quite accurately describes the types of surfaces where elastohydrodynamic films are formed.

The surfaces of the balls in a ball bearing theoretically make contact with the raceways at points, rollers of roller bearing make contact with the raceways along lines, and meshing gear teeth also make contact along lines. These types of surface are nonconforming. Under the pressures applied to these elements by the lubricating film, however, the metals deform elastically, expanding the theoretical points or lines of contact into discrete areas.

Since a convergent zone exists immediately before these areas of contact, a lubricant will be drawn into the contact area and can form a hydrodynamic film. This type of film is referred to as an elastohydrodynamic film, the "elasto" part of the term referring to the fact that elastic deformation of the surfaces must occur before the film can be formed, and this type of lubrication is elastohydrodynamic lubrication (EHL). EHL films are very thin, in the order of 10 to 50 μ in. (0.25 – 1.25 μm) thick. However, even with these thin films, complete separation of the contacting surfaces can be obtained.

Any material that will flow at the shear stresses available in the system may be used for fluid film lubrication. In most applications, petroleum derived lubricating oils are used. There are some applications for greases. Some materials not usually considered to be lubricants, such as liquid metals, water, gases, are also used in some applications. The following discussions are concerned mainly with fluid film lubrication with oils and greases, since some rather complex special considerations are involved when other materials are used.

Thick Hydrodynamic Films   The formation of a thick hydrodynamic fluid film that will separate two surfaces and support a load can be described with reference to Fig. 5-1. The two surfaces are submerged in a lubricating fluid. As the upper surface moves, internal friction in the fluid causes it to be drawn into the space between the surfaces. The force drawing the fluid into space A is equal to the force tending to draw it out, but since the cross-sectional area at the outlet section is smaller than at the inlet, the flow of fluid is restricted at the outlet. The moving surface tries to "compress" the fluid to force it through this restricted section with the result that the pressure in the fluid rises. This pressure rise tends to:

Fig. 5-1   Converging Wedge

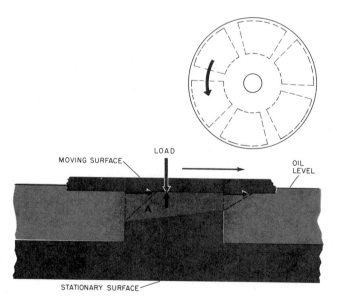

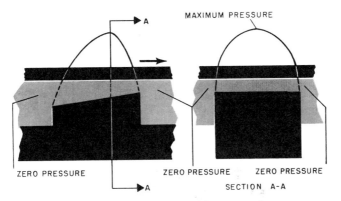

Fig. 5-2 Pressure Distribution in the Wedge Film The left view shows the pressure distribution in the direction of motion, while the right view shows the pressure distribution in the direction normal to motion.

1. Retard the flow of fluid into space A
2. Increase the flow at the restricted outlet section
3. Cause side leakage in the direction normal to the direction of motion

The most important effect, however, is that the pressure in this wedge shaped film enables it to support a load without contact occurring between the two surfaces.

In order for the elements of Fig. 5-1 to work as a bearing, the moving surface would have to be of infinite length. A more practical approach is to distribute a series of wedge shaped sections around the circumference of a disc as shown in the inset. The moving surface then becomes a disc to which load can be applied, and the complete assembly can be used as a thrust bearing.

The pressure distribution in the wedge shaped fluid film of Fig. 5-1 is shown in Fig. 5-2. In the direction of motion (left view), the pressure gradually rises more or less to a peak just ahead of the outlet section and then falls rapidly. In the direction normal to motion (right view), the pressure is at its maximum in the middle of the shoe but, because of side leakage, falls off rapidly toward each side.

In Fig. 5-2, the taper of the shoe and the thickness of the fluid film are greatly exaggerated for illustrative purposes. In a practical bearing of this type, the amount of taper might be on the order of 0.002 in. (50 μm) in a length of 6 in. (150 mm), and the film thickness at the outlet would be on the order of 0.001 in. (25 μm). With proper design and fluid viscosity, there would be no contact between the disc and shoes when the speed is above a certain minimum during normal operation. The bearing would not be frictionless, however, since a force would have to be applied to the disc to overcome the resistance to shear of the fluid films.

The same principles outlined in Fig. 5-2 can be used to generate a fluid film which will lift and support a shaft or journal in a bearing. If the moving surface and tapered shoe of Fig. 5-2 are "rolled up," the journal and bearing of Fig. 5-3 will result. A journal turning at a suitable speed will then draw fluid into the space between the journal and bearing and develop a load sup-

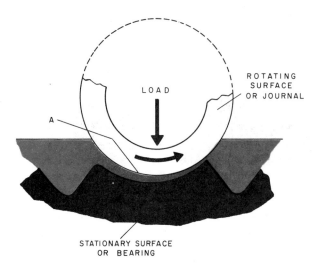

Fig. 5-3   Partial Journal Bearing

porting film. This bearing is called a partial bearing and is suitable for some applications where the load on the journal is always in one direction. However, bearings are more commonly of the full 360 degree type as shown in Fig. 5-4 to provide restraint in the event of load variations and to permit easier enclosure and sealing.

The stages in the formation of a hydrodynamic film in a full journal bearing are illustrated in Fig. 5-4. First, the machine is at rest with the oil shut off, and the oil has leaked from the clearance space. Metal-to-metal contact exists between the journal and bearing. When the machine has been started (left) and the oil supply turned on, filling the clearance space, the shaft begins to rotate counterclockwise and friction is momentarily high so the shaft tends to climb the left side of the bearing. As it does this, it rolls onto a thicker oil film so that friction is reduced and the tendency to climb is balanced by the tendency to slip back. As the journal gains speed (center),

Fig. 5-4   Development of Hydrodynamic Film in a Full Journal Bearing with Downward Load

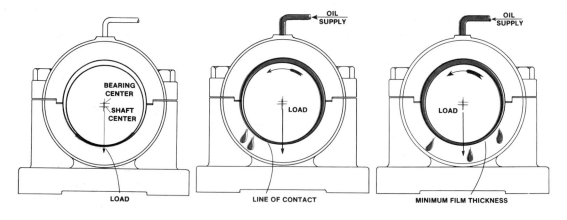

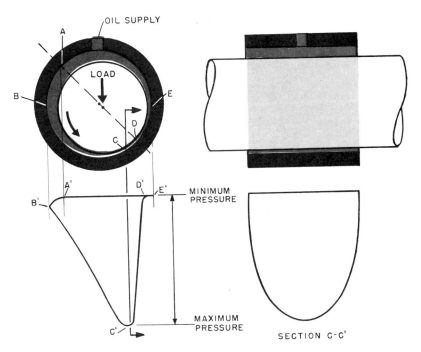

Fig. 5-5 Pressure Distribution in a Full Journal Bearing In the left view, from C the point of minimum film thickness, to D the pressure drops rapidly to the minimum which may be atmospheric, slightly above atmospheric, or slightly below atmospheric depending on such factors as bearing dimensions, speed, load, oil viscosity, and supply pressure. From A to C the pressure increases as the clearance decreases. Pressure distribution along the line of minimum film thickness is shown at the right. From a maximum value near the center, it drops rapidly to zero at the ends because of end leakage.

drawing more oil through the wedge shaped space between it and the bearing, pressure is developed in the fluid in the lower left portion of the bearing that lifts the journal and pushes it to the right. At full speed (right), the converging wedge has moved under the journal, which is supported on a relatively thick oil film with its minimum thickness at the lower right, on the opposite side to the starting position shown (left). Under steady conditions the upward force developed in the oil film just equals the total downward load, and the journal is supported in the slightly eccentric position shown. Variations in speed, load, and oil viscosity will cause changes in this eccentricity, and also in the minimum film thickness. The pressures developed in the oil film are illustrated in Fig. 5-5.

Thick hydrodynamic film lubrication is used in journal and thrust bearings for many applications. Characteristics and lubricant requirements of various applications are discussed more fully in later sections.

Elastohydrodynamic Films As the pressure on an oil increases, the viscosity increases (Fig. 5-6). Thus, as the lubricant is carried into the convergent zone approaching the contact area, elastic deformation of two surfaces such as a ball and its raceway or a roller and its raceway occurs due to the pres-

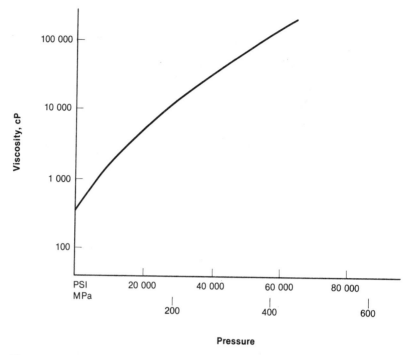

Fig. 5-6 Effect of Pressure on Viscosity

sure of the lubricant. As the viscosity increases, the pressure is further increased. This hydrodynamic pressure developed in the lubricant is sufficient to separate the surfaces at the leading edge of the contact area, and as the lubricant is drawn into the contact area the pressure on it increases further together with the viscosity. Due to this high viscosity and the short time required for the lubricant to be carried through the contact area, the lubricant cannot escape and separation of the surfaces can be achieved.

As noted, EHL films are very thin, on the order of about 10 to 50 μin. (0.25–1.25 μm). The thickness of the film increases as the speed is increased, the lubricant viscosity is increased, the load is decreased, or the geometric conformity of the mating surfaces is improved. Load has comparatively little effect on the film thickness because at the pressures in the contact area, the oil film is actually more rigid than the metal surfaces. Thus an increase in load, for example, mainly has the effect of deforming the metal surfaces more and increasing the area of contact rather than decreasing the film thickness.*

If the surfaces shown in Fig. 5-7 were rougher, that is if the asperities were higher, the opposing surface asperities would be more likely to make contact through the film. Thus the relationship between the roughnesses of the surfaces and the film thickness is important in the consideration of elas-

---

*There is some evidence that at high contact pressures, above 150,000 psi (1.0 GPa), load may have an increased effect on film thickness. The reasons for this are not yet fully defined.

tohydrodynamic lubrication. This had led to the introduction of the quantity specific film thickness (λ=lambda) in EHL considerations. The specific film thickness is the ratio of the film thickness (h) to the composite roughness (σ) of the two surfaces:

$$\lambda = h/\sigma$$

The composite roughness of the two surfaces can be calculated as the square root of the sum of the squares of the individual surface roughnesses:

$$\sigma = \sqrt{\sigma_1^2 + \sigma_2^2}$$

where σ_1 and σ_2 are the root mean square (RMS) roughness values of the two surfaces.*

The specific film thickness is related to the ability of the lubricating film to prevent or minimize wear or scuffing. Fig. 5-8 shows the effect of λ on the L_{10} fatigue life† of a series of cylindrical roller bearings. For values of λ of approximately 3 to 4, the L_{10} fatigue life is relatively constant, so is essentially independent of the oil film thickness. This region is labeled "full EHL" and corresponds to full fluid film conditions that exist in thick hydrodynamic film lubrication. In the region labeled "partial EHL," the film becomes progressively thinner, and more and more surface asperities penetrate the film.

*A number of calculations are given in the literature for the composite roughness of using centerline average (CLA) values. The relationship between RMS and CLA values is:

$$\sigma_{RMS} = 1.3\sigma_{CLA}$$

†The L_{10} fatigue life is the average time to failure of 10 percent of a group of identical bearings operating under the same conditions.

Fig. 5-7 Schematic of Contact Area between Ball and Inner Race

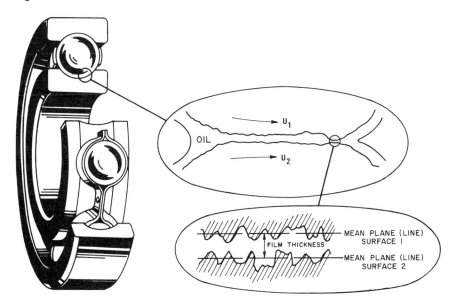

Fatigue life is decreased accordingly. This region is one of mixed film lubrication. When λ is below about 1 boundary lubrication prevails and the surfaces are in contact nearly all the time, fatigue failure is generally rapid unless lubricants with effective antiwear additives are used.

These ranges for λ are considered to be applicable to rolling element bearings. As will be discussed later, somewhat lower values of the calculated safe minimum film thickness may be applicable to gears.

Increased understanding of EHL film characteristics and the use of specific film thickness have led to improved understanding of the lubricant function in many applications where unit loads are high and to improved methods of selecting lubricants for such applications.

Hydrostatic Films In hydrostatic film lubrication, the pressure in the fluid film that lifts and supports the load is provided from an external source. Thus, relative motion between opposing surfaces is not required to create and maintain the fluid film. The principle is used in plain and flat bearings of various types where it offers such advantages as low friction at very low speeds or when there is no relative motion, more accurate centering of a journal in its bearing, and freedom from stick slip effects.

The simplest type of hydrostatic bearing is illustrated in Fig. 5-9. Oil under pressure is supplied to the recess or pocket. If the supply pressure is sufficient, the load will be lifted and floated on a fluid film. At equilibrium conditions, the pressure across the pad will vary approximately as shown. The total force developed by the pressure in the pocket and across the lands will be such that the total upward force is equal to the applied load.

Fig. 5-8 L10 Fatigue Life Relative to Book Value as a Function of Specific Film Thickness The data are for a series of cylindrical roller bearings.

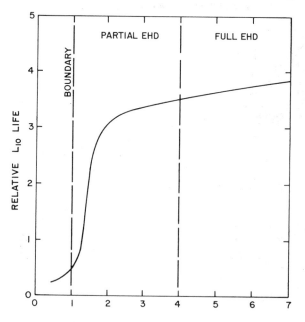

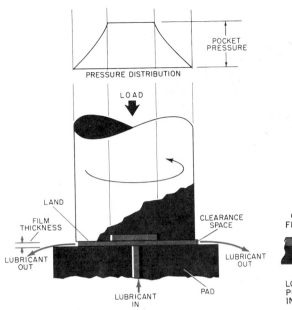

Fig. 5-9   Simple Hydrostatic Thrust Bearing

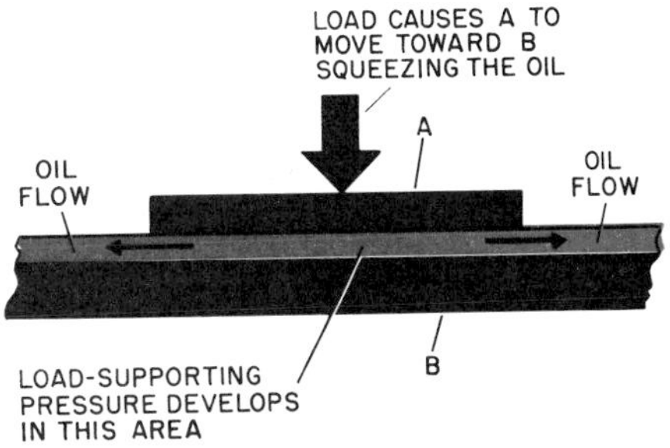

Fig. 5-10   Squeeze Film Principle

The clearance space and the oil film thickness will be such that all the oil supplied to the bearing can flow through the clearance spaces under the pressure conditions prevailing.

Squeeze Films   As discussed under EHL films, as the applied pressure on an oil increases, the viscosity of the oil increases. This fact contributes to the formation of what are called squeeze films.

The principle of the squeeze film is shown in Fig. 5-10, where application of a load causes plate A to move toward stationary plate B. As pressure develops in the oil layer, the oil starts to flow away from the area. However, the increase of pressure also causes an increase in the oil viscosity so the oil cannot escape as rapidly and a heavy load can be supported for a short time. Sooner or later, if load continues to be applied, all the oil will flow or be forced from between the surfaces and metal-to-metal contact will occur, but for short periods such a lubricating film can support very heavy loads.

One application where squeeze films are formed is in piston pin bushings (Fig. 5-11). At the left, the load is downward on the pin and the squeeze film develops at the bottom. Before the film is squeezed so thin that contact can occur, the load reverses (right view) and the squeeze film develops at the top. The bearing oscillates with respect to the pin, but this motion probably does not contribute much to film formation by hydrodynamic action. Nevertheless, bearings of this type have high load carrying capacity.

Thin Films   A copious, continuous, supply of lubricant is necessary to maintain fluid films. In many cases, it is not practical or possible to provide such an

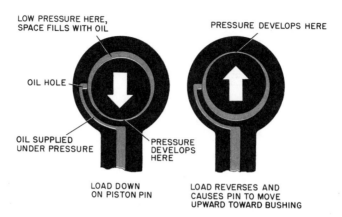

Fig. 5-11 Squeeze Film in Piston Pin Bushing

amount of lubricant to machine elements. In other cases, as for example during starting of a hydrodynamic film bearing, loads and speeds are such that fluid films cannot be maintained. Under these conditions, lubrication is by what are called thin films.

When surfaces run together under thin film conditions, enough oil is often present so that part of the load is carried by fluid films and part is carried by contact between the surfaces. This condition is often called *mixed film lubrication*. With less oil present, or with higher loads, a point is reached where fluid oil plays little or no part. This condition is often called *boundary lubrication*.

Nature of Surfaces   Machined surfaces are not smooth but are wavy because of inherent characteristics of the machine tools used. Minute roughnesses — "hills and valleys" — are superimposed on these waves, somewhat as shown in Fig. 5-12. The surfaces are not clean even when freshly machined, but are coated with incidental films — moisture, oxides, cutting fluids, rust preventives — and in service with lubricating films. These films must be pierced or rubbed away before contact between clean surfaces can occur.

Surface Contact   When rubbing contact is made between the surface peaks, known as asperities, a number of actions take place, as shown in Fig. 5-13, which represents a highly magnified contact area of a bearing and journal. These actions are:

Fig. 5-12 Magnified Surface Profile

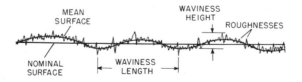

1. There is heavy rubbing as at A; surface films are sheared and elastic or plastic deformation occurs. Real (compared to apparent) areas of contact are extremely small and unit stresses will be very high.
2. The harder shaft material plows through the softer bearing material B, breaking off wear particles and creating new roughnesses.
3. Some areas C are rubbed or sheared clean and the clean surfaces weld together. The minute welds break immediately as motion continues, but depending on the strength of the welds may break at another section so that metal is transferred from one member to the other. New roughnesses are formed, some to be plowed off to form wear particles.

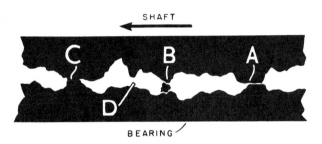

Fig. 5-13   Actions Involved in Surface Contacts

These actions account for friction and wear, boundary lubrication conditions. Under mixed film conditions, the force necessary to shear the oil films D also is part of the frictional force. However, the presence of partial oil films reduces friction and wear that result from the other actions.

The actions involved in rubbing contact result in the development of very high temperatures in the minute areas of contact. These local temperatures, which are of short duration, are known as flash temperatures. With light loads and low speeds, the small total amount of heat is conducted away through the surfaces without excessive rise in the general temperatures of the surfaces. However, with higher loads or speeds, or both, the number of contacts in a given area may increase to the point where the heat cannot be dissipated fast enough and the surfaces may run hot.

Solid Films   In many applications, oils or greases cannot be used, either because of difficulties in applying them, sealing problems, or other factors such as environmental conditions. A number of more or less permanently bonded lubricating films have been developed to reduce friction and wear in applications of this type.

The simplest type of solid lubricating film is formed when a low friction, solid lubricant such as molybdenum disulfide is suspended in a carrier and applied more or less in the manner of a normal lubricant. The carrier may be a volatile solvent, a grease, or any of several other types of materials. After the carrier is squeezed or evaporates from the surfaces, a layer of molybdenum disulfide remains to provide lubrication.

Solid lubricants are also bonded to rubbing surfaces with various types of resins, which cure to form tightly adhering coatings with good frictional properties. In the case of some plastic bearings, the solid lubricant is sometimes incorporated in the plastic. During operation, some of the solid lubricant may then be transferred to form a lubricating coating on the mating surface.

In addition to molybdenum disulfide, polytetrafluoroethylene (PTFE), polyethylene, and a number of other materials are used to form solid films. In some cases, combinations of several materials, each contributing some characteristics to the film, are used.

PLAIN BEARINGS

The simplest types of plain bearing are shown in Fig. 5-14. The bearing may be only a hole in a block (left), it may be split to facilitate assembly (center), or in some cases where the load to be carried is always in one direction, the bearing may consist of only a segment of a block (right). The part of the shaft within a bearing is called the journal, and plain bearings are often called journal bearings.

Plain bearings are designed for either fluid film lubrication or thin film lubrication. Most fluid film bearings are designed for hydrodynamic lubrication, but increasing numbers of bearings for special applications are being designed for hydrostatic lubrication.

Hydrodynamic Lubrication The primary requirement for hydrodynamic lubrication is that sufficient oil be present at all times to flood the clearance spaces.

The oil wedge formed in a hydrodynamic bearing is a function of speed, load, and oil viscosity. Under fluid film conditions, an increase in viscosity or speed increases the oil film thickness and coefficient of friction, while an increase in load decreases them. The separate consideration of these effects presents a complex picture that is simplified by combining viscosity Z,

Fig. 5-14 Elementary Plain Bearings

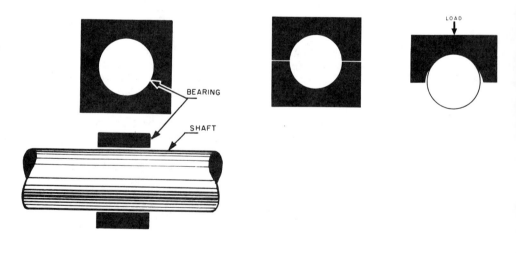

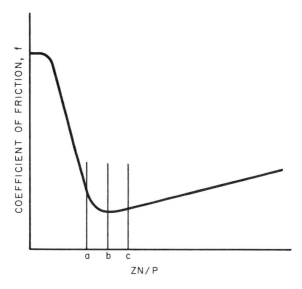

Fig. 5-15 Effect of Viscosity, Speed, and Load on Bearing Friction For each hydro-dynamic film bearing, there is a characteristic relation, such as that shown, between the coefficient of friction and the factor combining viscosity (Z), speed (N), and load (P).

speed N, and unit load P, into a single dimensionless factor called the ZN/P factor.* Although no simple equation can be offered that expresses the coefficient of friction f in terms of ZN/P, the relationship can be shown by a curve such as that in Fig. 5-15. A similar type curve could be developed experimentally for any fluid film bearing.

In Fig. 5-15, in the zone to the right of c, fluid film lubrication exists. To the left of a, boundary lubrication exists. In this latter zone, conditions are such that a full fluid film cannot be formed, some metallic friction and wear commonly occur, and very high coefficients of friction may be reached.

The portion of the curve between points a and c is a mixed film zone including the minimum value of f corresponding to the ZN/P value indicated by b. From the point of view of low friction, it would be desirable to operate with ZN/P between b and c, but in this zone any slight disturbance such as a momentary shock load or reduction in speed might result in film rupture. Consequently, good practice is to design† with a reasonable factor of safety so that the operating value of ZN/P is in the zone to the right of c. The ratio

*The expression ZN/P is dimensionless when all quantities are in consistent units, for example, Z in poises, N in revolutions per second, and P in dynes per square centimetre; or Z in pascal seconds, N in revolutions per second (reciprocal seconds), and P in pascals; or Z in reyns, N in revolutions per second, and P in pounds-force per square inch.

†Equations, procedures, and data for plain bearing design and performance calculations, are available in many technical papers and books. Among the latter are the following:
Bearing Design and Application, Wilcock and Booser, McGraw-Hill Book Company, N.Y., N.Y.; *Theory and Practice of Lubrication for Engineers*, Fuller, John Wiley & Sons; *Analysis and Lubrication of Bearings*, Shaw and Mack, McGraw-Hill Book Company.

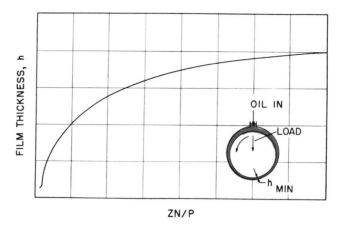

Fig. 5-16 Effect of Viscosity, Speed, and Load on Film Thickness

of the operating ZN/P to the value of ZN/P for the minimum coefficient of friction (point b) is called the bearing safety factor. Common practice is to use a bearing safety factor on the order of 5.

In an operating bearing, if it becomes necessary to increase the speed, ZN/P will increase and it may be necessary to decrease the oil viscosity to keep ZN/P and the coefficient of friction in the design range. An increase in load will result in a decrease in ZN/P, and it may be necessary to increase the oil viscosity to keep ZN/P and the coefficient of friction in the design range.

Film thickness can be related to ZN/P in the manner shown in Fig. 5-16. This curve is typical of large, uniformly loaded, medium speed bearings such as are used in steam turbines. In general, film thickness increases if ZN/P is increased, for example if the load is reduced while the oil viscosity and journal speed remain constant. With a proper bearing safety factor, the film thickness will be such that normal variations in speed, load, and oil viscosity will not result in the film thickness being reduced to the point where metal-to-metal contact would occur.

The work done against fluid friction results in power loss, and the energy involved is converted to heat. Most of the heat is usually carried away by the lubricating oil, but some of it is dissipated by radiation or conduction from the bearing or journal. The operating temperature is the result of a balance between the heat generated, overcoming fluid friction, and the total heat removal.

The effect of increasing temperature is to decrease oil viscosity. The reduction in viscosity results in a lower ZN/P and coefficient of friction. Also, there is less work required to overcome fluid friction, less heat is developed, and there is a tendency for the temperature to decrease. This has a stabilizing influence on bearing temperatures.

In general, if excessive temperatures develop even though load, speed, and oil viscosity are within the correct range, it may be that there is insufficient oil flow for proper cooling. It may then be necessary to provide extra grooving or increase the clearance in order to increase the flow of oil through the bearing.

Grease Lubrication   While the grease in a rolling element bearing acts as a two component system in which the soap serves as a sponge reservoir for the fluid lubricant, greases in plain bearings behave as homogeneous mixtures with unique flow properties. These flow properties are described by the apparent viscosity (see Chap. 3), that is, the observed viscosity under each particular set of shear conditions. As the rate of shear is increased, the apparent viscosity decreases, and at high shear rates approaches the viscosity of the fluid lubricant used in the formulation. In many plain bearings, the shear rate in the direction of rotation is high enough to cause the apparent viscosity of a grease to be in the same general range as the viscosities of lubricating oils normally used for hydrodynamic lubrication. As a result, fluid film formation can occur with a grease, and it is now considered that some grease lubricated plain bearings operate on fluid films, at least part of the time. In addition, hydrodynamic film bearings designed for grease lubrication are now being used in some applications.

The pressure distribution in a grease lubricated hydrodynamic film bearing is similar to that in an oil lubricated bearing (Fig. 5-5). However, toward the ends of the bearing, because of reduced pressure in the film, the shear stress is lower, the apparent viscosity of the grease remains high and end leakage is lower. This results in high pressures being maintained farther out toward the ends of the bearing; the average pressure in the film is higher, and the maximum pressure is correspondingly lower. The minimum film thickness for the same bearing load and speed will be greater. The coefficient of friction may be equal or less than that with an equivalent oil lubricated bearing, depending on such factors as the type of grease used and the viscosity of the oil component in the grease.

Fluid film bearings lubricated with grease have some advantages compared to those lubricated with oil. As a result of the lower end leakage, the amount of lubricant required to be fed to the bearing is lower, so grease lubricated bearings can be supplied by an all loss system with either a slow, continuous feed, or a timed, intermittent feed in conjunction with adequate reservoir capacity in the grooves of the bearing.

When a grease lubricated bearing is shut down for a period of time with the flow of lubricant shut off, the grease usually does not drain or squeeze out completely. Some grease remains on the bearing surfaces so that a fluid film can be established almost immediately when the bearing is restarted. Starting torque and wear during starting may be greatly reduced. During shutdown periods, retained grease also acts as a seal to exclude dirt, dust, water, and other environmental contaminants, and to protect bearing surfaces from rust and corrosion. If the grease provides a lower coefficient of friction, power consumption during operation will also be lower.

A disadvantage of grease lubrication for fluid film bearings is that greases do not provide as effective cooling as oils. This may be partially offset if the coefficient of friction is lower with a grease; however, it may be a limitation if speeds or loads are high.

Hydrostatic Lubrication

In a hydrostatic bearing, the oil feed system used must be such that the pressure available, when distributed across the pocket and land surfaces, is sufficient to support the maximum bearing load that may be applied. The system must also be designed to provide an equilibrium condition for loads below the maximum. Three types of lubricant supply are used to accomplish this—constant volume system, constant pressure system with flow restrictor, and constant pressure system with flow control valve.

Constant Volume System In this type of system, the pump delivers a constant volume of oil at whatever pressure is necessary to force that volume through the system. That is, if the back pressure increases, the pump pressure automatically increases sufficiently to maintain the flow rate. In most cases, the volume delivered by the pump actually decreases somewhat as the pressure increases, but this has relatively little effect on the way the system operates.

A constant volume system must have adequate pressure capability to support any applied load. Referring to Fig. 5-9, when the pump is turned on, oil will flow into the pocket and the pressure will increase until the load is lifted sufficiently to establish a clearance space through which the volume of oil flowing in the system will be discharged. The clearance space and oil film thickness will be a function of both the volume of flow in the system and the applied load.

If the load is then increased, the clearance space and film thickness will decrease, and the pump pressure will have to increase in order to discharge the same volume of oil through the reduced clearance space. Only small changes in clearance space and film thickness accompany fairly large variations in load, so the bearing is said to be very "stiff."

The disadvantage of the constant volume system is that it does not compensate for variations in the point of application of the load in multiple pocket bearings. In the two pocket bearing of Fig. 5-17 using a constant volume

Fig. 5-17 Two Pocket Hydrostatic Thrust Bearing When flow restrictors are used they are installed at the locations marked R.

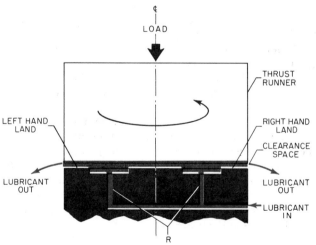

system, if the load is shifted to the right, the runner will tend to tilt. This will decrease the clearance at the right hand land and increase the clearance at the left hand land. Oil can then flow more freely out of the left pocket, the pressure in the system will decrease, and the load will sink until metallic contact might occur at the right side. This problem can be compensated for with either of the following systems.

Constant Pressure System with Flow Restrictor   A constant pressure system requires an accumulator or manifold to maintain the pressure at a relatively constant value. If this constant pressure is applied to the pockets of the bearing (Fig. 5-17) through flow restrictors, such as capillaries or orifices, a compensating force will be developed when the runner tends to tilt. The pressure drop across a flow restrictor increases as the flow through it increases, and decreases as the flow decreases. Thus the pressure in the pocket, being the difference between the system pressure and the pressure drop across the restrictor, decreases as flow increases and increases as flow decreases. With this system, if an off-center load causes the runner to tilt, the clearance at the right hand land will decrease, the flow to that pocket will decrease, the pressure drop across the restrictor will decrease, and the pocket pressure will increase. The opposite effect will occur in the left pocket. This will tend to lift the load over the right hand pocket and let it sink over the left hand pocket, restoring the runner to a more or less coaxial position.

In single pocket bearings, a constant pressure system with a flow restrictor is not as stiff as a constant volume system. As the load is increased with the constant pressure system, the pocket pressure must increase to support it. This requires that the foil flow decreases in order to reduce the pressure drop across the flow restrictor. With less flow and higher pressure, a considerably smaller clearance space will be required to discharge the oil. However, the compensating feature results in wide use of constant pressure systems in both thrust and journal bearings with multiple pockets.

Constant Pressure with Flow Control Valve   If a flow control valve is used as the restrictor in a constant pressure system, the system becomes essentially constant pressure and constant volume. The bearing will have the stiffness of a constant volume system and also have the compensating features of the constant pressure system.

Hydrostatic Bearing Applications   One of the more common applications of the hydrostatic principle is in "oil lifts" for starting heavy rotating machines, such as steam turbines, large motors in steel mills, and rotary ball and rod mills. Because metal-to-metal contact exists between the journal and the bearings when the journal is at rest, extremely high torque may be required to start rotation, and damage to the bearings may occur. By feeding oil under pressure into pockets machined in the bottoms of the bearings, the journal can be lifted and floated on fluid films (Fig. 5-18). The pockets are generally kept small in order not to interfere seriously with hydrodynamic film capacity of the bearings. When the journal reaches a speed sufficient to create hydrodynamic films, the external pressure can be turned off and the bearings will continue to operate in the hydrodynamic manner. The reverse procedure may be used during shutdown.

The low friction characteristics of hydrostatic film bearings at low

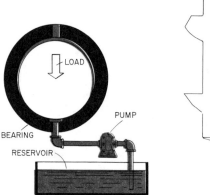
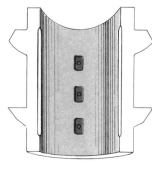

Fig. 5-18 Hydrostatic Lift The view at the right shows one type of shallow pocket through which the oil pressure can be applied.

speeds are being used in a variety of ways. One application is in "friction-less" mounts or pivots for dynamometers. Another is in the bearings for tracking telescopes where the relative motion is extremely slow but must be completely free of stick slip effects. Increasingly, the hydrostatic principle is being applied to the guides and ways of large machine tools, particularly where extremely precise movement and location of the ways is required.

The characteristic of controlled film thickness of hydrostatic film bearings is being used in high speed applications such as machine tool spindles for high precision work. Spindles of this type are equipped with multiple pocket bearings with a constant pressure system and a flow restrictor for each pocket. With this arrangement, any change in the lateral loading on the spindle as a result of a change in the cutting operation is automatically compensated for by changes in the pressures in the individual pockets. Lateral movement of the spindle is thus minimized, and very accurate control of the centering of the spindle in the bearings can be achieved.

Hydrostatic bearings are being used with fluids of low viscosity or poor lubrication characteristics, such as in pumps for liquid metals or water. Also, they are being used for extremely high speed and/or high temperatures where the lubricating fluid is often a compressible fluid such as air.

Thin Film Lubrication Many bearings are designed to operate on restricted lubricant feeds as the most practical and economic approach. The lubricant supplied to the bearings gradually leaks away and is not reused; thus this type of lubrication is generally referred to as "all loss" lubrication. Because of the restricted supply of lubricant, these bearings operate on thin lubricating films, either of the mixed film or boundary type. The simplest type of all loss lubrication is hand oiling (see Fig. 5-19). Hand oiling results in flooded clearances immediately after lubrication. This condition may permit formation of fluid films for a brief period of time; however, the oil quickly leaks away to an amount less than that considered acceptable for safe operation. In short, the bearing passes through the regime of mixed film lubrication and operates much of the time under boundary conditions.

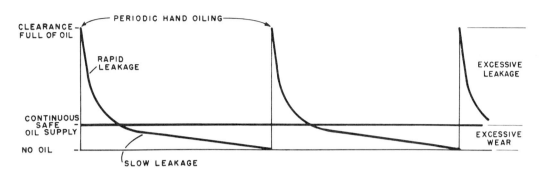

Fig. 5-19 Hand Oiling In this diagram, the condition of "feast or famine" that is always present with periodic hand oiling is compared with the safe continuous supply of oil that is closely approximated by devices which feed oil frequently in small quantities.

A closer approach to maintaining a safe oil supply may be accomplished with application devices such as wick feed oilers, drop feed cups, waste packed cups, bottle oilers, and central dispensing systems such as force feed lubricators or oil mist systems. These devices supply oil on either a slow, constant basis, or at regular, short intervals. With greases, leakage is not as serious a problem, but the use of centralized lubrication systems will provide a more uniform lubricant supply than grease gun application (see Chap. 6).

Even with regular application of small amounts of lubricant, thin film bearings require proper design and installation, and proper lubricant selection to control wear and provide satisfactory service life.

Wearing In of Thin Film Bearings In a new bearing, the journal normally will make contact with the bearing over a fairly narrow area (Fig. 5-20, left). Only light loads can be carried by such a bearing under thin film conditions, since unit loads beyond the ability of the oil film to prevent metallic contacts would probably exist. Under favorable conditions wear will occur, but will have the effect of widening the contact area (Fig. 5-20, right) until the load is distributed over a sufficiently large area that wear becomes practically negligible. Under unfavorable conditions, this initial wear may be so rapid that bearing failure occurs.

Large bearings are often fitted prior to operation by hand scraping, or by counterboring the loaded area to radius of the journal. Fitting of this type

Fig. 5-20 Contact Area before and after Wearing In of Thin Film Bearing

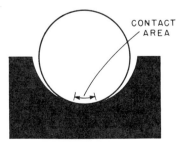
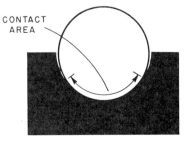

can be done only when the direction of loading is known and is constant. After fitting, the arc of contact should be in the range of 90 to 120 degrees.

Mechanical Factors   In plain bearings, there are several mechanical factors that affect lubrication. These include:

1. Ratio of the length of the bearing to diameter
2. Projected area of the bearing
3. Clearance between journal and bearing
4. Bearing materials
5. Surface finish

Length to Diameter Ratio   The diameter of a journal is determined by mechanical requirements which involve the torque transmitted, shaft strength, and shaft rigidity. The bearing length is governed by the load to be carried, the space available, and the operating characteristics of the particular type of machine, expecially with regard to retention of bearing alignment and shaft flexing. More and more, space limitations are becoming extremely important in the determination of bearing length.

As a general rule, the shorter the bearing the higher end leakage will be, and the more difficult it will be to develop load supporting lubricant films. On the other hand, with too long a bearing, metal-to-metal contact, high friction, high temperatures, and wear may occur at the bearing ends due to even moderate misalignment or shaft deflection under load. Length to diameter (L/D) ratios of as much as 4 have been used but the trend for many years has been to lower ratios. Most fluid film bearings now have L/D ratios of 1.5 or less, connecting rod bearings for internal combustion engines may have L/D ratios of 0.75 or less. Thin film bearings have L/D ratios in about the same range.

Projected Area   The axial length of a bearing multiplied by its diameter is called the projected area, and it is common to express unit loads in force per unit of projected area. Established practices pertaining to unit loads on this basis vary considerably for different classes of machinery and different bearing materials, ranging from as low as 15 psi (103 kPa) for lightly loaded line shafting to as high as 5000 psi (34.5 MPa) or more for internal combustion engine crankpins and wristpins. Most industrial bearings carrying constant loads — as in turbines, centrifugal pumps, and electrical machinery — fall in the range from 50 to 300 psi (345 – 2700 kPa) with most under 200 psi (1380 kPa). Heavier loads are encountered in bearings of reciprocating machinery and in other bearings subject to varying or shock loads. Peak hydraulic pressures within the oil films (Fig. 5-5) are usually 3 to 4 times these unit loads based on projected area.

Clearance   A full bearing must be slightly larger than its journal to permit assembly, to provide space for a lubricant film, and to accommodate thermal expansion and some degree of misalignment and shaft deflection. This clearance between journal and shaft is specifid at room temperature.

One of the principal factors controlling the amount of clearance that must be allowed is the coefficient of thermal expansion of the bearing material. The higher the coefficient of thermal expansion the more clearance must be allowed to prevent binding as the bearing warms up to operating temperature. Babbitts and bearing bronzes have the lowest coefficients of thermal expansion of common bearing materials. Clearances for these materials in general machine practice range from 0.1 to 0.2 percent of the shaft diameter. Many precision bearings will have less clearance than this, while rough machine bearing will have more. Because of their higher coefficients of thermal expansion, aluminum bearings require somewhat more clearance than babbitts or bronzes, and some of the plastic bearing materials require considerably more, in many cases as much as 0.8 percent of the diameter.

Bearing Materials During normal operation of a fluid film lubricated bearing under constant load, the most important property required in the bearing material is adequate compressive strength for the hydraulic pressures developed in the fluid film. When cyclic loading is involved, as in reciprocating machines, the material should have adequate fatigue strength to operate without developing cracks or surface pits. With shock loading, the material should be of such ductility that neither extrusion nor crumbling will occur. Under boundary lubrication conditions, the material also requires:

1. Scoring resistance, requiring appreciable hardness and low shear strength
2. The ability to conform to shaft irregularities and misalignment
3. The ability to embed abrasive particles

If operating temperatures are high, resistance to corrosion and softening may be important.

Although these properties are somewhat conflicting, numerous materials have been developed to obtain satisfactory bearings for the wide range of conditions encountered.

Bearing materials most often encountered in industrial machines are bronzes and babbitts. Suitable bronzes and babbitts are available for practically all conditions of speed, load, and operating temperature encountered in general practice. Steel and cast iron are used for a limited number of purposes, usually involving low speeds or shock loads. In recent years, there has been considerable growth in the use of plastic and elastomeric materials such as nylon, thermoplastic polyesters, laminated phenolics, polytetrafluoroethylene and rubber for bearings, particularly in applications where contamination of, or leakage from, oil lubricated bearings might result in high maintenance costs or short bearing life. Some of these materials can be lubricated with water or soluble oil emulsions in certain applications. Allowable unit loads for these bearing materials usually are lower, although in a number of cases filled nylon bearings have been used as direct replacements for bronze bearings.

For internal combustion engines, babbitt bearings are made with a very thin layer of babbitt over a backing of bronze or steel to increase the load carrying capacity. Even then, the loads may be greater than babbitts can handle, so a number of stronger bearing materials have been developed. These are usually fabricated in the form of precision inserts (Fig. 5-21), which are interchangeable and require no hand fitting or finish machining when installed.

Precision insert bearings, which are usually constructed of layers of different materials, provide:

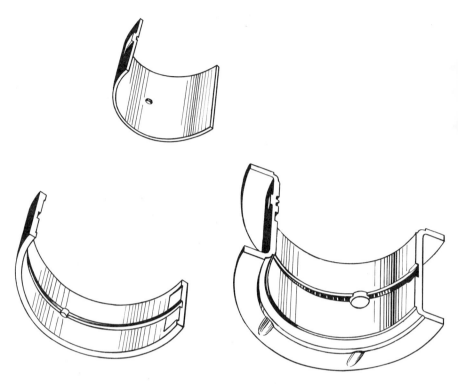

Fig. 5-21 **Precision Insert Bearings** These are typical main and connecting rod bearings used in internal combustion engines.

1. A thin surface layer (sometimes as little as 0.0003 in., 0.0075 mm) having good surface characteristics — such as low friction, scoring resistance, conformability, and resistance to corrosion
2. A thicker layer (0.008 – 0.025 in., 0.2 – 0.6 mm) of bearing material having adequate compressive strength and hardness, suitable ductility, and good resistance to fatigue
3. A still thicker (usually 0.05 – 0.125 in., 1.25 – 3.2 mm) back or shell of bronze or steel

Some of the more common combinations used with this type of construction are babbitt over leaded bronze over steel, lead alloy over copper-lead over steel, lead alloy over silver over steel, and tin over aluminum alloy over steel. These bearings all require smooth, hardened journals, rigid shafts, and minimum misalignment.

Surface Finish Machined surfaces are never smooth. The peak-to-valley depth of roughness in machined surfaces ranges from about 160 μin. (4 μm) for carefully turned surfaces to about 60 μin. (1.5 μm) for precision ground surfaces. Finer finishes, approximately 10 μin. (0.25 μm), can be obtained by other commercial methods.

Finely finished surfaces would, in general, be damaged less than rough surfaces by the metal-to-metal contact that occurs under boundary lubrication conditions. However, some degree of "wearing in" of new bearing surfaces always occurs. New surfaces, which may be relatively rough, tend under favorable conditions and careful wearing in to become smoother. In some cases, certain degrees of roughness aid this wearing in process.

Under fluid film conditions, the minimum safe film thickness is a function of the roughness of the surfaces, rougher surfaces requiring thicker films in order to prevent contact of surface asperities through the film. On the other hand, the finer the surface finish, the lower the minimum safe film thickness, and the less clearance is necessary. Since the film thickness decreases with increases in unit loading, if the minimum safe film thickness is lower as a result of finer surface finishes, the allowable unit loading is higher, all other factors being equal. Conversely, it can be said that bearings designed for high unit loads and small clearances must have finely finished surfaces. Tests show that fluid films may also be formed at a lower speed when starting up a bearing with a smooth finish than when starting one with a rough finish.

Grooving   In all plain bearings, some provision must be made to admit the lubricant to the bearing and distribute it over the load carrying surfaces. Lubricant is generally admitted through a hole or holes and then distributed by means of grooves cut in the bearing surface. The location of the supply hole and the type of grooving used depend on several factors, including the type of supply system, the direction and type of load, and the requirements of the bearings. Certain basic principles apply to all cases.

Grooving for Oil   The distribution of oil pressure in a typical fluid film bearing with steady load is shown in Fig. 5-5. Usually, oil should be fed to a bearing of this type at a point in the no load area where the oil pressure is low. When the shaft is horizontal and the steady load is downward, it is usually convenient to place the supply hole at the top of the bearing, as shown.

Generally, grooves should not be extended into the load carrying area of a fluid film bearing. Grooves in the load carrying area provide an easy path for oil to flow away from the area. Oil pressure will be relieved and load carrying capacity will be reduced. This effect for an axial and a circumferential groove is shown in Figs. 5-22 and 5-23. However, to provide increased oil flow for better cooling in certain force feed lubricated bearings, it is sometimes necessary to extend the grooves through the load carrying area. With variable load direction, it may also be necessary to extend the grooves through the load carrying area. Usually, this is done in precision insert bearings for internal combustion engines.

With constant load direction, a single oil hole may be sufficient. To increase oil flow or improve distribution an axial groove cut through the oil supply hole (Fig. 5-24) often is all that is required. Normally the groove should not extend to the ends of the bearing since that would allow the oil to flow out the ends rather than being carried into the oil wedge. An exception to this is in certain high speed bearings where carefully sized ports or orifices are provided at the ends of the groove to permit increased flow for cool-

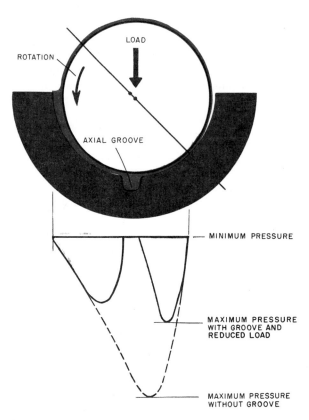

Fig. 5-22 Axial Groove Reduces Load Carrying Capacity An axial groove through the pressure area of a fluid film bearing provides an easy path for leakage and relief of oil pressure. In the lower figure, solid lines represent the approximate pressure distribution when the groove is present, and the dashed line represents approximately what it would be without the groove.

ing purposes. Also, in medium speed equipment, end bleeder ports frequently are provided to give a continuous flushing of dirt particles.

The grooving needed for distribution in a two part bearing usually is formed by chamfering both halves at the parting line (Fig. 5-25). Both sides are chamfered if it is necessary to provide for rotation in both directions. These chamfers should also stop short of the ends of the bearings. Frequently an oil inlet port and distribution groove are combined with the chamfer at the split. Where a top inlet hole is used, an overshot feed groove may be machined in the upper half as shown in Fig. 5-26.

If a stationary journal and a rotating bearing are used, oil may be fed through a hole and axial groove in the journal. Again, the groove should be placed on the no load side.

Where heavy thrust loads are to be carried, fluid film bearings of the tilting pad or tapered land type are often used. Tilting pad bearings require no grooving since the oil can readily flow out around the pad mountings. Tapered land bearings require radial grooves located just ahead of the point

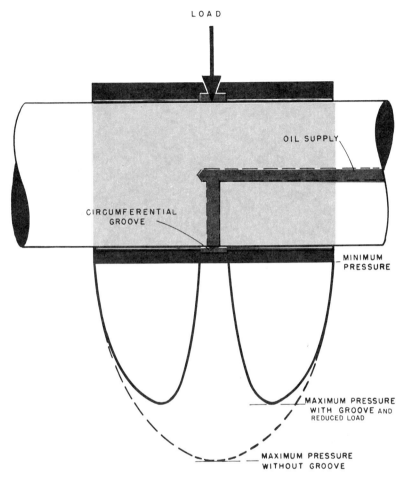

Fig. 5-23 Circumferential Groove Reduces Load Carrying Capacity As in the previous figure, the solid lines show the pressure distribution with the groove present, while the dashed line shows what it would be without the groove.

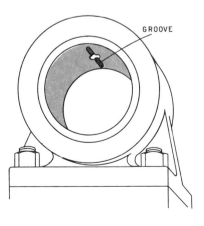

Fig. 5-24 Axial Distribution Groove in One Part Bearing

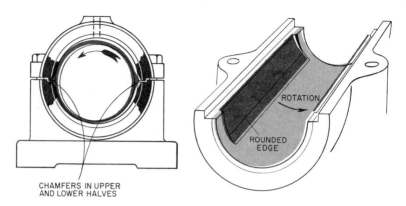

Fig. 5-25 Distribution Grooves in Two Part Bearing

where the oil wedge is formed. If thrust load is carried by one end face of a journal bearing, the axial groove or chamfers may be extended to the thrust end so that oil will flow directly to the thrust surfaces. The end of the bearing should be rounded or beveled to aid in the flow of oil between the end face and the thrust collar or shoulder.

Circumferential grooves are sometimes cut near one or both ends of a bearing to collect end leakage and drain it to the sump or reservoir. This oil might otherwise flow along the shaft and leak through the shaft seals. When collection grooves are used, they mark the effective ends of the bearing.

Vertical shaft bearings often require only a single oil hole in the upper half of the bearing in the no load area. The lower the supply pressure, the higher the hole should be, as a general rule. Sometimes a circumferential groove may be added near the top of the bearing to improve distribution (Fig. 5-27 left). If leakage from the bottom of the bearing is excessive, a spiral groove is sometimes cut in the bearing in the proper direction relative to shaft rotation so that oil will be pumped upward (Fig. 5-27 right).

Increased oil flow to cool a hot running bearing can be obtained by sim-

Fig. 5-26 Overshot Feed Groove and Chamfers

Fig. 5-27 Grooving for Vertical Bearing

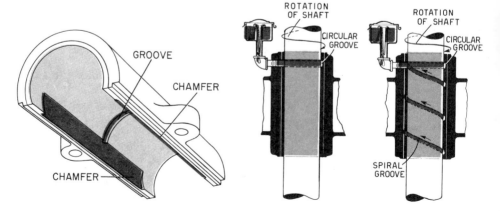

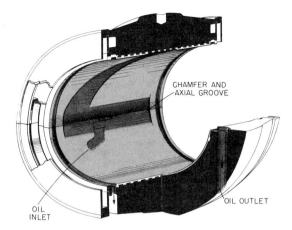

Fig. 5-28 Grooving to Increase Oil Flow for Cooling   This cutaway of a large turbine bearing shows a wide groove cut diagonally in the top (unloaded) half to permit a large flow of oil for cooling purposes. A relatively small part of the oil passing through this bearing would be needed for the fluid film.

ple forms of grooving. An axial groove on the no load side, for example, will increase oil flow by as much as 3 or 4 times compared to a single oil hole alone. Circumferential grooves also increase oil flow, but not as much as an axial groove. They also have the disadvantage of reducing the load carrying capacity of the bearing. Increased clearance can often be used in lightly loaded, high speed bearings to increase oil flow. When increased clearance might reduce load carrying capacity too much, extra grooving or clearance relief in the unloaded portion can be used to increase oil flow for cooling (see Fig. 5-28).

Where the direction of bearing load changes as in reciprocating machines, it is still essential that oil be fed into an unloaded or lightly loaded area. One way of doing this is with a circumferential groove. While this in effect divides the bearing into two shorter bearings of reduced total load carrying capacity, it may be the most effective alternative. Also, it may be desirable to provide a path for oil flow to other bearings; for example, as in many internal combustion engines (Fig. 5-29). An axial groove or chamfer may be used with a circumferential groove to improve oil distribution or increase oil flow (Fig. 5-30).

Ring oiled bearings are somewhat of a special case with regard to grooving. Where the direction of load is fixed and is toward the bottom of the bearing, a simple axial groove connecting with the ring slot (Fig. 5-31) is adequate. In a two part bearing, the axial groove may be formed by chamfering the bearing halves at the parting line. However, if loads are sideways because of belt pull or gear reaction, this type of groove can be blocked as shown in Fig. 5-31. To overcome this problem, grooving such as shown in Figs. 5-32 and 5-33 is often used. Groving of this type is used for electric motors, which may be belted or geared to the load, so that the direction of loading and shaft rotation are not known to the motor manufacturer.

Grooving for thin film lubricated bearings generally follows the same pattern as for fluid film lubricated bearings. Due to the restricted supply of oil to

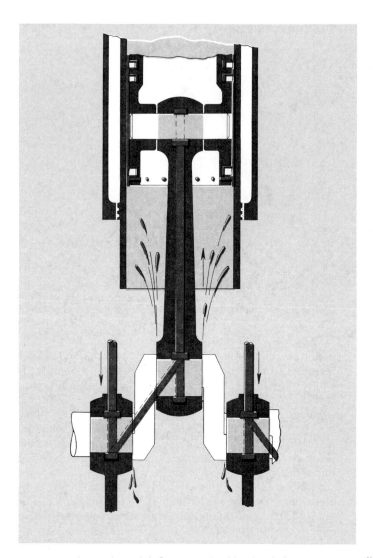

Fig. 5-29 Circumferential Grooves In this circulation system an oil pump, driven from the crankshaft, takes oil from the crankcase sump and delivers it under pressure to the crankpin and wristpin bearings through passages in the crankshaft and connecting rods.

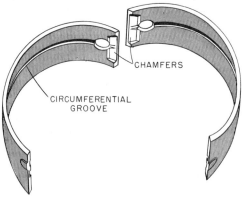

CHAMFERS

CIRCUMFERENTIAL GROOVE

Fig. 5-30 Circumferential Groove and Axial Chamfer An axial groove (also called a spreaded groove) is often used with a circumferential groove in precision bearings.

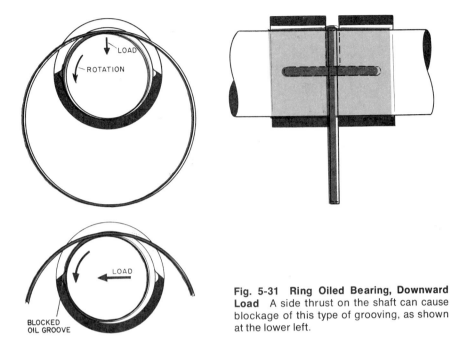

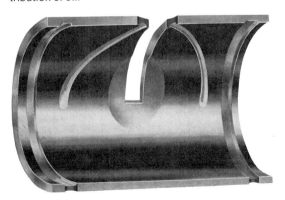

Fig. 5-31 Ring Oiled Bearing, Downward Load A side thrust on the shaft can cause blockage of this type of grooving, as shown at the lower left.

this type of bearing, it may be necessary to provide additionl reservoir capacity in the grooves, and to design the grooves to retain and distribute the oil (Figs. 5-34 to 5-36).

Grooving for Grease Grooving principles for grease are practically the same as for oil. Because grease has higher resistance to flow at the low shear rates in the supply system, grease grooves need to be made wider and deeper than oil grooves. Annular grooves cut near the ends of the bearing, similar to oil

Fig. 5-32 Section of Ring Oiled Bearing With X Grooves These grooves cross in the top of the bearing. The area around the end of the ring slot is relieved to aid in the distribution of oil.

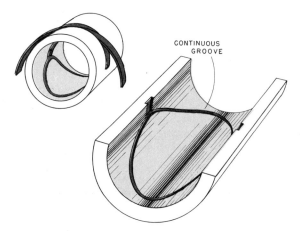

Fig. 5-33   Ring Oiled Bearing, Continuous Groove   With this pattern of grooving, oil feed cannot be blocked and there is little interruption of load carrying surface along any axial line.

collection grooves but without the drain hole, are often used to reduce grease leakage. Some of the grease forced into these grooves in the loaded zone flows back out in an unloaded area and enters the bearing for reuse.

Lubricant Selection   Lubricants for plain bearings must be carefully selected if the bearings are to give long service life with low friction and power loss.

Oil Selection   All the lubrication systems for fluid film bearings are essentially circulation systems; thus, the oil is reused many times over long periods.

Fig. 5-34   Auxiliary Groove   Where oil application is infrequent, extra storage capacity is sometimes needed in a bearing. This can be provided by means of an auxiliary groove cut just ahead of the load carrying area.

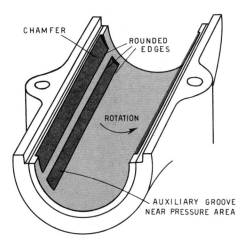

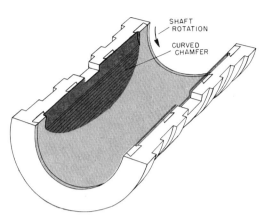

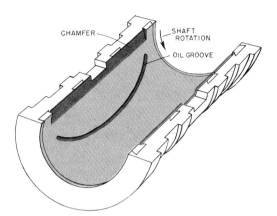

Fig. 5-35 Curved Chamfer Where clearance at the horizontal center line is large, end leakage of oil can be reduced by using a curved chamfer. This tends to guide a restricted supply of oil toward the center of the bearing so that it will be drawn into the load carrying area.

Fig. 5-36 Straight Chamfer and Curved Groove The chamfer provides oil storage capacity and aids distribution. The curved groove catches oil flowing toward the bearing ends and guides it back toward the center of the bearing.

Moisture and other contaminants may enter the system. These conditions require that the oil have:

1. Good chemical stability to resist oxidation and deposit formation in long term service
2. Provide protection against rust and corrosion
3. Separate readily from water to resist the formation of troublesome emulsions
4. Resist foaming

These properties are generally found in high quality circulation oils.

To select the proper viscosity of oil, three operating factors must be determined — speed, load, and operating temperature. Speed (rpm) is usually readily determined. Bearing loads may be calculated from the total weight supported by the bearings divided by the total projected area of the bearings. For simplicity, bearing loads of 200 psi (1380 kPa) can often be assumed for bearings operating under light to moderate loads with no shock loads, and 500 psi (3450 kPa) for heavily loaded bearings or where shock loads are present. Operating temperatures can be approximated by using the oil emergence temperature of the hottest bearing.

When these three factors are known, the optimum viscosity of oil can be determined from a chart such as that shown in Fig. 5-37. Since this chart gives the viscosity at the operating temperature, ASTM viscosity-temperature charts must then be used to determine the desired viscosity at one of the standard measuring temperatures (see discussion of viscosity in Chap. 2).

Where bearings are to be started or operated at low temperatures, the pour point and low temperature fluidity of the oil selected must be considered. This may sometimes necessitate the use of an oil with a viscosity lower than optimum in order to ensure circulation at start up.

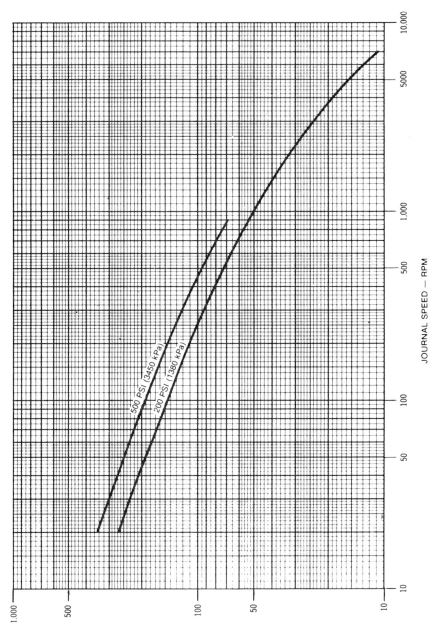

OIL VISCOSITY AT OPERATING TEMPERATURE — CENTISTOKES

JOURNAL SPEED — RPM

Fig. 5-37 Oil Viscosity Chart for Fluid Film Bearings To determine the viscosity required, follow up from the appropriate journal speed to either of the lines, then to the left to read the oil viscosity at the operating temperature. The 200 psi line should be used for bearings operating under normal loads without shock loads, and the 500 psi line where loads are heavy or shock loads are present.

In thin film lubricated bearings, the oil is normally not reused, so the important characteristics of the oil selected are that it distribute readily and form tough, tenacious films on the bearing surfaces. Generally, oils with antiwear and friction reducing properties are preferred.

With thin film lubricated bearings, it may be assumed that loads are high enough for the bearings to operate under boundary conditions at least part of the time. Therefore, only speed and operating temperature must be determined to select the correct viscosity of oil. Once speed and operating temperature are known, the oil viscosity may be obtained from a chart such as Fig. 5-38. Again, the viscosity must be converted to a value at one of the standard reporting temperatures by means of viscosity-temperature charts. Ambient temperatures must also be considered, and so must the type of application device since some of these devices are limited as to the maximum viscosity of oil they will dispense.

Grease Selection The greases used for fluid film and thin film bearings are essentially similar. Enhanced film strength is probably required more frequently in thin film bearings, but where heavy, shock, or vibratory loads are encountered in fluid film bearings, this property is usually also required.

Since the methods used to lubricate plain bearings are all loss systems, the grease used is not subjected to long term service that might cause oxidative breakdown. On the other hand, operating temperatures may be higher in grease lubricated bearings than in comparable oil lubricated bearings because of the poorer cooling ability of grease. Thus, there may be exposure to high temperatures while the grease is in the bearings. There is also severe mechanical shearing of the grease, particularly as it passes through the load carrying zone, which may cause softening and lead to increased end leakage.

The method of application has considerable influence on both the type of grease and the consistency selected for plain bearings. With centralized lubrication systems, the grease must be a type suitable for dispensing through such systems. Where bearings must be lubricated at low ambient temperatures, softer greases are usually needed, since they pump more easily at low temperatures.

These requirements generally dictate the use of greases with good mechanical stability, adequate dispensing and pumpability characteristics, corrosion protection properties, adequate film strength, and adequate high temperature characteristics for the operating temperatures. In addition, for fluid film bearings, the apparent viscosity of the grease must be such that fluid films can be formed at the shear rates prevailing, but not so high that friction losses will be excessive.

ROLLING ELEMENT BEARINGS

The term "rolling element bearings" is used to describe that class of bearings in which the moving surface is separated from the stationary surface by elements such as balls, rollers, or needles that can roll in a controlled manner. These bearings are often referred to as "antifriction" bearings.

The essential parts of a rolling element bearing (Fig. 5-39) are a stationary ring, a rotating ring, and a number of rolling elements. The inner ring fits

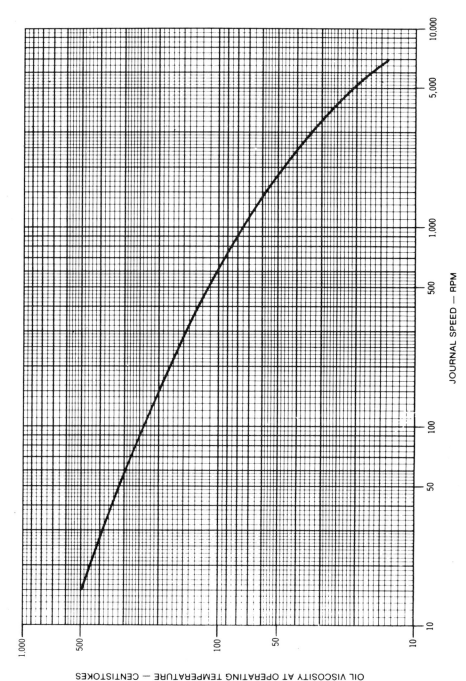

Fig. 5-38 Oil Viscosity Chart for Thin Film Bearings Since loads can be assumed to be heavy enough for the bearing to operate in a thin film regime, only journal speed and operating temperature affect oil viscosity selection.

JOURNAL SPEED — RPM

OIL VISCOSITY AT OPERATING TEMPERATURE — CENTISTOKES

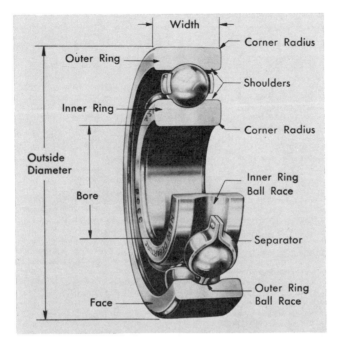

Fig. 5-39 **Ball Bearing Nomenclature**

the shaft or spindle, and the outer ring fits in a suitable housing. Shaped raceways are machined in the rings to confine and guide the rolling elements. These rolling elements are usually held apart from each other and their relative positions maintained to keep the shaft or spindle centered by the separator, which is also called a cage or retainer. In full complement bearings, the rolling elements completely fill the space between the rings and no separator is used. Rolling element bearings ranging in size from smaller than a pinhead to 18 ft. (6 m) or more in outside diameter have been manufactured.

Manufacturers now supply a wide variety of designs of rolling element bearings of which those in Figs. 5-40 and 5-41 represent only some of the more popular types. The bearing in Fig. 5-39 is a single row, deep groove bearing, which is usually the initial choice for any application. When the fatigue life of this design is inadequate; space is limited; self-alignment is required; thrust loads must be carried; or any of a variety of other conditions must be met, one of the other types can be selected. When one of these other types is selected the desired performance characteristics are very often obtained at the expense of higher cost, a lower speed limit, or more severe lubrication requirements.

When properly lubricated, the load capacity and life of a rolling element bearing are limited primarily by the fatigue strength of the bearing steel. Normal rolling action applies repeated compressive loading in the contact area with stresses up to about 350,000 psi (2.4 GPa). In addition to causing elastic deformation of the rolling element and raceway, this stress induces shearing stress in the steel in a zone approximately 0.002 to 0.003 in.

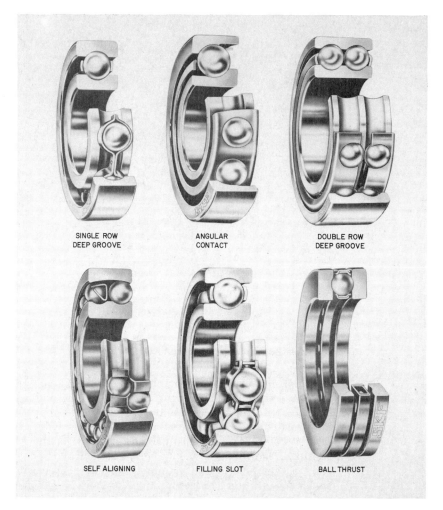

Fig. 5-40   Popular Types of Ball Bearings

(0.05 – 0.075 mm) below the surface. This shearing stress induces fatigue cracks in the steel, which gradually grow and intersect. Small surface areas loosen and break away, forming pits. The actual time required for this to occur depends on many factors, including load, speed, and continuity of service, as well as the fatigue strength of the bearing steel.

As discussed earlier under elastohydrodynamic films, inadequate lubrication can greatly reduce the fatigue life of rolling element bearings. In addition, current research has shown that chemical effects have a considerable influence on fatigue life. The chemical composition of the lubricant additives, the base stock, and the contact surface materials are the major chemical variables. The water content of the atmosphere and the lubricant may also contribute significantly to chemical effects. At operating temperatures below 212°F (100°C) most industrial oils will contain dissolved water. Because of the small size of the water molecules compared to oil molecules, they readily diffuse to the tips of the initial microcracks resulting from cyclic stressing. Although the precise mechanism by which this water accelerates fatigue cracking is not clear, there is evidence that the water in the microcracks breaks down and liberates atomic hydrogen. This causes hydrogen embrittlement of the adjacent metal. Research data indicate that water accelerated fatigue can reduce bearing life from 30 to 80 percent.

Need for Lubrication Operation of rolling element bearings always results in the development of a certain amount of friction which comes from two main sources:

1. The fluid friction which results from shearing of the lubricating films.
2. Displacement or churning of lubricant in the path of the rolling elements

Fig. 5-41 Popular Types of Roller Bearings

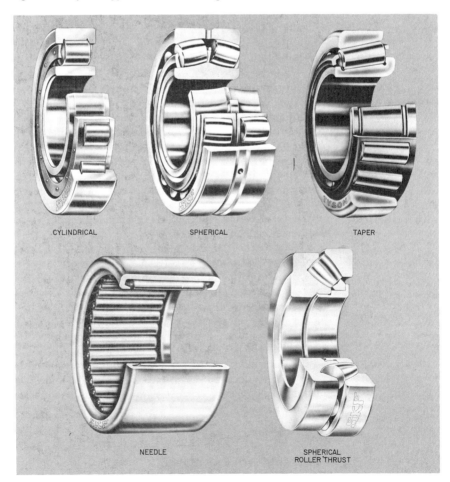

CYLINDRICAL   SPHERICAL   TAPER

NEEDLE   SPHERICAL ROLLER THRUST

3. Sliding between various elements of the bearing

The main areas where sliding occurs are:

1. Slipping between rolling elements and the raceways. Because of the elastic deformation of both the raceway and the rolling elements, true rolling does not occur and slight sliding results.
2. Sliding between the rolling elements and the separator.
3. Sliding between the ends of rollers and the raceway flanges of roller bearings. This type of sliding may be particularly severe in certain types of bearings designed for thrust loads.
4. Sliding between the shaft or spindle and contact type housing seals.
5. Sliding between adjacent rolling elements in full complement bearings.

Lubrication aims to maintain suitable films between all of these sliding parts—EHL films for the contacts between the rolling elements and raceways, or thin films for other sliding parts. These films must be adequate to minimize friction and protect against wear. In addition, lubrication is expected to protect against rusting or other corrosive effects of contaminants, and may provide part of the sealing against contaminants. In some cases, the lubricant also acts as a coolant to remove heat generated by friction, or heat conducted to the bearing from outside sources.

Factors Affecting Lubrication

Whether oil or grease is used for rolling element bearings depends on a number of factors. Where bearings are installed in machines that require oil for other elements, the same oil is often supplied to the bearings. In other cases, it may be more convenient to lubricate the bearings separately and grease is often used. Groups of similar bearings may be lubricated with oil by a circulation system or an oil mist system, or with grease in a centralized lubrication system. Many bearings are now "packed for life" with grease by the bearing manufacturer and need no further lubrication in service. The characteristics of these various methods of application have some influence on the oil or grease selected.

The thickness of the elastohydrodynamic films formed in the contact areas between rolling elements and the raceways is a function of the speed at which the surfaces roll together, the load, the oil viscosity, and the operating temperature. The film thickness increases with increases in speed or oil viscosity and decreases with increases in load or operating temperature (since higher temperatures reduce the oil viscosity). For any given set of operating conditions, the oil viscosity (or viscosity of the oil in a grease) should be selected to provide a safe minimum film thickness.

The lubrication requirements of the other sliding surfaces of a bearing must also be considered in the selection of a lubricant, but in general, the primary consideration in lubricant selection is the requirements of the EHL films.

Effect of Speed The speed at which the surfaces of a rolling element bearing roll together is a fairly complex calculation, so an approximation called the "bearing speed factor," nd_m, is usually used. The bearing speed factor is determined by multiplying the rotational speed in revolutions per minute n

Table 5-1   Bearing Speed Factors

Bearing Type	Oil Lubricated	Grease Lubricated
Radial Ball Bearings	500,000	340,000
Cylindrical Roller Bearings	500,000	300,000
Spherical Roller Bearings	290,000	145,000*
Thrust — Ball and Roller Bearings	280,000	140,000

*Grease lubrication is not recommended for spherical roller thrust bearings.

by the pitch diameter in millimetres d_m. For convenience, the pitch diameter is taken as one-half the sum of the bearing bore d and the bearing outside diameter D.* That is:

$$nd_m = \frac{n(d + D)}{2}$$

Manufacturers have established maximum bearing speed factors for both oil and grease lubricated bearings. These factors are shown in Table 5-1 for commercial grade bearings. Precision bearings are rated 5 to 50 percent higher. The higher speed factors allowed for oil lubrication generally reflect the better cooling available with oil, and the lower fluid friction of low viscosity oils.

Recent Mobil research into simplified solutions for EHL calculations determined that the relation of bearing outside diameter to speed gave the greatest accuracy in determining the correct oil viscosity for rolling element bearings.

Where bearings are lubricated with grease, the considerations relative to bearing speed are mainly concerned with the physical properties of the grease. At low to moderate speeds, the grease must be soft enough to slump slowly toward the rolling elements, but must not be so soft that excess grease gets into the path of the rolling elements. Excess grease increases shearing friction and can cause high operating temperatures. For high speeds, a relatively stiff grease is sometimes used but it must not be so stiff that the rolling elements, after cutting a channel through the grease, are unable to pick up and distribute sufficient amounts of grease to provide continuous replenishment of the lubricating films. Moreover, in order to keep shearing friction low and prevent leakage from housing seals, the grease must have good resistance to softening as a result of mechanical shearing.

Effect of Load   Under EHL conditions, the effect of load on film thickness is not as great as the effect of speed or oil viscosity. For example, while doubling the speed or oil viscosity might result in over a 50 percent increase in film thickness, doubling the load might result in only about a 10 percent de-

*The older DN factor (bearing bore in millimetres D multiplied by the speed in rpm N) is still found in some references although it has not been officially used by the bearing manufacturers' associations for many years. It is not a satisfactory factor to use because for a given bearing bore there are bearings with considerable variations in outside diameter and size of the rolling elements. Thus, for any given bore diameter there can be considerable variation in the speeds with which the elements roll together.

crease in film thickness. As a result, under steady load conditions the viscosity of oil can generally be selected on the basis of the bearing speed factor and operating temperature without regard to the load.

When shock or vibratory loads are present, somewhat greater film thickness is necessary to prevent metal-to-metal contacts through the film; thus, higher viscosity oils are usually required. Under severe shock loading conditions, lubricants with enhanced antiwear properties may be desirable.

Pressures between the rolling elements and separators are not high enough to form EHL films. Additionally, these pressures act more or less continuously on the same spots in the separator pockets. As a result, thin film lubrication exists and a higher viscosity oil can sometimes improve the resistance to wiping of the oil films and reduce wear of the separator. When grease is the lubricant, wear can only be reduced if the lubricant can penetrate the clearances between the rolling elements and separators. This is especially important with machined separators where clearances are small.

In tapered and spherical roller thrust bearings, heavy loads sometimes force the ends of the rollers against the flanges of the raceways with considerable pressure. Severe wiping of the lubricant films and wear of the rollers and flanges may occur. In these types of bearings, a lubricant with enhanced antiwear properties may be required, even though the rollers and raceways can be lubricated satisfactorily with conventional products.

Effect of Temperature   Since oil viscosity and grease consistency are both functions of temperature, the operating temperature of bearings must always be considered when selecting lubricants. Bearing operating temperatures may be increased above normal by heat that is conducted to the bearing from a hot shaft or spindle, or by heat radiated to the housing from a hot surrounding atmosphere. Excessive churning of grease resulting from overfilling can also raise bearing temperatures (Fig. 5-42). Higher than normal

Fig. 5-42 Temperature Rise in Grease Lubricated Bearing   As shown by the curves, the final operating temperature of a bearing is a function of the amount of grease packed in it.

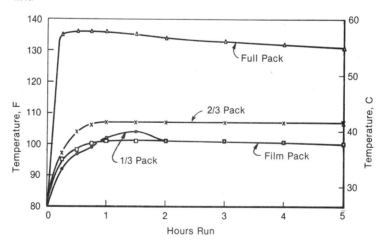

temperatures reduce oil viscosity so that film thickness may decrease below a safe level, and may soften grease so that excessive churning and further frictional heating occur.

High temperatures also increase the rate at which both oil and grease deteriorate due to oxidation. Oxidation may result in thickening of oil, and eventually could lead to deposits which will interfere with oil flow or bearing operation. Oxidation may also lead to deposits with greases, and in severe cases, to hardening such that the grease cannot feed and lubricate.

Under low temperature conditions, the lubricant must be such that the bearing can be started with the power available, and it must distribute sufficiently to prevent excessive wear before frictional heating warms the bearing and lubricant.

Contamination   Solid particles of any kind that are trapped between the rolling elements and raceways are the most frequent cause of shortened bearing life. Consequently, dirt should be kept out of bearings as much as possible, and lubricants should be changed before oxidation has progressed to the point where deposits begin to form. The use of oxidation inhibited lubricants can greatly extend the period of time that lubricants may be left in service without excessive oxidation.

Water that gets into a bearing tends to cause rusting which can quickly ruin a bearing. Water, when mixed with some greases, may cause them to soften so they may leak from bearings. Large quantities of water can wash the lubricant out. Sometimes fluids such as acids get into bearings and cause corrosion. Any of these conditions usually involves special precautions in the selection of the lubricant.

Lubricant Selection   Rolling element bearings generally require high quality lubricants, many of which are specially formulated for the purpose.

Oil Selection   Oils for the lubrication of rolling element bearings should have the following characteristics:

1. Excellent resistance to oxidation at operating temperatures to provide long service without thickening or formation of deposits that would interfere with bearing operation.
2. Proper viscosity at operating speeds and temperatures to protect against friction and wear. Two charts (Fig. 5-43), may be used to select the proper viscosity at the operating temperature from the type of bearing, the operating speed, and bearing outside diameter (D). The speed factor (SF) is determined for the speed and type of bearing from Fig. 5-43 (top). The bearing speed/size factor (BS/SF = SF × D) is used in Fig. 5-43 (bottom) to determine the correct oil viscosity for a given specific film thickness (λ). For steady loads, the minimum viscosity line, corresponding to a specific film thickness, $\lambda = 1.5$ (see discussion of EHL), will provide adequate film thickness. Where shock or vibratory loads are present, higher viscosity and thicker films are desirable. Therefore, the optimum viscosity line, corresponding to a specific film thickness, $\lambda = 4.0$, should be used. ASTM D341 "Standard Viscosity-Temperature Charts

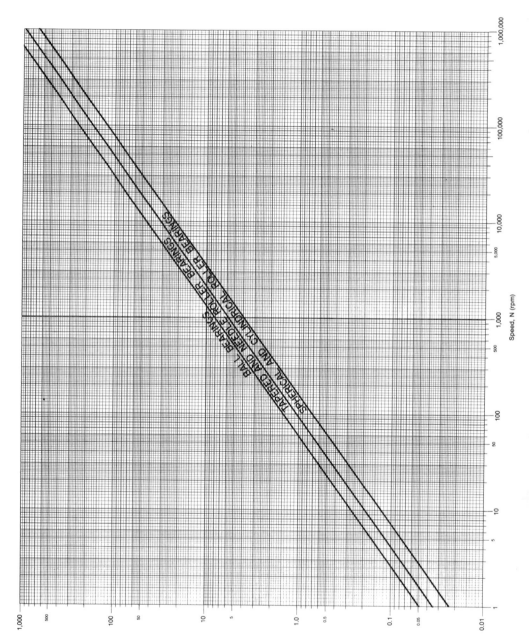

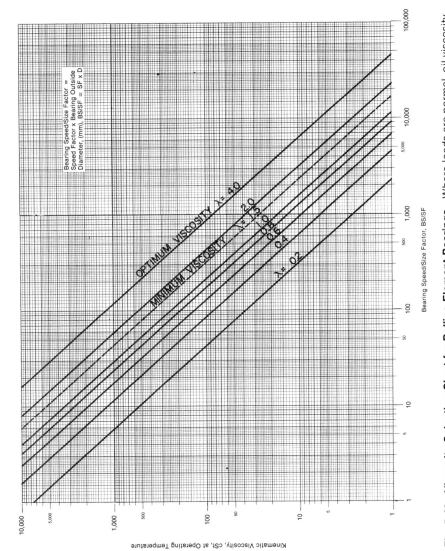

Fig. 5-43 Viscosity Selection Chart for Rolling Element Bearings Where loads are normal, oil viscosity selected with the minimum viscosity line is generally satisfactory. Where there are shock or vibratory loads, somewhat thicker oil films are required and the optimum viscosity line should be used.

for Liquid Petroleum Products"* should be used to translate this viscosity to the viscosity at the standard reporting temperatures.

3. Antirust properties to protect against rust when moisture is present.
4. Good antiwear properties where required because of heavy or shock loads, or where thrust loads cause heavy wiping between the ends of rollers and raceway flanges.

In addition to these characteristics, increasing emphasis is now being placed on the interrelationship between oil characteristics and fatigue. The specific lubricant properties that influence fatigue are not easily defined; but where average bearing life is lower than expected, special lubricants formulated to help minimize fatigue may be desirable.

Grease Selection Grease for the lubrication of rolling element bearings should have the following characteristics:

1. Excellent resistance to oxidation to resist the formation of deposits or hardening that might shorten bearing life. Greases used in packed for life bearings are expected to provide trouble free performance for the life of the bearing when operating temperatures are normal.
2. Mechanical stability to resist excessive softening or hardening as a result of shearing in service.
3. Proper consistency for the method of application and proper feedability at operating temperatures to maintain adequate lubricating films without excessive slumping that would increase friction and bearing heating.
4. Controlled oil bleeding, particularly for high speed bearings, to supply the small amount of oil needed to form EHL films.
5. Enhanced antiwear properties to resist the wiping action between roller ends and raceway flanges in bearings carrying heavy radial or axial thrust loads.
6. Ability to protect surfaces against rusting. When small quantities of water contaminate a bearing, the grease should absorb the water without appreciable softening or hardening. When large quantities of water are present, the grease must resist being washed out.

SLIDES, GUIDES, AND WAYS

This category of bearings, sometimes called flat bearings, includes all slides, guides, and ways used on forging and stamping presses; as crosshead guides of certain compressors, diesel and steam engines; and on metalworking machines such as lathes, grinders, planers, shapers, and milling machines. The service conditions under which these bearings operate vary widely. Crosshead guides operate at relatively high speeds under conditions which probably permit the formation of fluid lubricating films most of the time. The requirements for lubrication usually are not severe, and the guides usually are designed to operate on the same oil used for the main and connecting rod bearings of the machine.

*American Society for Testing and Materials, 1916 Rose Street, Philadelphia, Pa. 19103.

The lubrication of ways and slides of machine tools can present special problems. At low speeds and under heavy loads, the lubricant tends to be wiped off so that boundary lubrication prevails. While this results in higher friction, boundary films have the advantage that they are of almost constant thickness. On the other hand, with low loads and high traverse speeds if the oil viscosity is high, fluid films can be formed that will lift and float the slide. With variations in speed or load, these fluid films can vary in thickness enough to produce wavy surfaces on the parts being machined, or cause them to run offsize. Thus, precision machining generally requires that the slides and ways operate under boundary conditions at all times. This frequently means that friction reducing and antiwear additives must be included in the oil.

A phenonomen known as stick slip can be encountered in the motion of slides and ways. If the static coefficient of friction of the lubricant is greater than the dynamic coefficient, more force will be required to start the slide from rest than will be required to maintain it in motion after it has started. There is always some amount of free play in the feed mechanism, so when force is applied to start the slide in motion it will initially resist. When the force becomes high enough, the slide will begin to move. As soon as motion begins, the force required to maintain motion decreases so the slide will jump ahead until the free play in the feed mechanism is taken up. At low traverse speeds, this can be a continuous process, producing chatter marks on the workplace. With cross slides, stick slip effects can make it extremely difficult to set feed depths accurately. Stick slip effects can be overcome with additives that reduce the static coefficient of friction to a value equal to or less than the dynamic coefficient.

With vertical guides, there is a tendency for the lubricant to drain from the surfaces. To resist this tendency and secure adequate films, special adhesive characteristics are needed.

Film Formation When two parallel surfaces move with respect to each other, there is some tendency to draw oil in and form a fluid film between the surfaces. However, such a film does not have high load carrying capacity, and when the surfaces are long, the film is usually wiped or squeezed away rapidly. Transverse grooves in the slide to divide it into a series of shorter surfaces, with lubricant supplied at the grooves, can improve film formation and load carrying ability. In some cases, with precision machine tools, hydrostatic bearings are being used. Pockets are machined in the slide and supplied from a constant pressure system with a flow restrictor for each pocket. This arrangement eliminates stick slip effects, and if properly arranged, maintains the slide parallel to the way.

Lubricant Characteristics On the basis of the foregoing requirements, the characteristics of suitable lubricants for slides, guides, and ways can be summarized as follows:

1. Proper viscosity at operating temperature for ready distribution to the sliding surfaces and for forming the necessary boundary films
2. High film strength to maintain the required boundary films under heavy

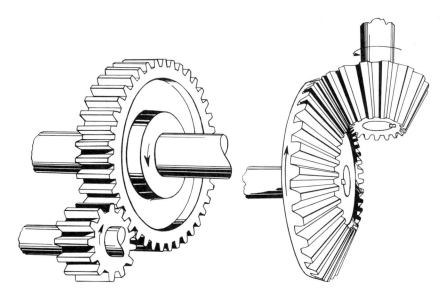

Fig. 5-44 **Spur Gears**

Fig. 5-45 **Bevel Gears** These gears transmit motion between intersecting shafts. The shafts need not meet at a right angle.

loads and antiwear capability to control wear under these boundary conditions
3. Proper frictional characteristics to prevent stick slip and chatter
4. Adequate adhesiveness to maintain films on intermittently lubricated surfaces, especially vertical surfaces, and to resist the washing of any cutting coolants used

Slides and guides in some machines such as open crankcase steam engines and forging machines may be exposed to considerable amounts of water. Lubricants for these applications must be specially formulated to resist being washed off.

GEARS

Gears are employed to transmit motion and power from one revolving shaft to another, or from a revolving shaft to a reciprocating element. With respect to lubrication and the formation and maintenance of lubricating films, gears can be classified as follows:

1. Spur (Fig. 5-44), bevel (Fig. 5-45), helical (Fig. 5-46), herringbone (Fig. 5-47), and spiral bevel (Fig. 5-48) gears
2. Worm gears (Fig. 5-49)
3. Hypoid gears (Fig. 5-50)

The difference in the action between the teeth of these three classes of gears have a considerable influence on both the formation of lubricating films and the properties of the lubricants required for satisfactory lubrication.

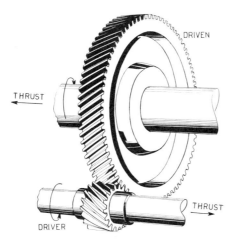

Fig. 5-46 **Helical Gear and Pinion**

Action Between Gear Teeth

As gear teeth mesh, they roll and slide together. The progression of contact as a pair of spur gear teeth of usual design engage is shown in Fig. 5-51. The first contact is between a point near the root of the driving tooth (upper gear) and a point at the tip of the driven tooth. In view A, these points are identified as 0-0 lying on the line of action. At this time, the preceding teeth are still in mesh and carrying most of the load. As contact progresses, the teeth roll and slide on each other. Rolling is from root to tip on the driver and from tip to root on the driven tooth. The direction of sliding at each stage of contact is as indicated by the small arrows.

In view B, contact has advanced to position 3-3, which is approximately the beginning of "single tooth" contact when one pair of teeth pick up all of

Fig. 5-47 **Herringbone Gear and Pinion**   Fig. 5-48 **Spiral Bevel Gears** These gears transmit motion between intersecting shafts.

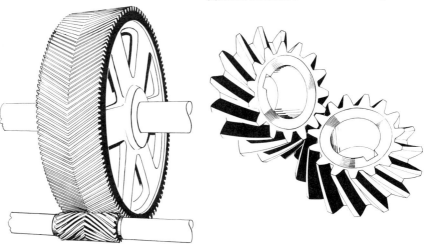

Fig. 5-49 Worm Gears In this drawing the worm is represented as an endless rack.

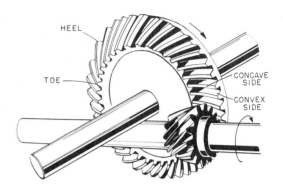

Fig. 5-50 Hypoid Gears These gears transmit motion between nonintersecting shafts crossing at a right angle.

the load. It will be seen that to reach this point of engagement, since the distance 0-3 on the driven gear is greater than the distance 0-3 on the driver, there must have been sliding between the two surfaces. View C, position 4-4, shows contact at the pitch line where there is pure rolling—no sliding. It should be noted, particularly, that the direction of sliding reverses at the pitch line. Also, sliding is always away from the pitch line on the driving teeth, and always toward it on the driven teeth. View D shows contact at position 5-5, which marks the approximate end of single tooth contact. As shown, another pair of teeth is about to make contact. In view E, two pairs of teeth are in mesh, but as shown at position 8-8, the original pair of teeth is about to disengage.

It will be seen that rolling is continuous throughout mesh. Sliding, on the other hand, varies from a maximum velocity in one direction at the start of mesh, through zero velocity at the pitch line, then again to a maximum velocity in the opposite direction at the end of mesh.

This combination of sliding and rolling occurs with all meshing gear teeth regardless of type. The two factors that vary are the amount of sliding in proportion to the amount of rolling, and the direction of slide relative to the lines of contact between tooth surfaces.

With conventional spur and bevel gears, the theoretical lines of contact run straight across the tooth faces (Fig. 5-52). The direction of sliding is then at right angles to the lines of contact. With helical, herringbone, and spiral bevel gears, because of the twisted shape of the teeth, the theoretical lines of contact slant across the tooth faces (Fig. 5-53). Because of this, the direction of sliding is not at right angles to the lines of contact, and some side sliding along the lines of contact occurs.

With worm gears, as with spur gears, the same sliding and rolling action occurs as the teeth pass through mesh. Usually, this sliding and rolling ac-

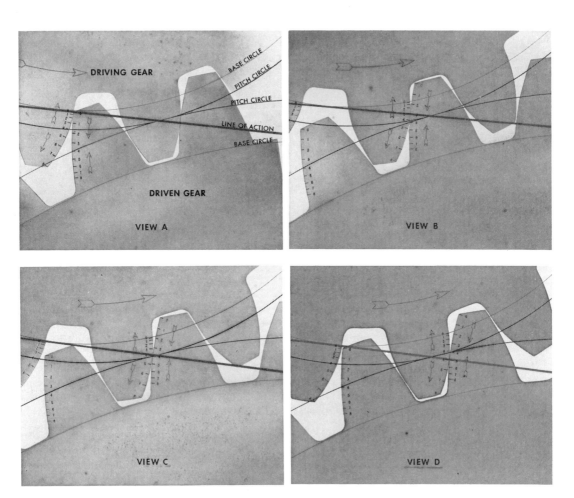

VIEW A

DRIVING GEAR

BASE CIRCLE
PITCH CIRCLE
PITCH CIRCLE
LINE OF ACTION
BASE CIRCLE

DRIVEN GEAR

VIEW B

VIEW C

VIEW D

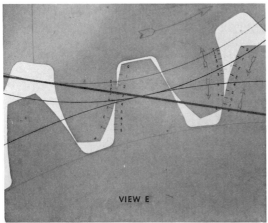

VIEW E

Fig. 5-51 Meshing of Involute Gear Teeth These photographs show the progression of rolling and sliding as a pair of involute gear teeth (a commonly used design) pass through mesh. The amount of sliding can be seen from the relative positions of the numbered marks on the teeth.

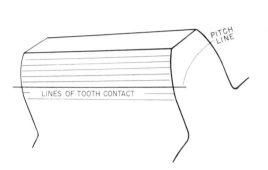

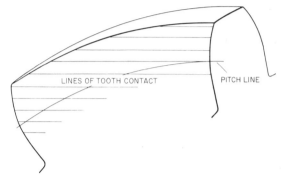

Fig. 5-52   A Spur Gear Tooth Showing Lines of Tooth Contact   On the driving tooth, this contact first occurs below the pitch line. As the gear turns, this contact progressively sweeps upward to the top of the tooth. The action is reversed on the driven tooth.

Fig. 5-53   Tooth Contact of Low Angle Helical Gear

tion is usually relatively slow because of the low rotational speed of the worm wheel. In addition, rotation of the worm introduces a high rate of side sliding. The combination of two sliding actions produces a resultant slide which in some areas is directly along the line of contact.

In addition to the usual rolling action, hypoid gears have a combination of radial and sideways sliding that is intermediate between worm gears and spiral bevel gears. The greater the shaft offset, the more nearly the sliding conditions approach those found in worm gears.

Film Formation
As pointed out earlier, the requirements of EHL film formation is that the pressure must be high enough to cause elastic deformation of the contracting surfaces. Thus, the theoretical lines or points of contact are expanded into areas, and a convergent zone exists ahead of the contact area. In the case of industrial gears, in all except very lightly loaded applications, tooth loading is high enough to produce elastic deformation along the line of contact. A convergent zone immediately ahead of the contact area also exists at all times — see Fig. 5-54, and the conditions for the formation of EHL films are met. The equations used for the calculation of EHL film thicknesses are complex, and are beyond the scope of this publication.* However, certain factors in these calculations are of importance in the general consideration of selection of lubricants for industrial gear drives.

The equations used do not consider the effect of tooth sliding action on the formation of th EHL films. The entraining velocity tending to carry the lubricant into the contact zone is considered to be the rolling velocity alone. The rolling velocity, for convenience, is usually calculated at the pitch line and considered representative for the entire tooth.

A recent study† indicates that the critical specific film thickness λ for gears is not only considerably lower than for rolling element bearings but is

---

*An excellent reference on this subject is the Mobil *EHL Guidebook*.
†Wellauer, E. J., and Holloway, G. A., Application of EHD Oil Film Theory to Industrial Gear Drives, Trans. ASME, *J. Eng. Ind.*, 98B(1):626 – 634 (May 1976).

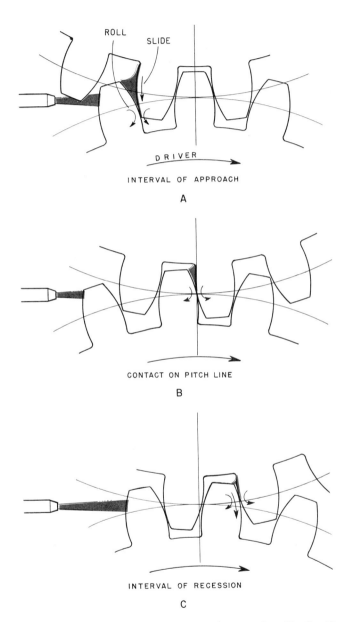

Fig. 5-54 Convergent Zone between Meshing Gear Teeth These drawings show that if oil is present between meshing gear teeth, it will be drawn into the convergent zone between the teeth, and that the point of this wedge shaped zone always points toward the roots of the driving teeth.

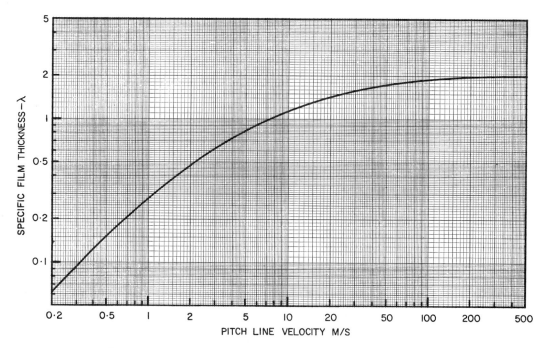

Fig. 5-55 Critical Specific Film Thickness for Gears This curve is based on a 5 percent probability of surface distress to define the target film thickness, which is adjusted to reflect the root mean square surface roughness [$\sigma = (\sigma_1^2 + \sigma_2^2)^{1/2}$].

also a function of the pitch line velocity. The curve of Fig. 5-55, developed from experimental data, shows that at low speeds values of λ as low as 0.1 or lower can be tolerated without surface distress in the form of pitting or wear. At higher speeds, values of λ of up to 2.0 or higher are required for equal freedom from tooth distress.

Currently, no analysis has been made of the reasons for these lower specific film thicknesses providing satisfactory results in gears. However, it is generally accepted that in the range below $\lambda = 1.0$, lubricants containing extreme pressure and antifatigue additives are required.

In the selection of lubricants for gears, tooth sliding is considered from two aspects:

1. It tends to increase the operating temperature because of frictional effects.
2. Sliding along the line of contact tends to wipe the lubricant away from the convergent zone; thus, it is more difficult to form lubricating films.

Factors Affecting Lubrication of Enclosed Gears The lubricant in an enclosed gear set, which represents the major portion of gear usage, is subjected to very severe service. It is thrown from the gear teeth and shafts in the form of a mist or spray. In this atomized condition, it is exposed to the oxidizing effect of air. Fluid friction and, in some cases, metallic friction generate heat which raises the lubricant temperature. The violent churning and agitation of the lubricant by the gears of splash lubricated sets also raises the temperature. Raising the temperature increases the rate of oxidation. Sludge or deposits, formed as a result of oil oxidation, can restrict oil flow, or interfere with heat flow in oil coolers or from the sides of the gear case. Restrictions in the oil flow may cause lubrication failure, while heat insulating deposits decrease cooling and cause further increases in the rate of oxidation. Eventually, lubrication failure and damage to the gears may result.

In selecting the lubricant for enclosed gear sets, in addition to the requirement for adequate oxidation resistance, the following factors of design and operation require consideration:

1. Gear type
2. Gear speed
3. Reduction ratio
4. Operating temperature
5. Transmitted power
6. Surface finish
7. Load characteristics
8. Drive type
9. Application method
10. Water contamination
11. Lubricant leakage

Gear Type   With spur and bevel gears the line of contact runs straight across the tooth face and the direction of sliding is at right angles to the line of contact. Both these conditions contribute to the formation of effective lubricating films. However, only a single tooth carries the whole load during part of the meshing cycle, resulting in high tooth loads. Additionally, if one tooth wears, there is no transfer of load to other meshing teeth to relieve the load on the worn tooth, and wear of that tooth will continue.

Helical, herringbone, and spiral bevel gears always have more than one pair of teeth in mesh. This results in better distribution of the load under normal loading. Under higher loading, the individual tooth contact pressures may be as high as in comparable straight tooth gears under normal loading. The sliding component along the line of contact, because of time and high viscosity of the lubricant in the contact area, has little or no effect on the EHL film in the contact area. In the convergent zone ahead of the contact area, the sliding component tends to wipe the lubricant sideways. Therefore, not as much lubricant is available to be drawn into the contact area, and the resultant pressure increase in the convergent zone may not be as great. These effects may contribute to a need for slightly higher viscosity lubricants, although, in general, oils for these types of gears are selected on the same basis as for straight tooth gears.

An additional factor with helical, herringbone, and spiral bevel gears is that if one tooth wears, the load is transferred simultaneously to other teeth in mesh. This relieves the load on the worn tooth and may make these types of gear somewhat less critical of lubricant characteristics. The important factor with all of these types of gear is that the lubricant have a high enough viscosity to provide effective oil films, but not so high that excess fluid friction will occur.

The high rate of side sliding in worm gears results in considerable frictional heating. Generally, the rolling velocity is generally quite low so the velocity tending to carry the lubricant into the contact area is low. Combined with the sliding action tending to wipe the lubricant along the convergent zone, this makes it necessary to use high viscosity lubricants. In order to help reduce the wiping effect and reduce friction, lubricants containing friction reducing materials or antiwear additives are usually used.

Hypoid gears are of steel to steel construction and are heat treated. They are designed to transmit high power in proportion to their size. Combined with the side sliding that occurs, these gears operate under boundary or mixed film conditions essentially all the time, and require lubricants containing active extreme pressure additives.

Gear Speed The higher the speed of meshing gears, the higher will be the sliding and rolling speeds of individual teeth. When an ample supply of lubricant is available, speed assists in forming and maintaining fluid films. At high speed, more oil is drawn into the convergent zone; in addition, the time available for the oil to be squeezed from the contact area is less. Therefore, comparatively low viscosity oils may be used, for despite their fluidity there is insufficient time to squeeze out the oil film. At low gear speeds, however, more time is available for oil to be squeezed from the contact area and less oil is drawn into the convergent zone; thus, higher viscosity oils are required.

Reduction Ratio Reduction ratio influences the selection of the lubricating oil because high ratios require more than one step of reduction. When the reduction is above about 3:1 or 4:1, multiple reduction gear sets are usually used and above about 8:1 or 10:1, they are nearly always used. In a multiple reduction set, the first reduction operates at the highest speed and so requires the lowest viscosity oil. Subsequent reductions operate at lower speeds so require higher viscosity oils. The low speed gear in a gearset is usually the most critical in the formation of an EHL film. In the case of a gear reducer, this would be the output gear. In very high speed gear reducers, both the lowest speed and highest speed gears should be checked to determine the more critical condition. In some cases, a dual viscosity system may be employed, using a lower viscosity oil for the high speed gears and a higher viscosity oil for the low speed gears. In some gear sets, this can be accomplished automatically by circulating the cool oil first to the low speed gears, and then, after it is heated and its viscosity decreased, to the high speed gears.

Operating Temperature The temperature at which gears operate is an important factor in the selection of the lubricating oil since viscosity decreases with increasing temperature and oil oxidizes more rapidly at high temperatures. Both the ambient temperature where the gear set is located and the temperature rise in the oil during operation must be considered.

When gear sets are located in exposed locations, the oil must provide lubrication at the lowest expected starting temperature. In splash lubricated units, this means that the oil must not channel at this temperature, while in pressure fed gear sets, the oil must be fluid enough to flow to the pump suction. At the same time, the oil must have a high enough viscosity to provide proper lubrication when the gears are at operating temperature.

During operation, the heat generated by metallic friction, between the tooth surfaces and by fluid friction in the oil, will cause the temperature of the oil to rise. The final operating temperature is a function of both this temperature rise in the oil and the ambient temperature surrounding the gearcase. Thus, a temperature rise of 90°F (50°C) and an ambient temperature of 60°F (15.6°C) will produce an operating temperature of 150°F (66°C), while the same temperature rise at an ambient temperature of 100°F (38°C) will produce an operating temperature of 190°F (88°C). In the latter case, an oil of higher viscosity and better oxidation stability would be required to provide satisfactory lubrication and oil life at the operating temperature. For gear sets equipped with heat exchangers in the oil system, both the ambient temperature and the temperature rise are less important since the operating temperature of the oil can be adjusted by varying the amount of heating or cooling.

Transmitted Power As noted in the discussion of EHL film formation, load does not have a major influence on the thickness of EHL films. However, it cannot be ignored. As load is increased, the viscosity of the lubricant must be increased to maintain the same thickness of oil films.

Load also has an influence on the amount of heat generated by both fluid and mechanical friction. Gears designed for higher power ratings will have wider teeth, or teeth of larger cross section, or both. Regardless, a greater surface area will be swept as the teeth pass through mesh, so mechanical and fluid friction will be greater. At the same time, the relative area of radiating surface in proportion to the heat generated is usually less in a large gear set than in a small one. As a result, larger gear sets, transmitting more power, tend to run hotter unless they are equipped with oil coolers. However, if the operating temperature of a gear set is properly taken into account in the selection of lubricant viscosity, the heating effects based on the amount of power transmitted will be taken care of.

Surface Finish As discussed, surface roughness has an important influence on the thickness of oil films required for proper lubrication. Rougher surfaces require thicker oil films to obtain complete separation, and higher viscosity oils. On the other hand, smoother surfaces can be lubricated successfully with lower viscosity oils. Since some smoothing of the surfaces results from running in, some authorities recommend using an estimated "run in" surface roughness rather than the "as finished" values in oil film thickness calculations and for selection of oil viscosity.

Load Characteristics The nature of the load on any gear set has an important influence on the selection of a lubricating oil. If the load is uniform, the torque (turning effort) and the load carried by the teeth will also be uniform. However, excessive tooth loads due to shock loads may tend to momentarily rupture the lubricating films. Therefore, where the shock factor has not been considered in the design or selection of the gear set, a higher than normal oil viscosity is required to prevent film rupture.

In some operations, the conditions may be still more severe due to overloads or to a combination of heavy loads and extreme shock loads, for instance, on rolling mill stands or where gears are started under heavy load. In such cases, it may be impossible to maintain an effective oil film. Hence, during a considerable part of mesh, boundary lubrication exists. This condition generally requires the use of extreme pressure (EP) oils.

Occasionally, due to lack of space or other limiting and unavoidable factors, gears are loaded so heavily that it is difficult to maintain an effective lubricating film between the rubbing surfaces. Such a condition is quite usual for hypoid gears in the automotive field. When operating under this condition of extreme loading, metal-to-metal contact is so severe that wear cannot be completely avoided. However, it can be controlled by the use of special extreme pressure lubricants containing additives designed to prevent welding and surface destruction under severe conditions. Only slow wear of a smooth and controlled character will then take place.

Drive Type Electric motors, steam turbines, hydraulic turbines, and gas turbines produce uniform torque. Therefore, when the power transmitted by gears is developed by one of these prime movers, gear tooth loading is uniform. Reciprocating engines, however, produce variable torque, so some variation in gear tooth loading results. When gears are driven by steam engines, diesel engines, and so forth, higher viscosity oils may be required to assure effective oil films. Higher viscosity oils may not be necessary when the type of drive has been considered and compensated for in the design or selection of the gear set.

Application Method When lubricating oil is applied to gear teeth by means of a splash system, the formation of an oil film between the teeth is less effective than when the oil is circulated and sprayed directly on the meshing surfaces. This is particularly true of low speed, splash lubricated units in which only a limited amount of oil may be carried to the meshing area. A higher viscosity oil is needed to offset this condition, since higher viscosity will result in more oil clinging to the teeth and being carried into the mesh.

When a gear set is lubricated by a pressure system rather than a splash lubrication system, there is better dissipation of heat. This is because the pressure tends to throw the oil against all internal surfaces of the gear case, and more heat is conducted away by these radiating surfaces. With a splash system, particularly a low speed unit, the oil may dribble over only a small part of the internal surface of the gear case, thus restricting heat dissipation. As a result, splash lubricated units usually run hotter and require higher viscosity oils.

Water Contamination Water sometimes finds its way into the lubrication systems of enclosed gears. This water may come from cooling coils, condensed steam, or condensation of moisture in the atmosphere. In the latter case, it is often an indication of inadequate venting of the gear case and oil reservoir. It is apt to occur in gear sets operated intermittently, with warm periods of operation alternating with cool periods of idleness. Where moisture contamination may occur, it is necessary to use an oil with good demulsibility, that is, an oil that separates readily from water.

Water and rust also act to speed up deterioration of the oil. Water separates slowly, or not at all, from oil that has been oxidized or contaminated with dirt. In this respect, iron rust is a particularly objectionable form of contamination. Water in severely oxidized oil or dirty oil usually forms stable emulsions. Such emulsions may cause excessive wear of gears and bearings by restricting the amount of oil flowing through pipes and oilways to the gears and bearings. The fact that oxidized oil promotes the formation of stable emulsions constitutes another reason for using an oxidation resistant oil in enclosed gears. Obviously then, to protect gear tooth surfaces and bearings, the oil must not only separate quickly from water when new but must also have the high chemical stability necessary to maintain a rapid rate of separation even after long service in a gear case.

Lubricant Leakage Although most enclosed gear cases are oil tight, extended operation or more severe operating conditions may result in lubricant leakage at seals or joints in the casing. Where the amount of leakage is high and cannot be controlled by other methods, special lubricants designed to resist leakage may be required.

Lubricant Characteristics for Enclosed Gears The necessary characteristics of lubricants for enclosed gears may be summarized as follows:

1. Correct viscosity at operating temperature to assure distribution of oil to all rubbing surfaces, and formation of protective oil films at prevailing speeds and pressures
2. Adequate low temperature fluidity to permit circulation at the lowest expected start up temperature
3. Good chemical stability to minimize oxidation under conditions of high temperatures and agitation in the presence of air, and to provide long service life for the oil
4. Good demulsibility to permit rapid separation of water and protect against the formation of harmful emulsions
5. Antirust properties to protect gear and bearing surfaces from rusting in the presence of water, entrained moisture, or humid atmospheres
6. Noncorrosive to prevent gears and bearings from being subjected to chemical attack by the lubricant
7. Foam resistance to prevent the formation of excessive amounts of foam in reservoirs and gear cases

In addition to these characteristics, many modern gear sets operating under severe service conditions or where loads are heavy or shock loads are present require lubricants with extreme pressure (EP) properties to minimize scuffing and destruction of gear tooth surfaces. Worm gears usually require lubricants with mild wear and friction reducing properties, or mild extreme pressure properties. Some other gear sets operating under severe service conditions require lubricants with leak resistant properties to minimize lubricant loss from the gear case. In some cases where heavy water contamination is unavoidable, lubricants that are heavier than water may be required. Heavy lubricants will displace the water from the gear cases.

AGMA Specifications for Lubricants for Enclosed Gear Drives AGMA (American Gear Manufacturers Association) Standard 250.03 provides specifications for rust and oxidation (R&O) and mild extreme pressure (EP) gear lubricants for industrial enclosed gearing, and tables of viscosity grade recommendations for various types of gearing. The viscosity grade ranges correspond to those in ASTM D2422, "Standard Recommended Practice for Viscosity System for Industrial Fluid Lubricants" (see Chap. 2), although the AGMA grades in the current issue of the standard (1972) are based on viscosities in Saybolt Universal Seconds measured at 100°F. A later AGMA Standard, 251.02 "Lubrication of Open Gearing," indicates that until the changeover to 40°C for viscosity determinations is fully implemented, lubricants falling within the stated viscosity tolerances at either 100°F or 40°C are acceptable. The AGMA grades and the corresponding ISO Viscosity Grades are shown in Table 5-2.

These AGMA lubricant numbers are recommended for the lubrication of enclosed industrial gearing as shown in Tables 5-3 and 5-4.

Factors Affecting Lubrication of Open Gears

In contrast to enclosed gears flood lubricated by splash or circulation systems, there are many gears for which it is not practical or economical to provide oil tight housings. These so-called open gears can only be sparingly lubricated and perhaps only at infrequent intervals.

Gears of this type are lubricated by either a continuous or an intermittent method. The three most common continuous methods are splash, idler immersion, and pressure. In the first two the lubricant is lifted from a reservoir or sump (sometimes referred to as a slush pan) by the partially submerged gear or an idler. Pressure systems require a shaft or independently driven pump to draw oil from a sump and spray it over the gear teeth. Most gears lubricated in any of these ways are equipped with relatively oil tight enclosures.

Many different intermittent methods of application are used. Some are arranged for automatic timing, while others must be controlled manually. Methods used include automatic spray, semiautomatic spray, forced feed lubricators, gravity or forced drip, and hand application by brush. When grease type lubricants are used, hand or power grease guns or a centralized lubrication system can be used.

With the continuous methods of application, lubrication of open gears is similar to that with enclosed gears. Since the gears are usually large and relatively slow moving, higher viscosity lubricants are required.

With the intermittent methods of application, fluid films may exist when lubricant is first applied to the gears. However, these films quickly become thinner as the lubricant is squeezed aside until only extremely thin

Table 5-2   AGMA Viscosity Grades for Enclosed Gearing

AGMA Lubricant No.	Viscosity Range SUS @ 100°F	Corresponding ISO Viscosity Grade
1	193–235	46
2, 2 EP	284–347	68
3, 3 EP	417–510	100
4, 4 EP	626–765	150
5, 5 EP	918–1122	220
6, 6 EP	1335–1632	320
7 comp* 7 EP	1919–2346	460
8 comp 8 EP	2837–3467	680
8A comp	4171–5098	1000

*Oils marked "comp" are compounded with 3 to 10 percent fatty or synthetic fatty oils.

Table 5-3 AGMA Lubricant Number Recommendations for Enclosed Helical, Herringbone, Straight Bevel, Spiral Bevel, and Spur Gear Drives

Type of Unit	Size of Unit Main Gear Low Speed Centers	Other Lubricants, Ambient Temperature, °F		AGMA Lubricant Number Ambient Temperature, °F	
		−40 to 0	−20 to +25	15 to 60	50 to 125
Parallel shaft, single reduction	up to 8 in.	Automatic Transmission Fluid (or similar product—see note 5)	SAE 10W/30 or 10W/40 Motor Oil (or similar product—see note 5)	2–3	3–4
	over 8 in. and up to 20 in.			2–3	4–5
	over 20 in.			3–4	4–5
Parallel shaft, double reduction	up to 8 in.			2–3	3–4
	over 8 in. and up to 20 in.			3–4	4–5
	over 20 in.			3–4	4–5
Parallel shaft, triple reduction	up to 8 in.			2–3	3–4
	over 8 in. and up to 20 in.			3–4	4–5
	over 20 in.			4–5	5–6
Planetary gear units	O.D. housing up to 16 in.			2–3	3–4
	O.D. housing over 16 in.			3–4	4–5
Spiral or straight bevel gears units	Cone distance up to 12 in.			2–3	4–5
	Cone distance over 12 in.			3–4	5–6
Gearmotors and shaft mounted units				2–3	4–5
High speed units (See note 4)				1	2

1. Pour point of the lubricant selected should be at least 10°F lower than the expected minimum ambient starting temperature. If ambient starting temperature approaches lubricant pour point, oil sump heaters may be required to facilitate starting and ensure proper lubrication.
2. Ranges are provided to allow for variations in operating conditions such as surface finish, temperature rise, loading, speed, etc.
3. AGMA viscosity number recommendations listed above refer to R&O gear oils shown in Table 5-2. EP gear lubricants in the corresponding viscosity grades may be substituted where deemed necessary by the gear drive manufacturer.
4. High speed units are those operating at speeds above 3600 rpm or pitch line velocities above 500 fpm. Refer to Standard AGMA 421.06, *Practice for High Speed Helical & Herringbone Gear Units*, for detailed lubrication recommendations.
5. When they are available, good quality industrial oils having similar properties are preferred over the automotive oils. The recommendation of automotive oils for use at ambient temperatures below +15°F is intended only as a guide pending widespread development of satisfactory low temperature industrial oils. Consult gear manufacturer before proceeding.
6. Drives incorporating overrunning clutches as backstopping devices should be referred to the clutch manufacturer as certain types of lubricants may adversely affect clutch performance.

Source: Extracted from AGMA Specification — *Lubrication of Industrial Enclosed Gear Drives* (AGMA 250.03), with the permission of the publisher, the American Gear Manufacturers Association, 1901 North Fort Meyer Drive, Arlington, Va. 22209.

Table 5-4   AGMA Lubricant Number Recommendations for Enclosed Cylindrical and Double-Enveloping Worm Gear Drives

Worm Centers	Type of Worm	Worm Speed Up to rpm	AGMA Lubricant No. Ambient Temperature, °F		Worm Speed[b] Above rpm	AGMA Lubricant No. Ambient Temperature, °F	
			15–60[a]	50–125		15–60[a]	50–125
Up to 6 in. inclusive	Cylindrical	700	7 comp	8 comp	700	7 comp	8 comp
	Double-Enveloping	700	8 comp	8A comp	700	8 comp	8 comp
Over 6 in. centers up to 12 in. centers	Cylindrical	450	7 comp	8 comp	450	7 comp	7 comp
	Double-Enveloping	450	8 comp	8A comp	450	8 comp	8 comp
Over 12 in. centers up to 18 in. centers	Cylindrical	300	7 comp	8 comp	300	7 comp	7 comp
	Double-Enveloping	300	8 comp	8A comp	300	8 comp	8 comp
Over 18 in. centers up to 24 in. centers	Cylindrical	250	7 comp	8 comp	250	7 comp	7 comp
	Double-Enveloping	250	8 comp	8A comp	250	8 comp	8 comp
Over 24 in. centers	Cylindrical	200	7 comp	8 comp	200	7 comp	7 comp
	Double-Enveloping	200	8 comp	8A comp	200	8 comp	8 comp

[a]Pour point of the oil used should be less than the minimum ambient temperature expected. Consult gear manufacturer on lube recommendations for ambient temperatures below +15°F.

[b]Wormgears of either type operating at speeds above 2400 rpm or 2000 ft./min, rubbing speed may require force feed lubrication. In general, a lubricant of lower viscosity than recommended in the above table may be used with a force feed system.

Note: Worm gear drives will also operate satistactorily using R & O gear oils, sulfur phosphorous EP oils, or lead naphthenate EP oils. These oils should be used, however, only upon approval by the manufacturer.

Source: Extracted from AGMA Specification — Lubrication of Industrial Enclosed Gear Drives (AGMA 250.03), with the permission of the publisher, the American Gear Manufacturers Association, 1901 North Fort Meyer Drive, Arlington, Va. 22209.

films remain on the metal surfaces. During much of the time, therefore, these gears operate under conditions of boundary lubrication. Under boundary conditions, the extremely thin film must resist being rubbed or squeezed off the surfaces of the teeth. The ability of the film to resist the rubbing action depends both on its viscosity and the action that takes place between the lubricant and the metal surfaces. The lubricating film must bond so strongly to the tooth surfaces that metal-to-metal contact (and resultant wear) is minimized. With a properly selected lubricant applied at sufficiently frequent intervals, wear may be kept to a negligible amount.

Temperature Whether or not it is possible to maintain adequate boundary films under the pressures existing between meshing teeth depends, among other things, on the operating temperature to which the gears are exposed. Heat causes the lubricant to thin out and drop off or be thrown off the gear teeth, decreasing the amount of lubricant remaining on the rubbing surfaces. This results in thinner films. Temperature also decreases the resistance of the lubricant to being rubbed off the surface. Thus, high temperatures require more viscous lubricants. Conversely, when gears operate at low temperatures, the lubricant must not become so viscous or hard that it will not distribute properly over the tooth surfaces. At the same time, the lubricant must be such that it does not harden and chip or peel from the teeth at the lowest temperatures encountered.

Dust and Dirt Many open gears, whether operating outdoors or indoors, are exposed to dusty and dirty conditions. Abrasive dust, adhering to oil wetted surfaces, will form a lapping compound that causes excessive wear of the teeth. When viscous lubricants are used, the dirt may pack in the clearance space at the roots of the teeth, forming hard deposits. Packed deposits between gear teeth tend to spread the gears and overload the bearings.

Water Open gears operating outdoors are often exposed to rain or snow, and outdoor and indoor gears to the splash of process fluids. The lubricant must resist being washed off the gears by these fluids in order to protect the gear tooth surfaces against wear or rust and corrosion.

Method of Application The method of application must be considered when selecting a lubricant for open gears. If the lubricant is to be applied by drip, force feed lubricator, or spray, it must be sufficiently fluid to flow through the application equipment. For brush application, the lubricant must be sufficiently fluid to be brushed evenly on the teeth. In any case, during operation, the lubricant should be viscous and tacky to resist squeezing from the gear teeth. Some types of very viscous lubricants can be thinned for application by heating, or diluent type products may be used. These latter products contain a nonflammable diluent that reduces the viscosity sufficiently for application. Shortly after application, the diluent evaporates, leaving the film of viscous base lubricant to protect the gear teeth. When open gears are lubricated by dipping into a slush pan, the lubricant must not be so heavy that it channels as the gear teeth dip into it, and it must offer excessive drag. When open gears are lubricated by grease type materials, the consistency and pumpability must permit easy application under the ambient conditions prevailing.

Table 5-5 AGMA Viscosity Grades for Open Gear Lubricants

AGMA Lubricant No.	Viscosity Range			Corresponding ISO Viscosity Grade
	SUS @ 100°F	SUS @ 210°F	cSt @ 210°F	
4, 4 EP	626–765	—	—	150
5, 5 EP	918–1122	—	—	220
6, 6 EP	1335–1632	—	—	320
7, 7 EP	1919–2346	—	—	460
8, 8 EP	2837–3467	—	—	680
9, 9 EP	6260–7650	—	—	1500
10, 10 EP	13,350–16,320	—	—	—
11, 11 EP	19,190–23,460	—	—	—
12, 12 EP	28,370–34,670	—	—	—
13, 13 EP	—	850–1000	182–214	—
Residual Compounds*				
14R	—	2000–4000	429–858	—
15R	—	4000–8000	858–1715	—

*Residual Compounds — Diluent Type, commonly known as solvent cutbacks, are heavy bodied oils containing a volatile, nonflammable, diluent for ease of application. The diluent evaporates, leaving a thick film of lubricant on the gear teeth. Viscosities listed are for the base compound without diluent.

Caution: These lubricants may require special handling and storage procedures. Diluents can be toxic and irritating to the skin. Consult lubricant supplier's instructions.

Load Characteristics Open gears are often heavily loaded and shock loads may be present. These conditions usually require the use of lubricants with enhanced antiwear properties.

AGMA Specifications for Lubricants for Open Gearing The AGMA standard 251.02 contains specifications for three types of open gear lubricants: R&O gear oils, EP gear oils, and residual gear oils. The viscosity ranges for the various grades are shown in Table 5-5. As in the case of the AGMA grades for enclosed gearing, products meeting the ranges at either 100°F or 40°C are acceptable.

The recommendations for the use of these various grades are based on the method of application, ambient temperature, gear speed, and character of operation. Tables 5-6 and 5-7 contain these recommendations.

CYLINDERS

Cylinders and their pistons are essential elements of a great variety of machines, including compressors, steam engines, internal combustion engines, hydraulically operated machines, some types of air motors, percussion air tools, and linear actuators that operate a great variety of control mechanisms. In general, where gases are involved, cylinders are lubricated with meager quantities of oil. Nevertheless, mechanical wear due to inadequate lubrication is seldom a problem. Wear that does occur usually is the result of abrasive contaminants or, in some cases, corrosion.

Generally, the main problems with cylinder lubrication result from conditions imposed on the oil after it has left the cylinder, or, as in internal

Table 5-6 Recommended AGMA Lubricants for Industrial Open Gearing Continuous Methods of Application

Ambient Temperature[a] °F	Character of Operation	Pressure Lubrication Pitch Line Velocity		Splash Lubrication Pitch Line Velocity		Idler Immersion Pitch Line Velocity
		Under 1000 ft/min	Over 1000 ft/min	Under 1000 ft/min	1000 to 2000 ft/min	Up to 300 ft/min
15–60[b]	Continuous	5 or 5 EP	4 or 4 EP	5 or 5 EP	4 or 4 EP	8–9 8 EP–9 EP
	Reversing or Frequent "Start Stop"	5 or 5 EP	4 or 4 EP	7 or 7 EP	5 or 6 EP	8–9 8 EP–9 EP
50–125[b]	Continuous	7 or 7 EP	6 or 6 EP	7 or 7 EP	6 or 6 EP	11 or 11 EP
	Reversing or Frequent "Start Stop"	7 or 7 EP	6 or 6 EP	9–10[c] 9 EP–10 EP	8–9[d] 8 EP–9 EP	11 or 11 EP

[a]Temperature in vicinity of the operating gears.
[b]When ambient temperatures approach the lower end of the given range, lubrication systems must be equipped with suitable heating units for proper circulation of lubricant and prevention of channeling. Check with lubricant and pump suppliers.
[c]When ambient temperature remains between 90 and 125°F at all times, use 10 or 10 EP.
[d]When ambient temperature remains between 90 and 125°F at all times, use 9 or 9 EP.

Source: Extracted from AGMA Specification — *Lubrication of Industrial Open Gearing* (AGMA 251.02), with the permission of the publisher, the American Gear Manufacturers Association, 1901 North Fort Meyer Drive, Arlington, Va. 22209.

Table 5-7   Recommended AGMA Lubricants for Industrial Open Gearing
Intermittent Methods of Application — Limited to 1500 ft./min Pitch Line Velocity

| Ambient Temperature[a] °F | Mechanical Spray Systems[b] | | Gravity Feed or Forced Drip Method Using Extreme Pressure Lubricant |
	Extreme Pressure Lubricant	Residual Compound[c]	
15 to 60	—	14R	—
40 to 100	12 EP	15R	12 EP
70 to 125	13 EP	15R	13 EP

[a]Ambient temperature is temperature in vicinity of the gears.
[b]Greases are sometimes used in mechanical spray systems to lubricate open gearing. A general purpose EP grease of number 1 consistency (NGLI) is preferred. Consult gear manufacturer and spray system manufacturer before proceeding.
[c]Diluents must be used to facilitate flow through applicators.
[d]Feeder must be capable of handling lubricant selected.

Source: Extracted from AGMA Specification — *Lubrication of Industrial Open Gearing* (AGMA 251.02), with the permission of the publisher, the American Gear Manufacturers Association, 1901 North Fort Meyer Drive, Arlington, Va. 22209.

combustion engines, on conditions imposed by processes that occur within the cylinder. This interrelationship between the lubrication of cylinders and the machines of which they are part, makes it more practical to consider cylinder lubrication in later sections as part of the overall machine lubrication.

FLEXIBLE COUPLINGS

When two rotating shafts are to be connected, it is unavoidable that there will be some degree of misalignment, either because of static effects such as deflection of the shafts, thermal effects causing expansion and contraction, or dynamic loads causing the shafts to change positions in their supporting bearings. Misalignment may be angular where the shafts meet at an angle that is not 180 degrees, parallel where the shafts are parallel but displaced laterally, or axial such as results from end play. In some cases, all three types of misalignment may be present. To accommodate misalignment, flexible couplings of various types are used to connect shafts together.

In addition to protecting the machines against stresses resulting from misalignment, flexible couplings transmit the torque from the driving shaft to the driven shaft, and may help to absorb shock loads.

Universal joints and constant velocity joints may properly be considered as flexible couplings. Both types of joint will accommodate relatively large amounts of misalignment, and are used to some extent in industrial applications. However, since they are used to a greater extent in automotive applications; see Chap. 13 for a detailed discussion. In this chapter, only those types of flexible couplings used in industrial applications, where the amount of misalignment is relatively small, are considered.

Several types of lubricated flexible couplings are in use. Gear type couplings have hubs with external gear teeth (or splines) keyed to the shafts. A shell or sleeve with internal gear teeth at each end meshes with the teeth on the hubs and transmits the torque from the driving shaft to the driven shaft. Misalignment is accommodated in the meshing gears. Where large amounts of misalignment must be accommodated, special tooth forms that permit greater angular motion are used.

Chain couplings have sprockets connected by a chain wrapped around and joined at the ends. The chain may be of either the roller or silent type. Roller chain couplings depend on the relative motion between the rollers and sprockets for their flexibility. Barrel shaped rollers and special tooth forms may be used where large amounts of misalignment must be accommodated. Silent chain couplings depend on the clearance between the chain links and sprocket teeth for their flexibility.

Spring laced couplings, also called steel grid or spring type couplings, are made of two serrated flanges laced together with metal strips. Flexibility is obtained from movement of the metal strips in the serrations.

There are two types of sliding couplings. Sliding block couplings have two C shaped jaw members connected by a square center member. The jaws are free to slide on the surfaces of the block to accommodate misalignment. Sliding disc couplings have flanged hubs with slots machined in the flanges. These slots engage jaw projections on a center disc. The floating center disc allows relative sliding movement between the flange slots and the disc jaws.

Slipper type couplings, used mainly on rolling mills, are actually a type of universal joint. A C-shaped jaw attached to one shaft is fitted with pads that can turn in their recesses. A tongue on the other shaft fits between the pads. The tongue can move sideways in the pads to accommodate lateral misalignment, and the pads can rotate in their recesses to accommodate angular misalignment.

Flexing member couplings compensate for misalignment both by flexing of the components and sliding between them. The radial spoke coupling uses thin steel laminations to connect the coupling flanges. Clearance in the slots of the outer flange and flexing of the laminations allows for misalignment. Axial spoke couplings have flexible pins mounted in sintered bronze bushings connecting the flanges. The pins can flex or move in and out in their bushings to accommodate misalignment.

Lubrication of Flexible Couplings

All motion in flexible couplings is of the sliding type. Motion is more or less continuous when a coupling is revolving but the amount of motion may be small, varying with the amount of misalignment between the shafts. Contact pressure may be high, particularly when there is considerable misalignment, since misalignment tends to reduce the surface area over which the load is distributed. The thickness of the lubricating films formed and the type of lubricant required depend to a considerable extent on the type of coupling.

Gear Couplings Gear couplings can transmit more torque than any other type of flexible coupling of equal size. Since they rotate continuously, a supply of lubricant is usually contained in the assembly. As the unit revolves, centrifugal force causes the lubricant to be thrown outward forming an annulus of lubricant which should completely submerge the gear teeth. The relative motion between the gear teeth may be sufficient to create and maintain appreciably thick lubricating films.

Either oil or grease may be used for gear couplings. Greases are often used where speeds are normal to high and temperatures are moderate. For normal speeds, a soft grease made with a high viscosity oil, often with extreme pressure properties, is usually used. For higher speeds, a stiffer grease may be required.

Oil is generally preferred for gear couplings operating at normal to high speeds and high temperatures when the coupling is large enough to hold a reasonable supply of oil. Couplings may be oil filled, in which case they require filling only with enough oil to cover the gear teeth during operation, or supplied continuously. Continuously supplied couplings may be designed to recirculate the oil, or to collect it so it can be drained and disposed of after use. Continuous supply systems generally are required where operating temperatures are high. Regardless of the method of applying the oil, quite high viscosity oils with good antiwear properties are required.

Chain Couplings Chain couplings are usually grease lubricated. Couplings without dust covers are lubricated by brushing with grease periodically. The grease must resist throwoff and must penetrate into the chain joints to provide lubrication. Generally, soft greases with good adhesive properties are usually required. Couplings with covers are packed with grease, usually of a type similar to those used in geared couplings.

Spring Laced Couplings These couplings are grease packed. Again, greases similar to those used in geared couplings are usually required.

Sliding Couplings Sliding block couplings may have the block formed of an oil impregnated sintered metal. In other cases the block contains a reservoir. In the latter case, high viscosity oils with good adhesiveness are required to minimize throwoff. Sliding disc couplings require similar oils.

Slipper Joint Couplings Slipper joint couplings are usually large and carry both heavy loads and shock loads so the lubrication requirements are severe. All loss systems are usually used to apply the lubricant. High viscosity oils with enhanced film strength are required.

Flexible Member Couplings Axial spoke couplings have prelubricated bushings for the pins. When enclosed in a dust cover, a soft, adhesive grease is required for packing. Radial spoke couplings require a high viscosity oil or semifluid grease.

DRIVE CHAINS

Drive chains are used to transmit power in a wide variety of applications. Chains fall into two general categories:

1. Machined surface chains suitable for high speed, precision drives
2. Cast or forged link chains which are usually made without machined surfaces and are suitable for lower speeds and powers or where environmental conditions dictate the use of low cost drives

Silent and Roller Chains These chains are precision mechanisms that, when properly lubricated, will transmit high power at high speeds and will give long service life. Silent chains, also called inverted tooth chains, operate in such a manner that the links of the chain almost completely fill the clearance space in the sprocket throughout the arc of contact, greatly reducing backlash and minimizing noise. Roller chains are available in various designs, and with up to 20 strands or more for use where high powers must be transmitted.

With silent chains, the lubricant must penetrate between the links and distribute along the seat pins to provide lubricating films on the rubbing surfaces. With roller chains, the lubricant must coat the outsides of the rollers to lubricate the contact surfaces of the rollers and sprockets and must also penetrate and distribute along the rubbing surfaces between the rollers and bushings. As a result, in both types of chain, the viscosity of the lubricating oil used is extremely important. Oils with enhanced film strength are also generally desirable, both to improve protection against wear and to permit the use of lighter oils with better penetrating properties.

Silent and roller chains usually are installed in oiltight housings where the lubricant is applied by force feed, oil mist, slinger, or oil ring, or by the chain or sprocket dipping into a bath of oil. Occasionally these chains are used without housing for low speed, low power drives. For applications of this type, the lubricant is applied by one of the all loss methods such as drip, wick feed oiler, brush, oil mist, or pour can. The characteristics of the method of application must be taken into account in lubricant selection, as is the requirement for controlling drip and throw off when the chain is not enclosed.

Normally drive chains of these types are not grease lubricated. Under certain environmental conditions, such as corrosive atmospheres, a properly selected grease may provide better protection than oil.

Cast Off Forged Link Chains   These include detachable link, pintle, fabricated and cast roller types. Detachable link chains may be of cast malleable iron, or medium carbon steel hardened to improve strength and wear characteristics. Pintle links may be either cast or forged.

These chains usually operate at low speeds transmitting relatively low power, although fabricated chains suitable for speeds up to 1000 ft. (300 m) per minute are available. The chains may be open, or fitted with dust covers. Under most conditions, they require viscous, tacky oils, but under dusty or dirty conditions lower viscosity oils may reduce dirt pick up. Occasionally, open gear lubricants that have a dry, rubbery surface after application may provide better performance.

Viscosity Selection   The American Chain Association has proposed lubricant viscosities for silent and roller chains as shown in Table 5-8. Where drip or throwoff may be a problem, higher viscosity oils or oils with special adhesive properties may be desirable. Where water washing occurs, oils compounded to resist water washing and protect against rusting are required.

CAMS AND CAMS FOLLOWERS

There are many cases in machines where rotary motion must be converted to linear motion. Where the linear motion required is comparatively short, a

Table 5-8 Viscosity of Oil for Silent and Roller Chains

Ambient Temperature		Recommended Viscosity
°F	°C	
−20−+20	−30−−7	SAE 10
20−40	−7−4	SAE 20
40−100	4−38	SAE 30
100−120	38−49	SAE 40
120−140	49−60	SAE 50

common method of accomplishing this is with a cam and cam follower combination — in some applications called a "tappet." Probably the best known application of these mechanisms is for operating poppet valves in internal combustion engines, but many other applications also exist. Since the cam will only lift the cam follower, some arrangement must be made to keep the cam follower in contact with the cam. This may be accomplished by transmitted load from the mechanism being actuated, or directly with a loading mechanism such as a spring.

Depending on such factors as loads, speeds, and the complexity of the cam shape, three general types of cam follower are:

1. Flat surfaced cam followers are used in the valve systems of most internal combustion engines.
2. Roller cam followers may be used where extremely high loads are involved.
3. Spherical cam followers may be used where precise conformity to a complex cam is required.

Of these, the roller generally has the least severe lubrication requirements, since the rolling contact tends to develop and maintain oil films.

In many cases with flat cam followers, the cam is tapered and the cylindrical cam follower, which is free to rotate, is located with its axis toward one side of the cam. With this arrangement, the cam follower will be rotated as it is forced upward by the cam. This spreads the wear more uniformly over the surface of the cam follower, and may promote the formation of oil films.

In heavily loaded applications, both flat and spherical cam followers operate on EHL films at least part of the time. Even then, lubricants with antiwear additives are often necessary if rapid wear and surface distress are to be avoided. In the case of four cycle gasoline automobile engines, for example, one of the functions of the zinc dithiophosphates, now used almost universally, is to provide antiwear activity for the camshaft and valve lifters (cam followers).

Most cam and cam follower machine elements are included in lubrication systems supplying other machine elements as well. For this reason, lubricants are rarely selected to meet the requirements of cams and cam followers alone. Generally, the lubricant is selected to meet the basic requirements of the system, and if special requirements exist for the lubrication of the cams and followers, as in the case of a gasoline automobile engine, these must be superimposed on the basic requirements.

WIRE ROPES

Wire rope, or cable as it is sometimes called, is used for a great variety of purposes ranging from standing (stationary) service, as guys or stays and suspension cables, to service involving drawing or hoisting heavy loads. In these various services, all degrees of exposure to environmental conditions are encountered, from clean, dry conditions in applications such as elevator ropes in office buildings, to full exposure to the elements on outdoor equipment, immersion in water, for example, on dredging equipment, and exposure to corrosive environments such as the acid water found in many mines. These and other operating factors require that wire ropes be properly lubricated to provide long rope life and maximum protection against rope failure where the safety of people is involved.

A wire rope consists of several strands laid (helically bent, not twisted) around a core. The core is usually a rope made of hemp or other fiber, but may be an independent wire rope or strand. Each strand consists of several wires laid around a core, which usually consists of one or more wires but may be a small fiber rope. The number of wires per strand ranges from 7 to 37 or more.

Need for Lubrication A number of factors contribute to the need for proper lubrication of wire ropes.

Wear Each wire of a wire rope can be in contact with three or more wires over its entire length. Each contact is theoretically along a line, but this line actually widens to a narrow band because of a deformation under load. As load is applied and as a rope bends or flexes over rollers, sheaves, or drums, stresses are set up which cause the strands and individual wires to move with respect to each other under high contact pressures. Unless lubricating films are maintained in the contact areas, considerable friction and wear result from these movements.

Fatigue One of the principle causes of wire rope failure is metal fatigue. Bending and tension stresses, repeated many times, cause fatigue. Eventually, individual wires break and the rope is progressively weakened to the extent that it must be removed from service. If lubrication is inadequate, the stresses are increased by high frictional resistance to the movement of the wires over one another, fatigue failures occur more quickly, and rope life is shortened.

Corrosion Another principle cause of rope failure is corrosion, a term which covers both direct attack by corrosive fluids, such as acid water that may be encountered in mines and various forms of rusting. To protect against corrosion, lubricant films that resist displacement by water must be maintained on all wire surfaces.

Core Protection   Wear, deterioration, and drying out of the core result in reduction of the core diameter and loss of support for the strands. The strands then tend to overlap, and severe cutting or nicking of the wires may occur. The lubricant applied in service must be of a type that will penetrate through the strands to the core to minimize friction and wear at the core surface, seal the core against water, and keep it soft and flexible.

Lubrication in Manufacture   During manufacture, wire rope cores are saturated with lubricant. A second lubricant, designed to provide a very tenacious film, is usually applied to the wires and strands as they are laid up. These lubricants protect the rope during shipment, storage, and installation.

Lubrication in Service   Much of the core lubricant applied during manufacture is squeezed out when the strands are laid, and additional lubricant is lost from both the core and strands as soon as load is applied to a rope. As a result, in-service lubrication must be started almost immediately after a rope is placed in service.

Proper lubrication of wire ropes in service is not easy to accomplish. Some of the types of lubricants required for wire ropes may not be easy to apply, and often wire ropes are somewhat inaccessible. Various methods of applying lubricants are used, including brushing, spraying, pouring on a running section of the rope, drip or force feed applicators, and running the rope through a trough or bath of lubricant. Generally, the method of application is a function of the type of lubricant required to protect a rope under the conditions to which it is exposed.

Lubricant Characteristics   Wire rope lubricants should have the following characteristics:

1. Form a durable, adhesive coating that will not be thrown or wiped from the rope as it operates over pulleys or drums
2. Penetrate between adjacent wires in order to lubricate and protect them against wear and to keep the rope core from drying out and deteriorating
3. Provide lubrication between pulleys and sheaves and the wire rope
4. Resist being washed off by water
5. Protect against rusting or corrosion by acid, alkaline, or salt water
6. Form nonsticky films so that dust and dirt will not build up on ropes
7. Remain pliable and resist stripping at the lowest temperatures to which the rope will be exposed
8. Resist softening or thinning out to the extent that throwoff or drippage occurs at the highest temperatures at which the rope will operate
9. Be suitable for application to the rope under the service conditions encountered

These requirements necessitate some compromises. Wire rope lubricants may be formulated with asphaltic or petrolatum bases and usually contain rust preventives and materials to promote metal wetting and penetration. Diluent products are used in some cases for ease of application. Grease products containing solid lubricants such as graphite or molybdenum disulfide are now being used to an increasing extent.

Bibliography

Mobil Technical Books
Plain Bearings, Fluid Film Lubrication
Gears and Their Lubrication
Mobil EHL Guidebook

Mobil Technical Bulletins
Rolling Element Bearings — Care and Maintenance
Way Lubrication — Machine Tools
Open Gear Lubrication

6

Lubricant Application

After selecting the proper lubricant for an application, it must be delivered to the elements that require lubrication. Two categories of lubricant application are prevalent—*all loss* methods, where a relatively small amount of lubricant is applied periodically and after use leaks or drains away to waste; and *reuse* methods, where the lubricant leaving the elements is collected and recirculated to lubricate again. Various centralized application systems, not to be confused with circulation systems, are actually specialized all loss systems, and are discussed separately. Reuse application systems are preferred because they conserve lubricant and minimize waste control and environmental pollution.

ALL LOSS METHODS

Some open gears and wire ropes; many drive chains and rolling element bearings; and some cylinders, bearings, and enclosed gears are lubricated by all loss methods. Nearly all grease lubrication except sealed, packed rolling element bearings is an all loss method. Only relatively small amounts of grease are applied, mainly to replenish the lubricating films, but in some cases to flush away some or all of the old lubricant and contaminants.

Oiling Devices The oldest known method of applying lubricants is by using an oil can. With high viscosity lubricants such as are used on open gears and some wire ropes, a paddle, swab, brush, or caulking gun may be required in place of an oil can. These are all variations of hand oiling.

While still widely used, there are several disadvantages to hand oiling. Immediately after application there is usually an oversupply of oil, and excessive leakage or throwoff occurs. Then follows a period when more or less the proper quantity of oil is present; and finally, depending on the frequency of application, there is usually a period when too little oil is present. During this latter period, wear and friction may be high. Also, with hand application, lubrication points may be neglected, either because they are overlooked or because they are difficult or hazardous to reach. Oil leakage onto machine parts, floors, or goods being processed can be hazardous and costly in terms of safety and/or materials wasted. Hand oiling is costly both in terms of labor and because of lost production if machines must be shut down for the purpose. Many devices have been developed and are in wide use to overcome some of the disadvantages of hand oiling. The objective of these devices is to feed oil continuously or at regular, frequent intervals in small amounts with as little attention as possible.

Drop Feed and Wick Feed Cups These devices (Figs. 6-1 and 6-2) are often used to supply the small amount of oil required by high speed rolling element bearings, thin film plain bearings and slides, and some open gears. The rate of oil feed from the drop feed cup can be adjusted with a needle valve, while the wick feed can be adjusted by changing the number of strands in the wick. Both have the disadvantage that they must be started and stopped by hand when the machine is started or stopped. Some drop feed cups are controlled by solenoid operated valves which elminate the problems of manual actuation.

Bottle Oilers In a typical bottle oiler (Fig. 6-3), the spindle of the oiler rests on the journal and is vibrated slightly as the journal rotates. This motion results in a pumping action which forces air into the bottle, causing minute amounts of oil to feed downward along the spindle to the bearing. The oil feed is more or less continuous but stops and starts when the machine is stopped or started.

Wick and Pad Oilers In one type of wick oiler (Fig. 6-4), a felt wick is held against the journal by a spring. The wick draws oil up from the reservoir by capillary action, and the turning journal wipes oil from the wick. No oil is fed when the journal is not turning. In another variation (Fig. 6-5), the wick carries oil up to the slinger, which throws it into the bearing in the form of a fine spray. In both cases, oil leaking out along the shaft drains back to the reservoir so the devices have some elements of a reuse system.

Pad oilers are sometimes used for long, open bearings. One end of the pad rests in an oil reservoir, while the other end rests along the journal. Capillary action draws oil up the pad where it is wiped off by the turning journal.

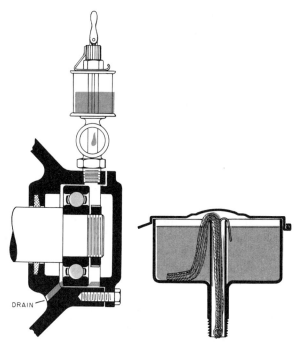

Fig. 6-1 **Drop Feed (left) and Wick Feed Cups** In the drop feed application, the oil drops fall on the lock nut, which throws a spray of oil into the bearing.

Fig. 6-2 **Drip Oiling System**

Fig. 6-3 **Bottle Oiler**

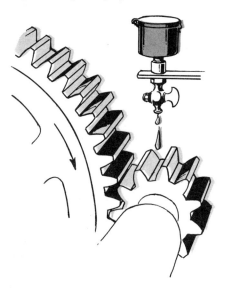

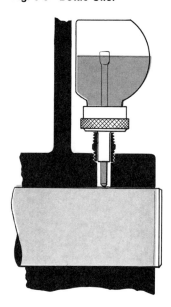

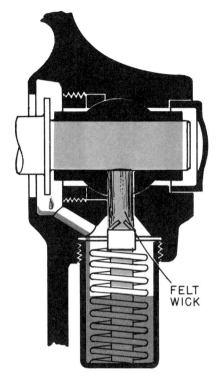

Fig. 6-4   One Type of Wick Oiler

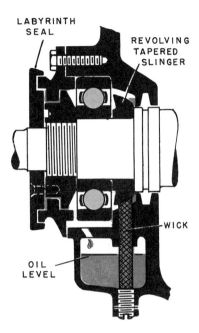

Fig. 6-5   Wick Fed Spray Oiler

Mechanical Force Feed Lubricators   Force feed lubricators are used in applications requiring positive feed of lubricants under pressure. A variety of force feed lubricators are in use. In the type shown in Fig. 6-6, oil is drawn from the reservoir in the base on the downstroke of the single plunger pump and forced under pressure on the upstroke, through the liquid filled sight glass to the delivery line. The pump is operated by an eccentric cam and lever which can be driven from a shaft on the machine, or can be operated by a hand crank. The stroke of the pump can be regulated to adjust the oil feed rate, which can be estimated by counting the drops as they pass through the liquid filled sight glass.

Some modern additive oils can cloud the liquid in the sight glasses of this type of lubricator, so lubricators with dry sight glasses have been developed. One of these is shown in Fig. 6-7. The pump shaft is cam shaped in cross section and causes the plunger to move up and down. Specially shaped surfaces *(A)* at right angles to the pump shaft axis also act on the eccentric head of the plunger, causing it to oscillate about its axis. On the upstroke of the plunger, the head is turned so that the suction slot registers with the suction channel port, while the delivery slot, which is 90° away, is blanked off. Oil is drawn up the suction tube and over to the sight glass where the falling drops may be observed. From the sight glass, the oil flows to the space below the plunger. On the downstroke of the plunger, the head rotates to align the de-

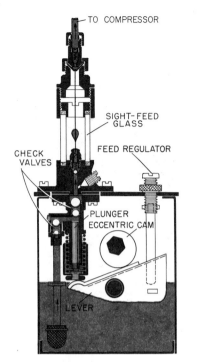

Fig. 6-6 Force Feed Lubricator
with Liquid Filled Sight Glass

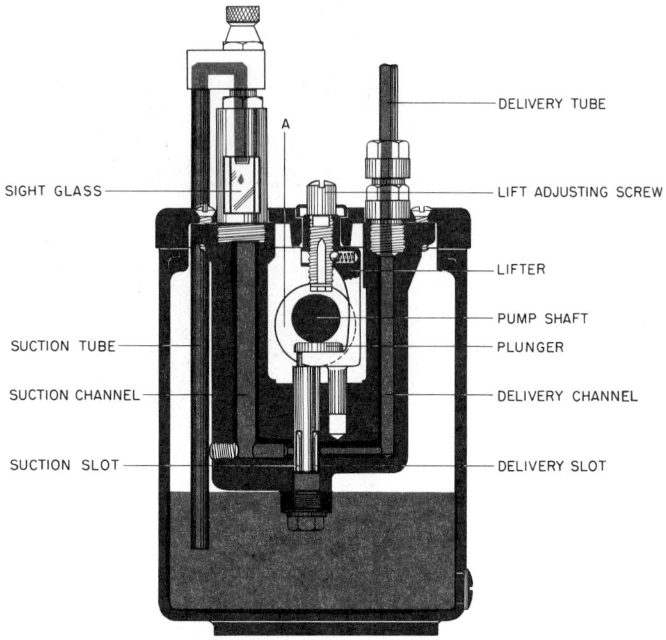

Fig. 6-7   Force Feed Lubricator with Air Drop Sight Glass

livery slot with the delivery channel port; the suction slot is blanked off. As the plunger is pushed downward, it forces oil out through the delivery tube. The plunger stroke, and amount of oil pumped, can be adjusted by means of the lift adjusting screw. The lubricator can be driven by the machine or turned by a hand crank.

Mechanical force feed lubricators are used on large stationary and marine diesel engines, gas engines, and crosshead reciprocating compressors to supply lubricant to the cylinders. One pumping unit is used for each cylinder feed, and all of the pumping units are usually mounted on a single reservoir.

Air Line Oilers*   Air powered cylinders and tools are often lubricated by "lubricating" the compressed air supply. Pneumatic cylinders used for actuating parts of machines may be lubricated by means of an oil fog lubricator such as that in Fig. 6-8. The flow of air through the lubricator creates an air-oil fog which carries sufficient oil to lubricate the cylinders.

Air tools such as rock drills and paving breakers are often lubricated by means of air line oilers. The air line oiler consists of an oil reservoir which contains a device for feeding a metered amount of atomized oil into the air stream. It is coupled into the air hose a short distance from the drill, and a

*Air line oilers are specialized variations of oil mist lubrication, which is discussed at the end of this chapter.

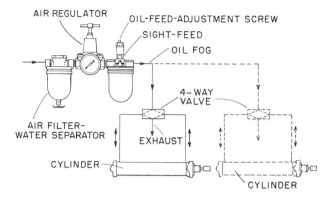

Fig. 6-8 Fog Lubricator for Pneumatic Cylinders

fine spray of oil is carried from it to lubricate the drill wearing surfaces that are reached by the air. A method of varying the rate of oil feed is provided so that the oil feed can be adjusted to the drill and drilling conditions.

In the air line oiler (Fig. 6-9), air for rock drill actuation passes through the center tube, and line pressure is applied to the oil reservoir via port A and the vertical drilled passages shown in the left view. As the result of a Venturi effect, a reduced pressure prevails at port B. Because of the difference in pressure, oil feeds through the valve (right view) and port B into the air stream.

Air Spray Application* Some open gear and wire rope lubricants, including some grease type materials, are applied by means of hand operated air spray equipment using either external mixing nozzles or airless atomizing equipment, but this has the same disadvantages as hand oiling. A number of auto-

*Air spray lubrication is a specialized variation of oil mist lubrication, which is discussed at the end of this chapter.

Fig. 6-9 Air Line Oiler Oil is drawn into the air stream through the port B by Venturi effect. The enlarged section around the valve acts as a pendulum to swing the valve and keep the oil intake submerged in oil regardless of the position of the outer housing.

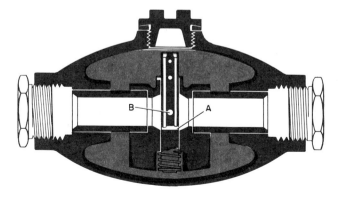
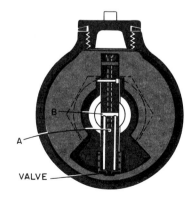

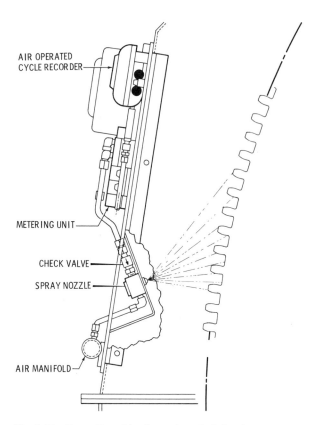

AIR OPERATED
CYCLE RECORDER

METERING UNIT

CHECK VALVE

SPRAY NOZZLE

AIR MANIFOLD

Fig. 6-10   Spray Panel for Open Gear Lubrication

matic or semiautomatic units such as that shown in Fig. 6-10 have been developed to overcome some of these problems.

Grease Application

Greases are used mainly for rolling element bearings, flexible couplings, and thin film plain bearings and slides. Because of the lower leakage tendencies of greases, periodic application with a hand or air powered grease gun does not have many of the disadvantages of hand oiling; but, as with any manual lubrication method, fittings will be overlooked and the cost of application is usually high. Therefore, central grease lubrication systems are being used more frequently and a number of bearings are being "packed for life."

Many rolling element bearings are manually packed, at the time they are installed and periodically thereafter, as is the case with automotive wheel bearings. Grease is forced into the spaces between the rolling elements by hand or with a bearing packer, and a moderate amount of grease placed in the hub. The housing must not be filled completely. If the housing is full there will be no place for the grease between the raceways to go when the bearing is operated. Excessive churning of the grease and overheating will then result.

Bearings intended for power gun lubrication must have provision for pressure relief to prevent pressure buildup and permit purging of old grease. This is sometimes accomplished by allowing sufficient clearance in the seals, but can also be achieved with a relief valve or a drain plug such as that in Fig. 6-11. When a bearing of this latter type is relubricated, the drain plug should be removed and left out until after the bearing has been operated long enough to expel all excess grease.

Packed for life bearings are used for many light to medium duty services. These bearings have seals and are packed at the factory with a special long life grease intended to last for the life of correctly applied bearings. In some cases, the service range of bearings of this type is extended by making provision for relubrication, by one of the following methods: removing one seal and applying grease through the seal by means of a gun with a hollow needle, or forcing grease through holes in the outer ring.

REUSE METHODS

Reuse methods of oil application include circulation systems supplying lubricant for one or more machines and self-contained systems such as bath, splash, flood, and ring oiling.

Circulation Systems The term "circulation system" generally refers to a system in which oil is delivered from a central reservoir to all bearings, gears, and other elements requiring lubrication. All the oil, disregarding minor leakage, drains back to a central sump and is reused. Two principal variations of this type of system

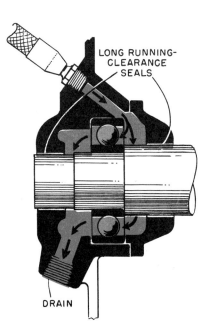

Fig. 6-11 Free Purging Housing Design
This design permits good purging of old grease from both sides of the bearing when new grease is applied.

are used—namely, pressure or gravity feed. In pressure feed systems, a separate sump and reservoir may be used, or the two may be combined (Fig. 6-12). The oil is pumped directly to the parts requiring lubrication. Where a separate sump is used, it may be either "wet," in which case the drain is located so that a certain amount of oil remains in the sump at all times, or "dry," in which case the drain is located and sized so that the sump remains essentially empty at all times. In gravity feed systems, oil is pumped to an overhead tank and then flows under gravity head to the elements requiring lubrication (Fig. 6-13).

In either type of system, very often the rate of flow is determined primarily by what is needed for cooling. This amount of oil usually will be more than is needed for lubrication. Means of cooling and purifying the returned oil are often included in circulation systems.

Although the many types of applications for circulation systems require considerable variations in size, arrangement, and complexity, in a general way, circulation systems can be considered in one of three groups:

1. Systems comprising a compact arrangement of pump, reservoir, and oil passages built into the housing of the lubricated parts such as that in Fig. 6-12.
2. Systems employing a multicompartment tank combining reservoir and purification facilities (Fig. 6-14)

Fig. 6-12   Pressure Feed Circulation System for Horizontal, Duplex Two Stage Compressor   The illustration shows a cutaway view through the main and crankpin bearings of the righthand frame. The spiral gear oil pump draws oil from the reservoir through a strainer and forces it through a fine mesh screen and the hollow pump arm to the crankpin bearing. Excess oil is bypassed to the reservoir through a relief valve (not shown). From the crankpin bearing, oil flows under pressure through internal passages to the main bearing, to the crosshead pin bearing and crosshead guides.

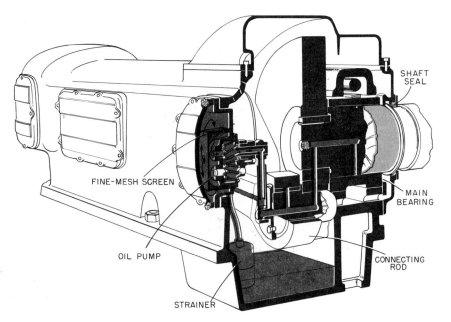

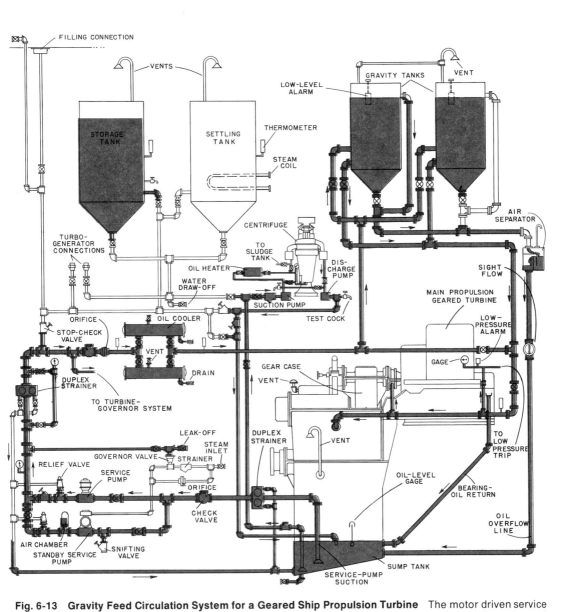

Fig. 6-13 Gravity Feed Circulation System for a Geared Ship Propulsion Turbine   The motor driven service pump takes oil from the sump tank and delivers it through oil coolers to the gravity tanks which are located about 30 ft. (10 m) above the main turbine. From the gravity tanks, oil flows directly to the turbine gears and bearings, and from these drains to the sump tank. Excess oil pumped to the gravity tanks overflows and drains directly to the sump tank. In this way, a constant pressure head is maintained on the supply for the bearings and gears. Standby pumps, storage and settling tanks each capable of holding a complete charge for the system, and a centrifuge with pumps and supply piping, are also provided.

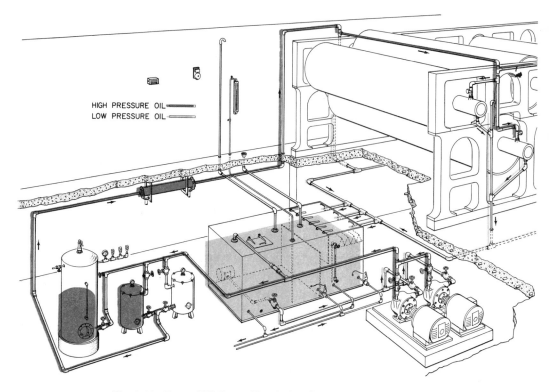

HIGH PRESSURE OIL
LOW PRESSURE OIL

Fig. 6-14 Paper Mill Dryer Circulation System

3. Systems comprising an assembly of individual units including a reservoir, an oil cooler, pumps, purification equipment, etc.

The system illustrated in Fig. 6-15 is fairly typical of the third type. Returning oil drains to a settling compartment, entering the reservoir at or just above the oil level. Water and heavy contaminants settle, and the sloping bottom of the reservoir helps to concentrate these impurities at a low point from which they can be drained. Partially purified oil overflows a baffle to the clean oil compartment. In some systems, especially where large reservoirs are used, baffles may be omitted. The clean oil pump takes oil, usually through a suction strainer, and pumps it to a cooler and then to bearings, gears, and other lubricated parts. The pressure desired in the oil supply piping is maintained by means of a relief valve, which discharges to the reservoir at a point below the oil level. A continuous bypass purification system is shown. The pump takes 5 to 15 percent of the oil in circulation from a point above the maximum level of separated water in the reservoir and pumps it through a suitable filter back to the clean oil compartment. The following discussion of good practices in circulation system design refers primarily to systems of this type, but the ideas presented are fundamental to most systems.

Oil Reservoirs   The bottom of an oil reservoir should slope more than 1 in 25 (4 percent) toward a drain connection, which should be located at the lowest point in the reservoir. This construction promotes the concentration of water and settled impurities and permits their removal without excessive loss of oil. In addition, it permits more complete removal of flushing oil or solvents used in cleaning operations prior to startup or during an oil change. Where oil is taken directly from the reservoir for purification, it should be removed close to the low point but above the water and settled impurities. An opening or openings above the oil level, adequate for inspection and cleaning, should be provided. Large reservoirs should have an opening large enough for a person to enter. These and any other openings should have well secured, dust tight covers.

   A connection should be provided at the highest point in the reservoir for ventilation. Proper ventilation results in the removal of moisture laden air and thereby reduces condensation on cooler surfaces above the oil level and subsequent rusting of these surfaces. Ventilation fixtures should be designed with air filters to prevent entrance of airborne contaminants. Instead of natural ventilation, where a source of water contamination is common—

Fig. 6-15   Circulation System Diagram   The elements illustrated include one type of pressure circulation system, and several commonly accepted practices.

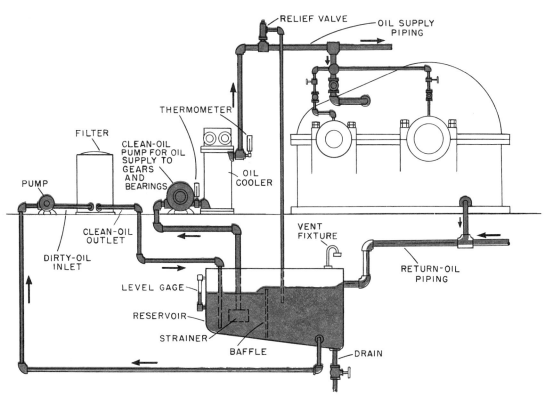

for example, large steam turbine systems — medium and large size reservoirs should be provided with a "vapor extractor," or exhauster, capable of maintaining a slight vacuum in the air space above the oil level. Too high a vacuum should be avoided, however, since it may have the effect of pulling plant atmospheric contaminants into the lubrication system.

The main oil return connection should be located at or slightly above the oil level and remote from the oil pump suction. Returning oil should not be permitted to drop from a considerable height directly into the oil body, since this action tends to whip air into the oil and cause foaming or to hold water and contaminants in suspension. Instead, the fall of oil should be broken and dispersed by means of a baffle, sheet metal apron, or fine screen. Where a line may carry any air, it should never discharge below the oil level. A connection for the return to the reservoir of a "solid" stream of oil, as from a pressure relief valve, should be placed about 6 in. (150 mm) below the oil level.

It is convenient to consider reservoir size in terms of the oil volume flowing in the system. The reservoir should be large enough so that oil velocity in it will be low and the oil will have sufficient rest time to assure adequate separation of water and entrained solids, separation of entrained air, and collapse of any foam that may exist. In practice, reservoir sizes range from a minimum of 2 times system flow per minute to a maximum of about 40 times. Many representative systems have reservoirs of 5 to 10 times system flow per minute.

Because of the importance of oil level in reservoirs, level indicators are commonly provided. Low oil level alarms are often provided on large reservoirs in addition to visual indicators.

Pump Suction The clean oil pump suction opening should be above the bottom of the reservoir to avoid picking up and recirculating settled impurities. However, it must be below the lowest oil level that may occur during operation. Where there is considerable variation of the oil level in the reservoir and it is desired to take oil at or near the surface, a floating suction may be used. Floating suction is frequently used in reservoirs of systems exhibiting constant, extreme water contamination. The oil supplied to the circulation systems from the top of the reservoir will have the least water contamination.

Bearing Housings The floors of bearing housings should have a slope of about 1 in 50 (2 percent) toward the drain connection. The design should be such that there are no pockets to trap oil and prevent complete drainage. Shaft seals should be adequate to prevent loss of oil or the entrance of liquid or solid contaminants. Any type of breather or vent fixture on a bearing housing should be provided with an air filter to keep out dust and dirt.

Return Oil Piping Return oil piping should be sized to operate about half full under normal conditions, and it should have a slope of at least 1 in 60 (1.7 percent) toward the reservoir. Any unavoidable low spots should have provision for periodic water removal.

Circulation System Metals Exclusive of bearings, the parts of a circulation system should preferably be made of cast iron or carbon steel. Fittings of bronze are acceptable. Stainless steel tubing is very good. Aluminum alloy tubing is acceptable as far as chemical inertness with oil is concerned, but it may not have sufficient structural strength for high pressure lines. No parts should be galvanized. As a general rule, no parts, exclusive of bearings, should be made of zinc, copper, lead, or other materials that may promote oil oxidation and deterioration. Copper tubing may be used for oil lines in some installations, but should not be used in systems such as those for steam turbines where extremely long oil life is desired. The use of copper should be confined to those applications where the oil is formulated specifically to inhibit the catalytic effects of copper.

Oil Coolers Oil coolers should be located so that all connections and flanged covers are accessible, and the tube bundles can be removed conveniently for cleaning. Cooler capacity should be adequate to prevent oil temperature from rising above a safe maximum during the hottest conditions. Means should be provided to control oil temperature at all times by regulating the flow of cooling medium.

Temperature Measurement Provision should be made so that oil temperatures can be measured on the discharge side of main bearings and other main operating components, and at the inlet and outlet of oil coolers.

OTHER REUSE METHODS

In addition to circulation systems, a number of other methods of oil application involve more or less continuous reuse of the oil. These are differentiated from integral circulation systems primarily in that pumps are not used to lift the oil.

Splash Oiling Splash oiling is encountered mainly in gear sets or in compressor or steam engine crankcases. Gear teeth, or projections on connecting rods, dip into the reservoir and splash oil to the parts to be lubricated or to the casing walls where pockets and channels are provided to catch the oil and lead it to the bearings (Fig. 6-16). In some systems, oil is raised from the reservoir by means of a disc attached to a shaft, removed by a scraper, and led to a pocket from which it is distributed (Fig. 6-17). This variation may be called a flood lubrication system. In either case, the oil returns to the reservoir for reuse after flowing through the bearings or over the gears. Accurate control of the oil level is necessary to prevent either inadequate lubrication or excessive churning and splashing of oil.

Bath Oiling The bath system is used for the lubrication of vertical shaft hydrodynamic thrust bearings and for some vertical shaft journal bearings. The lubricated surfaces are submerged in a bath of oil which is maintained at a constant level. When necessary, cooling coils are placed directly in the bath. The bath system for a thrust bearing may be a separate system or may be connected into a circulation system.

Ring, Chain, and Collar Oiling In a ring oiled bearing, oil is raised from a reservoir by means of a ring which rides on and turns with the journal (Fig. 6-18). Some of the oil is removed from the ring at the point of contact with the journal and is distributed by suitable grooves in the bearing. After flowing through the bearing, the oil drains back to the reservoir for reuse.

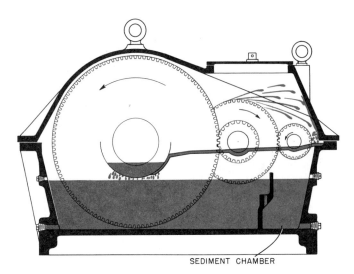

SEDIMENT CHAMBER

Fig. 6-16 Splash Oiling System The gear teeth carry oil directly to some gears and splash it to others and to collecting troughs which lead it to bearings not reached by splash.

Ring oiling is applied to a wide variety of medium speed bearings in stationary service. At high surface speeds, too much slip occurs between ring and journal and not enough oil is delivered. Also, at high speeds, in large, heavily loaded bearings, not enough cooling may be provided.

Oil rings are usually made about 1.5 to 2 times journal diameter, and where bearings are more than about 8 in. (200 mm) long, two or more rings are usually required. The oil level in reservoirs is usually maintained so that the rings dip less than one-quarter their diameter. The oil level, within a given range, is not usually critical, but too low a level may result in inadequate oil supply, and too high a level, because of excessive viscous drag, may cause ring slip or stalling. As a result, too little oil may reach the bearing and "flats" may wear on the rings to the extent that they will no longer perform satisfactorily.

Chains are used sometimes instead of rings in low speed bearings since they have greater capacity for lifting oil at low speeds.

Where oils of very high viscosity are required for low speed, heavily loaded bearings, a collar that is rigidly attached to the shaft may be used instead of a ring or chain. A scraper is required at the top of the collar to remove the oil and direct it to the distribution grooves in the bearing.

CENTRALIZED APPLICATION SYSTEMS

A number of factors have contributed to the growing use of centralized lubricant application systems. Among these are improved reliability, reduced cost of labor for lubricant application, reduced machine downtime required for lubrication, and, generally, a reduction in the amount of lubricant used through reduction of waste and more efficient use of lubricants.

Central Lubrication Systems

A number of types of central lubrication systems have been developed. Most can apply either oil or grease, depending on the type of reservoir and pump used. Greases generally require higher pump pressures since greater pressure losses occur in lines, metering valves, and fittings. Pump and reservoir capacities vary depending on the number of application points to be served, ranging from small capacities (Fig. 6-19), to units that install on standard drums (Fig. 6-20), and systems that operate directly from bulk tanks or bins where large volumes of lubricant are required.

In some systems, called *direct* systems, the pump serves to pressurize the lubricant and also to meter it to the application points. In *indirect* systems, the pump pressurizes the lubricant but valves in the distribution lines meter it to the application points.

Two basic types of indirect systems are in common use, and in turn each type has two variations. In *parallel* systems, also called header or nonprogressive systems, the metering valves or feeders are actuated by bringing the main distribution line up to operating pressure. All of the metering valves operate more or less simultaneously. This type of system has the disadvantage that if one valve fails, no indication of failure is given at the pumping station. However, all of the other application points will continue to receive lubricant. In *series*, or progressive, systems the valves are "in" the main distribution line (Fig. 6-21, right). When the main distribution line is brought up to pressure the first valve operates. After it has cycled, flow passes through it to the second valve, and then to each succeeding valve in turn.

Fig. 6-17 Flood Lubrication System In this single cylinder, water cooled compressor, running parts are lubricated by the splash and flood method. Crankshaft counterweights and an oil splasher dip into the oil reservoir and throw oil to all bearings. Oil pockets over the crosshead and over each double row tapered roller main bearing assure an ample supply for these parts.

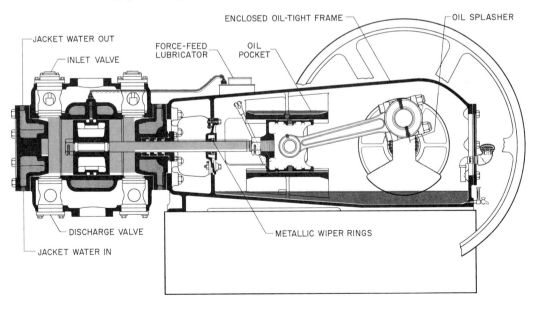

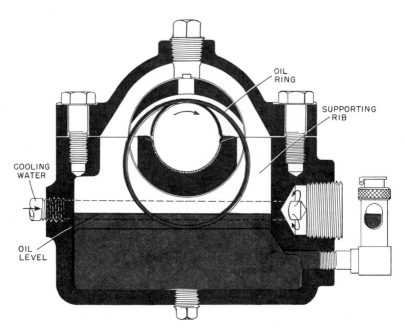

Fig. 6-18   Ring Oiled Bearing for Small Mechanical Drive Steam Turbine   The plug in the cap can be removed to observe the turning of the oil ring. The running oil level is indicated. This would be raised during shutdown as a result of oil draining from the bearing and walls of the housing. The bearing is cooled by means of water passages through supporting ribs on either side of the oil ring slot. The spring cap on the fill and level gage fitting (right) helps keep out dirt.

Thus, if one valve fails, all fail, and the pressure rise at the pump can be used to signal that a failure had occurred. The two variations of each of these basic types are as follows:

Two Line System   In this variation of the parallel system, two supply lines are used (Fig. 6-22). The four way valve directs pressure alternately to the two lines, at the same time relieving pressure in the line that is not receiving flow from the pump. The valve can be operated manually from the machine, cycled by a timer, or by a counter that measures the volume delivered by the pump. The valves are designed to deliver a charge of lubricant to the application point each time the flow in the lines is reversed.

Single Line Spring Return   In this variation of the parallel system, only a single distribution line is used. The layout of this type of system is shown in Fig. 6-23. As with the fourway valve of the previous system, the threeway valve may be operated manually from the machine, by a timer, or by a counter measuring pump output. The valves deliver a charge of lubricant when system pressure is applied to them, and reset themselves when the system pressure is relieved.

Series Manifolded System   In this type of system, a single supply line is used. No relief valve is required since the manifold type valves used automatically reset themselves and continue cycling as long as pressure is applied to them through the supply line. The system can be cycled by starting and stopping the pump.

Series System, Reversing Flow   This type of series system uses a single supply line with a fourway valve to reverse the flow in it (Fig. 6-24). The valves are designed to deliver a charge of lubricant, then permit lubricant flow to pass through to the next valve. When the flow in the supply line is reversed, the valves cycle again in sequence in the reverse order.

Mist Oiling Systems   In oil mist lubricators, oil is atomized by low pressure (10–50 psi, 70–350 kPa) compressed air into droplets so small that they float in the air forming a practically dry mist, or fog, that can be transported relatively long distances in small tubing. When the mist reaches the application point it is condensed, or coalesced, into larger particles that wet the surfaces and provide lubrication. Condensing can be accomplished in several ways. Oil mist systems have proven their reliability in an increasing variety of applications. They are used in all types of industries—from the very light duty service of lubricating dental handpieces to the heavy duty service of lubricating steel mill backup rolls. In the past, the systems were usually built onto or adapted to

Fig. 6-19   Motor Driven Pump and Reservoir   This unit is designed for continuous feeding of lubricant and would be suitable for a series manifolded system. The reservoir is filled at the top through a cone screen, and a spring loaded follower plate can be used. The gage registers the pressure in the transmission line.

Fig. 6-20   Drum Mounted Pump and Control Unit   A standard 400 lb drum serves as the reservoir. The pump can be of the air, electric, or hydraulic type, and develops 3500 psi (24 MPa) pressure. The unit is designed for a single line, spring return lubrication system.

Fig. 6-21   Parallel and Series Lubrication System   In the parallel system on the left the valves are "off" the supply line, while in the series system on the right the valves are "in" the supply line.

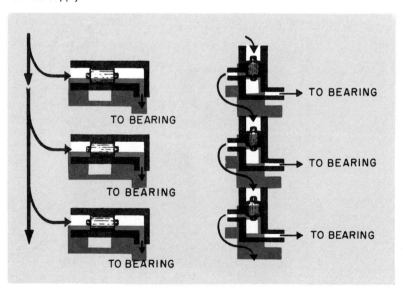

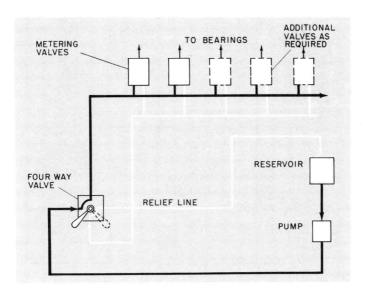

Fig. 6-22 Two Line Parallel System The four way valve, operated manually or automatically, alternately directs pump pressure to one line and then the other. When one line is pressurized, the other line is relieved.

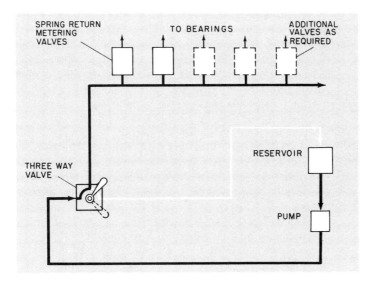

Fig. 6-23 Single Line Spring Return System The three way valve, operated manually or automatically, either directs pump pressure into the supply line or relieves the pressure in the line to permit the spring return valves to reset.

existing equipment. Machine tool builders are now designing them into their newer machines to provide greater reliability and productivity.

An oil mist lubrication system is simply a means to distribute oil of a required viscosity from a central reservoir to various machine elements.

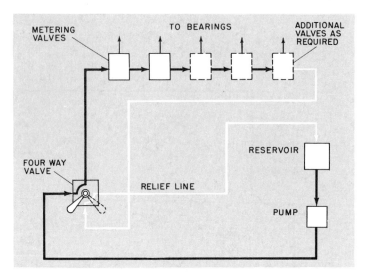

Fig. 6-24 Series Reversing Flow System The four way valve, operated manually or automatically, directs pump pressure to one end of the closed loop supply line while relieving the pressure at the other end.

A true oil mist is a dispersion of very small droplets of oil in smoothly flowing air. The size of these droplets averages from 1 to 3 μm (1 μm = 0.000039 in.) in diameter. In comparison, an ordinary airline lubricator produces an atomized mixture of droplets up to 100 μm in diameter, which are suspended (temporarily) in turbulent air flowing at high velocity and pressure. In an airline lubricator system, the air is a working fluid that is transmitting power, whereas in an oil mist system, air is used only as a carrier to transport the oil to points where it is required.

The droplet size is a very important consideration in the proper design of an oil mist system. The larger the droplets, the more likely they are to wet out and form an oily film at low impingement velocities. At practical, low flow rates the size limit is taken to be 3 μm. Droplets over this size will wet or spread out on surfaces quite readily while particles less than this diameter will not.

A dispersion of droplets less than 3 μm in diameter will form a stable mist and can be distributed for long distances through piping. At the points requiring lubrication, these drops can be made to wet metal surfaces by inducing a state of turbulence, causing small droplets to collide and form into large diameter drops. These larger drops wet metal surfaces to provide the necessary lubricant film.

This formation of larger drop sizes that will wet metal surfaces is referred to as condensation, although terms such as reclassification, condensing, coalescing, etc., are also used.

Different degrees of condensation may be achieved by using different adaptors at the points requiring lubrication. These adaptors are usually classified as: mist nozzles; spray, or partially condensing nozzles; and completely condensing, or reclassifying nozzles. When high speed rolling element bearings create sufficient turbulence in the bearing housing to cause the droplets to join and wet out, a mist type nozzle may be used. When gears are being lubricated, it is usually necessary to partially condense the oil mist so the limited amount of agitation within the gear housing will cause the droplets to coalesce and wet out. When slow moving slides or ways are being lubricated, it is usually necessary to completely condense the oil mist into a liquid which is then applied to the bearing surface.

In a typical oil mist system (Fig. 6-25) compressed air enters through a water separator, a fine filter, and an air regulator to the mist generator (a). From the generator the mist is carried to a manifold (b) and then to the various application points (c).

To produce an oil mist, liquid oil is blasted with air to mechanically break it up into tiny particles. Droplets over 3 μm are screened or baffled out of the flow and returned to the sump or reservoir. The resultant dispersion, containing oil droplets averaging 1 to 3 μm in diameter, is the oil mist to be fed into the distribution system.

The size of the Venturi throat, oil feed line, and pressure differentials impose a physical limit on the viscosity of oil that can be misted. By the judicious use of oil heaters in the reservoir, and in some designs air line heaters to heat incoming air, the viscosity of normally heavy bodied oils can be lowered to make misting possible. Systems without heaters can usually handle oils up to approximately 800 to 1000 SUS at 100°C (173 – 216 cSt at 38°C – at room temperature). If ambient temperatures are much below 70°F (21°C), then heat will likely be needed to reduce the oil's effective viscosity. Also, oils of over 1000 SUS at 100°F (216 cSt at 38°C) will usually require heating

Fig. 6-25   Oil Mist Lubrication System

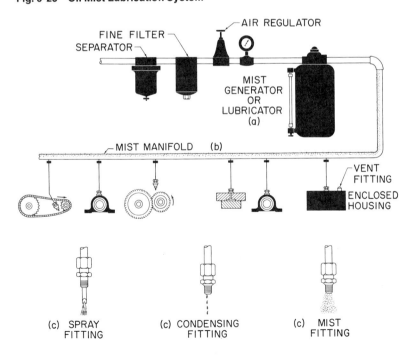

to lower their effective viscosity and make the formation of a stable oil mist possible.

If the immersion elements in the oil reservoir are not properly adjusted, additional heating of the oil by the heated air will raise the bulk oil temperature to the point where it will oxidize quite readily, and varnish or sludge may form in the generator. When heated air is being used, it should be no hotter than necessary to allow easy misting (usually below a maximum temperature of 175°F or 80°C). Also, the oil reservoir immersion elements used in conjunction with heated air should be used primarily at start up and later, only if necessary to maintain oil temperature during operation. Naturally, the immersion elements should be designed for oil heating and have a low watt density (for example, 6 W/square in.^{2}, or 0.9 W/cm^2).

In addition to lubrication, mist systems provide other benefits: (1) the flow of air through bearing housings provides some cooling effect; and (2) the outward flow of air under positive pressure helps prevent the entrance of contaminants. Proper venting must be provided to permit this air to flow out of housings.

Bibliography

Mobil Technical Books

Compressors and Their Lubrication
Gears and Their Lubrication
Plain Bearings, Fluid Film Lubrication

Mobil Technical Bulletins

Handling Storing and Dispensing Industrial Lubricants
Oil Mist Lubrication
Rock Drill Lubrication
Rolling Element Bearings, Care and Maintenance

7

Internal Combustion Engines

The term "internal combustion" describes engines that develop power directly from the gases of combustion. This class of engines includes the reciprocating piston engines, used in a wide variety of applications; the rotary engines, such as the Wankel; and most gas turbines. However, since closed system gas turbines are not truly internal combustion engines, gas turbines are discussed separately. The rotary engines (Wankel) are not discussed since they are not in general use. The following discussion is concerned primarily with reciprocating piston engines.

Piston engines range in size from the fractional horsepower units used to power model equipment, such as model airplanes, to engines for marine propulsion and industrial use that develop in the order of 50,000 hp or more. While this wide range of engine sizes and the types of applications of the engines present a variety of lubrication problems, certain factors affecting lubrication are more or less common to all reciprocating engines.

The primary objectives of lubrication of reciprocating engines are the prevention of wear and the maintenance of power producing ability and efficiency. These objectives require that the lubricant function effectively to lubricate, cool, seal, and maintain internal cleanliness. How well these factors can be achieved depends on the engine design, fuel, combustion, operating conditions, the quality of maintenance, and the engine oil itself.

DESIGN AND CONSTRUCTION CONSIDERATIONS

Among the design and construction features that affect lubrication are the:

1. Combustion cycle—whether two stroke or four stroke
2. Mechanical construction—whether trunk, piston, or crosshead type
3. Supercharging—whether the engine is supercharged or naturally aspirated
4. General characteristics of the lubricant application system

Combustion Cycle   In a reciprocating engine, the combustion cycle in each cylinder can be completed in one revolution of the crankshaft, that is, one upstroke* and one downstroke of the piston; or in two revolutions of the crankshaft, that is, two upstrokes and two downstrokes. The first engine is referred to as a two stroke cycle, or more simply, a two cycle engine, while the second is referred to as a four stroke cycle, or four cycle engine. Either cycle can be used for engines operating with spark ignition (gasoline or gas) or compression ignition (diesel). Since it is more widely used, the four stroke cycle will be described first.

Four Stroke Cycle   The sequence of events in the four stroke cycle is illustrated in Fig. 7-1. On the inlet or intake stroke at A, the intake valve is open and the piston is moving downward. Air (in a diesel engine) or an air-fuel mixture (in a spark ignition engine) is drawn in through the intake valve filling the cylinder. As the piston starts moving up at B, the intake valve closes and the air (or charge) is compressed in the cylinder. Near the top of this compression stroke, fuel is injected or a spark is passed across the spark plug. The fuel ignites and burns, and as it expands, it forces the piston down on the power stroke. Near the bottom of this stroke, the exhaust valve opens so that on the next upward stroke of the piston the burned gases are forced out of the cylinder. The assembly is then ready to repeat the cycle.

Most four cycle engines are equipped with poppet valves in the cylinder head for both intake and exhaust. Various arrangements are used to operate these valves. In the conventional arrangement, a camshaft is located along the side of the cylinder block. It is driven from the crankshaft by gears, a silent chain, or in some small modern passenger cars engines by a toothed belt. Cam followers (often called valve lifters) of either the roller, solid, or hydraulic type ride on the cams and operate push rods which in turn operate the rocker arms to open the valves. Valve closing is accomplished by springs surrounding the valve stems. This type of arrangement results in some mechanical lag in valve operation at high speeds; thus, some high speed automotive engines have the camshafts located above the cylinder

*Since not all reciprocating engines have vertical cylinders, this term is not strictly accurate; however, it does describe the stroke in which the piston approaches the head end of the cylinder.

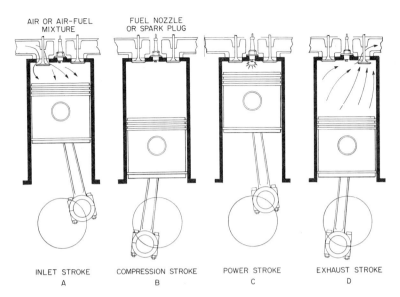

AIR OR AIR-FUEL
MIXTURE

FUEL NOZZLE
OR SPARK PLUG

INLET STROKE
A

COMPRESSION STROKE
B

POWER STROKE
C

EXHAUST STROKE
D

Fig. 7-1 Four Stroke Cycle At A the piston is moving downward drawing in a charge of air or air-fuel, mixture. At B the valves are closed and the piston compresses the charge as it moves upward. At C fuel has been injected and ignited by the high-temperature compressed air, or a spark is passed across the spark plug igniting the air-fuel charge, and the piston is pushed downward on the power stroke by the expanding hot gases. At D the burned gages are forced out of the cylinder through the open exhaust valve.

head so that the cams bear directly on the valve stems or on short rocker arms. This arrangement is called overhead camshaft construction. This design has been gaining acceptance for small modern engines. The cam drive for many of these utilizes a toothed rubber belt. Some large, medium, and low speed diesel engines now are equipped for direct operation of the valves in a somewhat similar manner, and fully hydraulic valve actuation is also used on a few engines. The complete valve operating mechanism is often referred to as the valve train.

Loading on the rubbing surfaces in the valve train may be high, particularly in high speed engines where stiff valve springs must be used to ensure that the valves close rapidly and positively. This high loading can result in lubrication failure unless special care is taken in the formulation of the lubricant.

Two Stroke Cycle The sequence of events in the two stroke cycle is illustrated in Fig. 7-2. Near the bottom of the stroke, the exhaust valves open and the piston uncovers the intake ports, allowing the scavenge air to force the exhaust gases from the cylinder. As the piston starts on the upstroke at B, the exhaust valves close and the piston covers the intake ports so that air (or charge) is trapped in the cylinder and compressed. Near the end of this compression stroke, the fuel is injected and begins to burn (or the charge is ignited). The expanding gases then force the piston down on the power stroke.

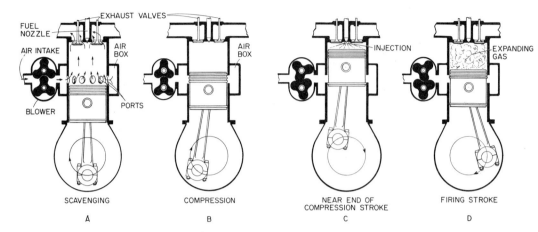

Fig. 7-2 Two Stroke Cycle At A air from the blower is driving exhaust gas from the cylinder. At B the charge of fresh air is trapped in the cylinder and compressed as the piston moves upward. At C fuel is injected. The fuel is ignited when the piston is close to the top of its stroke and the hot expanding gases of combustion force the piston down on the power stroke at D.

Two cycle engines are usually built with either ports for intake and valves for exhaust, or with ports for both intake and exhaust. With the combination of ports and valves, (see Fig. 7-2), the scavenge air and exhaust gases flow more or less straight through the cylinder, so this is referred to as *uniflow* scavenged. With ports for both intake and exhaust, if the intake ports are on one side of the cylinder and the exhaust ports are on the other side, the engine is referred to as *cross* scavenged. The scavenge air flows more or less directly across the cylinder. Where the exhaust ports are located on the same side of the cylinder as the intake ports, the engine is referred to as loop scavenged. The scavenge air must flow in a loop into the cylinder and then back to the exhaust ports. Most small, two cycle gasoline engines are of the cross scavenged type, while all three types of scavenging are used for two cycle diesel engines. In large engines, the type of scavenging has some influence on piston temperatures and the type of piston cooling that can be used — oil or water; therefore, indirectly it has an influence on lubricant selection.

Since the two stroke cycle does not have a full positive exhaust stroke to rid the cylinder of combustion gases, the scavenging process must be assisted by pressure developed outside the cylinder. This can be accomplished by means of a separate blower or compressor, driven either by the engine or an outside source, or by what is known as crankcase scavenging. With crankcase scavenging, the section of the crankcase under each cylinder is sealed, except for a check valve to admit air (or charge) on the upstroke of the piston, and a transfer passage to permit the air (or charge) to be delivered to the intake ports as the piston approaches and goes through its bottom dead center position. A variation of this principle is used on some large, crosshead diesel engines. The lower end of the cylinder is closed and a packing gland installed where the piston rod passes through. With the addition of a check valve and transfer piping to carry the air to the intake ports, the piston then will pump air for scavenging or supercharging purposes. This may be referred to as pulse charging.

In theory — for the same bore, stroke, and rational speed — a two stroke cycle engine will develop twice the power of a four stroke cycle engine. As a result, more fuel is consumed per unit of time, and cylinder and average engine temperatures tend to be higher, and more contaminants may find their way into the oil. On many engines, the oil is mixed with the fuel.

Mechanical Construction

Most piston engines, including all automotive engines, are of the trunk piston type. Large, low speed diesel engines for marine propulsion and industrial applications are usually of the crosshead type.

In a trunk piston engine, the piston is connected directly to the connecting rod by a piston pin or wristpin. Thus, the side thrust from the crankpin and connecting rod must be carried through the piston and rings to the cylinder wall. This may result in high rubbing pressures on one side of the piston and on the mating section of cylinder wall.

In a crosshead type of engine, the piston is connected rigidly to a piston rod, which is connected to the crosshead containing the wristpin. The crosshead also has a sliding guide bearing to absorb the side thrust from the crankshaft and connecting rod. This bearing is usually generously proportioned so that these thrust loads are readily carried by the lubricating films. No side thrust is carried by the piston. The cylinder assembly is usually completely separated from the crankcase, either by means of a diaphragm containing a packing gland, or by having the lower end of the cylinder closed so that it can be used for pulse charging.

Crosshead type engines are also built with double acting pistons, or with two pistons acting in opposite directions in the same cylinder — what is called an opposed piston engine. One of the chief disadvantages of crosshead construction is that it results in an engine that has a greater overall height than a trunk piston engine of the same horsepower. However, most engines with bores above 600 mm (24 in.) are of crosshead type.

Supercharging

One of the factors that limits the power developed by an internal combustion engine is the amount of air it can "breathe." It would be easy to supply more liquid fuel, but it would be undesirable to exceed the quantity that the available air can burn more or less completely. Supercharging is a method of increasing the available combustion air by supplying air at higher pressure, thus making it possible to burn more fuel.

The air for supercharging is provided by a blower, which may be engine driven, motor driven, or exhaust gas turbine driven. The latter is probably most frequently used now. With some large engines, supercharging by the blower may be supplemented by pulse charging.

Since supercharging increases the amount of air in the cylinders, and thus the amount of fuel that can be burned, it tends to raise combustion temperatures and to increase the deteriorating influences on the lubricating oil. To increase supercharging efficiency, the compressed air is sometimes cooled between compression and engine intake.

Methods of Lubricant Application

Small and medium size trunk piston engines usually are lubricated by a combination of pressure and splash. Oil under pressure is fed to the main and crankpin bearings and, from the crankpin bearings, through drilled passages to the wristpin bearings. Oil under pressure is also fed to an oil gallery from which it is distributed to the valve lifters and rocker arms. Oil is splashed to the lower cylinder walls to lubricate the rings and cylinders and to the undersides of the pistons for cooling. Eventually, all of this oil, neglecting the small amount that finds its way into the combustion chambers and is burned, drains back to the crankcase and is reused.

In some cases, the crankcase is arranged for dry sump operation with a reservoir, and possibly an oil cooler for the main supply. Where this arrangement is used, cylinder lubricant is supplied from the wristpin supply.

Trunk piston engines with bores approximating 250 mm (10 in.) usually have separate cylinder lubricators of the mechanical force feed type. Any excess cylinder lubricant then drains back into the crankcase supply, carrying contaminants with it. Crosshead engines are equipped with at least two separate lubrication systems. One system supplies the lubricant for the cylinders, while a second system supplies the main and crankpin bearings and crossheads, and usually the piston rod packing glands. A third system may be provided to lubricate the supercharger, or the supercharger may be lubricated from the crankcase system. Piston cooling oil is also supplied from the crankcase system in engines with oil cooled pistons. There is little or no contamination of the crankcase oil by excess cylinder oil, other than the minor amounts that may leak past the packing glands. However, any used cylinder oil that is collected above the diaphragms may present a disposal problem since it is generally unsuitable for further use and is extremely difficult to recondition satisfactorily.

A complicating factor in high output, late model, crosshead engines with oil cooled pistons is that piston temperatures may be high enough to cause thermal cracking of the oil while it is in the piston cooling spaces. This can lead to buildup of deposits in the piston cooling spaces, overheating of the pistons, and eventual cracking or other mechanical failure.

Most small two cycle gasoline engines, such as are used for marine outboard and utility purposes, are arranged for crankcase scavenging. In order to control the amount of oil carried into the cylinders with the charge, these engines are designed for dry sump operation. A small proportion of oil is premixed with the fuel, so the charge consists of a mixture of air, fuel, and oil. Some of the oil condenses on the crankcase surfaces where it provides lubrication of the main, crankpin, and wristpin bearings. Also, some of it is carried into the combustion chambers where it is burned with the fuel. Incomplete combustion of this oil, or the use of the wrong type of oil, can cause spark plug fouling and combustion chamber and port deposit buildup.

An important consideration with any internal combustion engine is the volume of oil in the system in proportion to the amount of contaminants that must be carried by the oil. Generally, large engines in marine and stationary applications can be arranged so that the volume of oil is large, and since operation is at essentially constant speed, combustion efficiency can be kept high. The rate of contaminant buildup is then quite low, and usually can be controlled to some extent by means of sophisticated purification equipment. With automotive engines, the amount of oil that can be carried into the system is proportionally much lower, for a variety of reasons. With automotive gasoline engines, particularly, it is desirable to restrict the volume of oil to promote more rapid warmup. In addition, the lower oil volume is more economical when frequent periodic changes must be performed. This helps to evaporate both the water of condensation and that diluting fuel, which result from start-stop type operations and cold starting. However, the variable speed conditions under which automotive engines operate, along with frequent starts and stops, tend to reduce combustion efficiency with the result that the rate of contaminant buildup in the oil is relatively high. The smaller oil quantity also leads to higher operating temperatures for the oil under extended operation so the exposure of the oil to oxidizing conditions may be more severe. Most modern engines are equipped with an oil filter.

FUEL AND COMBUSTION CONSIDERATIONS

When a hydrocarbon fuel is burned, certain products and residues are formed. One of the most important of these is water. In the case of liquid fuels, the volume formed is slightly greater than the volume of fuel burned. If the engine is fully warmed up, virtually all of this water passes out of the exhaust as vapor. If the engine is relatively cold, or under certain pressure and temperatures conditions in the cylinders, some of the water may condense and eventually find its way into the crankcase where it will mix with the engine oil. The amount of condensation that occurs and the other residues that are formed are largely dependent on the type of fuel and the operating conditions of the engines it is burned in. High temperature thermostats in the cooling system decrease warmup time.

Gasoline Engines   Most of the four cycle gasoline engines have automotive or similar applications. In these applications, the engine is generally started cold, and may operate much of the time at below normal temperatures, as a result of either short stop-and-go type operation, or carrying a light load. Under these conditions, a significant amount of the water formed from combustion of the gasoline may condense on the cylinder walls, and eventually reach the crankcase by way of blowby past the pistons and rings. In doing so, it tends to wash the lubricant films off the cylinder walls and promote rusting or corrosion. The rust formed is scuffed off almost immediately, but in the process metal is removed, so that what appears to be mechanical wear occurs. In the crankcase and in other areas where temperatures are low, the water can combine with the oil and other contaminants to form sludge that is usually of a soft, sticky consistency.

When a gasoline engine is cold, the mixture must be enriched by choking in order to provide enough vaporized fuel for starting. Some of the excess liquid fuel is blown out the exhaust, but a proportion of it drains past the pistons and into the crankcase, usually carrying with it some partially burned components, fuel soot, and solid residues from the lead antiknock agent (if used). As the engine warms up, much of the gasoline is evaporated off, but some of the heavier ends, some of the partially decomposed materials, and any solid residues remain in the oil. Even when the engine is fully warmed up, some of these fuel decomposition products blow by into the oil.

Fuel soot and partially burned fuel and water, when present in the oil, can lead to the formation of varnish, sludge, and deposits. Rusting of ferrous surfaces and corrosion of bearings can be promoted, particularly by the residues. The solid residues, such as fuel soot, when combined with small amounts of oil residue in the combustion chambers, can form deposits that adhere to piston tops and combustion chamber surfaces. Deposits of this type can significantly increase the fuel octane number requirement of an engine.

The use of unleaded gasolines required for those engines equipped with catalytic converters may have the effect of reducing the amount of rust and corrosion that occurs. In some engines, the absence of the lead antiknock may result in valve wear, called valve recession. Generally, this relates more to metallurgy than to lubrication. Somewhat higher operating temperatures result from the emission controls, but these have not presented a major problem from the lubrication point of view. The use of exhaust gas recirculation (EGR) to control nitrogen oxides NO_x emissions, along with other emission control changes, may result in a greater tendency to form valve stem deposits.

Two stroke cycle gasoline engines have such applications as outboard motors, snowmobiles, chain saws, and utility purposes such as pumps and lighting plants. These engines are lubricated by oil mixed with the fuel, so some oil is always present in the charge. Under low speed, low load conditions, such as trolling a boat, poor combustion can result in heavy buildup of port and piston crown deposits.

Diesel Engines

Diesel engines in trucks generally operate at more nearly constant speed and higher load factors than gasoline engines. Thus, combustion conditions are more nearly optimum, and water condensation and fuel dilution are not as serious problems. Where excessive fuel dilution occurs, it is usually the result of a mechanical problem, such as a faulty injector. Diesel fuel is not as readily evaporated from the engine oil as is gasoline; thus, if a problem exists, the concentration of diesel fuel will tend to increase steadily. This can lead to deposits and reduction of the oil viscosity to a sufficient extent that mechanical wear occurs.

One of the major problems with diesel engines is the sulfur in the fuel. Sulfur is present in diesel fuels in a considerably higher concentration than in gasolines. When it burns, it forms sulfur dioxide, part of which may be further oxidized to sulfur trioxide. In combination with water, these sulfur oxides form strong acids that are not only corrosive in themselves but also have a strong catalytic effect on oil degradation. Since piston temperatures are also high, this may result in heavy deposits of carbon and varnish on the pistons and in the ring grooves. Under severe conditions, deposits may build up in the ring grooves to the point where the rings cannot function properly, and high wear, blowby, and loss of power result. Manufacturers of diesel engine passenger cars are particularly concerned about soot deposits and recommend more frequent oil drains — more than twice those required for gasoline fueled car engines.

Many large diesel engines in marine and industrial service are operated on residual type fuels with sulfur content in the 2 to 4 percent range. The strong acids formed from the combustion of these high sulfur content fuels can be extremely corrosive to rings and cylinder liners, with the result that metal removal may be rapid and wear rates excessive.

Gaseous Fueled Engines
Engines burning clean gaseous fuels, such as liquified petroleum gas (LPG — propane, butane, and so forth) or natural gas, are comparatively free of the contaminating influences encountered in liquid fueled engines. Although, water is formed from combustion, most of it passes out the exhaust as vapor. Products of partial combustion that blow by to the crankcase tend to polymerize with the engine oil and cause an increase in viscosity and, eventually, may result in varnish and lacquer formation.

OPERATING CONSIDERATIONS

The operating conditions in an engine have a major influence on the severity of the service the lubricating oil is exposed to in the process of performing its functions of minimizing wear, assisting in cooling and sealing, and controlling deposits.

Wear
Normally, most engine wear is that experienced by cylinders, piston rings, and pistons. The three principal causes are:

1. Abrasion
2. Metal-to-metal contact
3. Corrosion

Much of the dust and dirt carried into the cylinders with the intake air is hard and abrasive in character. Some diesel fuels (particularly the residual fuels used in many large engines) may also contain abrasive materials. Abrasive particles carried by the oil onto load supporting surfaces of the cylinder walls and other areas can wear the rings and cylinder no matter how persistent the lubricating oil films. Fortunately, wear due to abrasives can be held to almost negligible values by the use of effective air, fuel, and oil filters. Proper maintenance of these filters is important, since a ruptured or damaged filter can allow free passage of abrasives.

As a result of the conditions of boundary lubrication that exist on the upper cylinder walls, metal-to-metal contact cannot be avoided entirely. The rate of wear from this cause depends to a large degree on the suitability of the lubricating oil. When the oil is of the correct viscosity and has adequate antiwear characteristics, wear due to metal-to-metal contact is kept at a low rate. However, reduced rates of oil flow and poor oil distribution, caused by deposits, can contribute to increased rates of metallic wear.

In large, highly supercharged, trunk piston engines, the side thrust on the cylinder walls is high. It has been found that some of these engines require oils with high load carrying ability in order to keep metallic wear rates to an acceptable level.

Corrosion and corrosive wear result from water or a combination of water and corrosive end products of combustion. When operating temperatures are low, either during the warmup period or as a result of low load or stop-and-go operation, condensation is increased and corrosive wear may be rapid. The use of high sulfur diesel fuels also promotes corrosive wear. Generally, it has been found that the use of alkaline additives in the lubricating oil acts to retard this type of wear. These materials neutralize the acidic materials so that their corrosive and catalytic effects are reduced.

Some hard alloy type bearings are susceptible to corrosive attack by certain oxidation products resulting from oil degradation. Oils for engines where corrosion susceptible bearings are used are formulated with additives that provide protection against these oil oxyproducts.

Metallic wear may be encountered on highly loaded valve train parts and fuel pump cams of some engines. This problem is most often encountered in passenger car engines. Antiwear agents are incorporated in oils for these engines in order to minimize wear. With this exception, metallic wear is not generally a problem with engine bearings as long as sufficient oil of the proper viscosity is available, and there is no interruption of oil flow.

Cooling Engine cooling is necessary to avoid engine damage and failure through overheating and thermal distortion. This is primarily a function of the cooling system, but the engine oil also has an important role to play in cooling, especially in large diesel engines where forced oil cooling of the pistons is used. Heat picked up by the oil is dissipated by natural radiation from the walls of the crankcase, or by means of an oil cooler where the oil is relied on for substantial engine cooling.

The specific heat of all petroleum oils is essentially the same and, of itself, has no relevance in the choice of a lubricant. Of importance is chemical stability and ability to resist the formation of deposits that might interfere with heat transfer, either from the engine parts to the oil or from the oil to its cooling medium. Under the severe conditions encountered in large diesel engines with oil cooled pistons, thermal stability is also important, that is, the ability to resist cracking of oil and deposit formation at high temperatures.

Sealing Effective cylinder sealing is necessary to minimize blowby and thereby maintain power and economy. Blowby, which cannot be entirely prevented, is a function of engine design (speed, ring, and cylinder conformity, and provision for cylinder lubrication) and oil. The greater responsibility lies with the rings and their ability to adjust themselves to the varying cylinder contours throughout the length of ring travel, but the oil has an important complementary role. Generally, as far as the oil is concerned, there will be maximum contribution if the ring grooves are clean and unobstructed so that the rings are free to move as required, and if there is no excessive removal of oil from the cylinder walls by the oil control rings. Too little oil on the cylinder walls will not only result in poor sealing but also may result in rapid wear. This, in turn, leads to even poorer sealing. Too much oil on the cylinder walls, on the other hand, results in more of the oil being exposed to combustion conditions so that oil consumption and rate of contaminant buildup in the oil may be high.

Deposits Control of deposits in an engine is a fundamental need in order to realize efficient engine performance. Deposit formation is affected by engine design, operating conditions, maintenance, fuel combustion, and the performance of the oil. In turn, deposits affect engine power output, noise, smoothness, economy, life and maintenance cost.

Two important sources of engine deposits have been discussed briefly, namely, dirt coming in with the fuel and combustion air and the fuel combustion process. Dirt in air or fuel causes abrasive wear, as already noted. In addition, it contributes to deposits on piston crowns, in ring grooves, and on valves. Deposits in these areas are often referred to as "carbon," but as this is a loose term. Usually, analysis will show that the deposits consist of dirt; solid combustion residues, such as lead salts and fuel soot; lubricating oil in various stages of decomposition; and residues from lubricating oil additives.

Temperatures in the combustion zone are high; thus, continued exposure of oil reaching this area causes it to oxidize, crack, and polymerize to heavier hydrocarbons. Large amounts of deposits in the combustion zone may be caused by excess amounts of oil reaching the upper cylinder area. Other possible causes are maladjustments in the fuel system, or improper fuels that cause smoky combustion.

Diesel engines are not particularly sensitive to deposits in the combustion zone, but gasoline engines are, especially those with high compression ratios. Depending on the character of the deposits, continued exposure to the combustion process may cause them to glow, which may provide a source of ignition in addition to the spark. Undesirable combustion phenomena such as knock, preignition, rumble, and run-on can result. Use of a higher octane fuel is usually beneficial in alleviating these problems, as is use of an oil that has less tendency to form adhering deposits in the combustion chambers.

Deposits in ring grooves are similar in origin to those in combustion chambers, but since they are not exposed to the direct flame of combustion, they may be more carbonaceous. Toward the bottoms of the pistons they also tend to have a higher oil content. In severe cases, groove deposits can be packed so tightly in the clearance spaces behind the compression rings that the rings cannot operate freely. Compression pressure cannot then be maintained, and blowby becomes excessive. Also, oil ring slots may become plugged; thus, oil control is lost.

Piston varnish is also formed from fuel and oil decomposition products. It may vary from a smooth, shiny, almost transparent coating often called lacquer, to a dark, opaque coating that becomes progressively more carbonaceous with continued operation.

Valve stem deposits are sometimes observed; see Fig. 7-3. These are generally the decomposition products of both fuel and oil, formed somewhat in the same manner as combustion chamber deposits. Intake valves may show more deposits than exhaust valves, particularly when they run cool because of low load operation. Under this condition, there is a greater opportunity for droplets of gasoline to form gums on the valve stems. Dirt and solid residues will then adhere to these gum deposits. In some designs of overhead valve engines, oil flow down the valve stems is excessive and this speeds up deposit formation, as well as being wasteful of oil.

The other major type of deposit is the emulsion or sludge formed by water, fuel decomposition residues, and solid residues. Sludge generally depos-

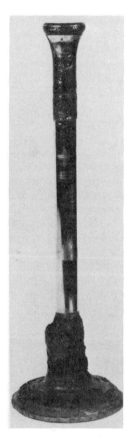

Fig. 7-3 Engine Valve Deposits

Fig. 7-4 Top Deck (Cylinder Head) Sludging This photograph shows heavy sludge buildup resulting from cold engine operation combined with an inadequate quality oil.

its on cooler engine surfaces, such as the bottom of the crankcase pan, the valve chambers, and the top decks; see Fig. 7-4. The main problem with this type of deposit is that it can be picked up by the oil and carried to areas such as the oil pump inlet screen or oil passages where it can obstruct oil flow and cause lubrication failure.

Although the physical appearance of deposits varies greatly throughout an engine, basically these deposits result from the combustion process and lubricating oil deterioration. The exact chemical and physical nature of deposits depend on the area where they are formed, the duration of exposure, and any relative motion present. In the case of sludge, water is an essential factor, and any condition that encourages the entrance and retention of water in an engine oil promotes sludge formation.

The contribution of the engine oil to the control of deposits is discussed more fully in the following section.

MAINTENANCE CONSIDERATIONS

The quality of maintenance of engine components that affect the lubrication process is of considerable importance. It is of course necessary to have clean combustion and crankcase ventilation air, which means that air cleaners must be serviced regularly. A clogged air cleaner, while it may be effective in cleaning the air, can restrict the volume of air reaching the engine to such an extent that power output is reduced significantly. In order to maintain power, the natural tendency is to open the throttle, which only adds to the difficulty since the extra fuel is not burned. Diesel engines are quite sensitive to such a condition and react by smoking and rapid buildup of engine deposits and lubricating oil contamination. In addition to keeping the filters serviced, it is important also that the piping connecting the filter to the engine be unobstructed and not have any leaks. This is particularly important in installations where the air filter is located a considerable distance from the engine.

The carburetor and ignition systems of spark ignition engines and the fuel injection systems of both gasoline and diesel engines should be in good working order and proper adjustment to assure as clean and complete combustion as possible. Malfunction or incorrect adjustment of these systems can result in increased amounts of unburned fuel in the cylinders, and dilution or more rapid buildup of contaminants in the oil.

Proper engine oil drain schedules require that the oil be drained before the contaminant load becomes so great that the oil's lubricating function is impaired, or heavy deposition of suspended contaminants occurs. In principle, the correct drain interval for a given engine in a given service is best established by means of a series of used oil analyses and inspections of engine condition. As a practical matter, this is only an economical approach for very large engines or fleets of similar engines in similar service. For most smaller engines, such as those used in passenger cars, drain intervals are usually established by the manufacturer on the basis of experience accumulated over the years with similar engines operating under typical conditions.

Drain intervals can vary considerably depending on the engine and service. For example, a large diesel engine in central station use, with a relatively large crankcase oil supply, may operate for many thousands of hours between oil changes. Such engines are usually in good adjustment, temperatures are moderate, and the contamination rate is low in comparison to the volume of oil in the system. On the other hand, a passenger car engine may require an oil change every few thousand miles. Such engines are physically small with relatively small crankcase capacity, and operate under conditions conducive to rapid oil contamination. Load factors are often low, they may be engaged mostly in short runs and start-stop service, and operate over wide ranges of ambient temperature, all of which favor the accumulation of oil contaminants and the risk of deposits.

The presence of an oil filter does not necessarily permit an extension of the oil drain interval. Filters do not remove oil soluble contaminants and water, which are important factors in deposit formation. Regular filter changes are, however, important in keeping the filter operable so that it can perform its function of removing insoluble contaminants from the oil.

ENGINE OIL CHARACTERISTICS

Some of the more important characteristics of engine oils are discussed in the following sections. The reader is also referred to the discussions of physical and chemical characteristics, additives, and evaluation and performance tests in Chap. 1.

Viscosity, Viscosity Index

In reciprocating engines, viscosity of the engine oil is extremely important. It has a bearing on wear prevention, sealing, oil economy, frictional power losses, and deposit formation. For some engines, particularly vehicle engines, it also is a factor in cranking speed and starting ease. Too high a viscosity may cause excessive viscous drag, reduction in cranking speed, and increased fuel consumption after the engine is started.

The viscosities of engine oils are usually reported according to the SAE "Viscosity Numbers for Crankcase Oils." While this system was originally intended only for automotive engine oils, its use has now been extended to include most oils for internal combustion engines. Engine manufacturers normally specify the viscosities of oils for their engines, according to ambient temperature and operating conditions, by SAE grade.

Viscosity index is important in engines that must be started and operated over a wide temperature range. In these cases, all other factors being equal, oils with higher viscosity indices will give less viscous drag during starting, and will provide thicker oil films for better sealing and wear prevention, and lower oil consumption, at operating temperatures. Viscosity index is of little importance for an oil to be used in engines not subject to frequent cold starting. Instead, the main problem is to assure sufficient viscosity at normal operating temperatures to provide suitable lubricating films.

Low Temperature Fluidity

When an oil is to be used in engines operating at low ambient temperatures, the oil must have adequate low temperature fluidity for easy dispensing from the containers to the engine crankcase, and to permit immediate flow to the oil pump suction when the engine is started. The pour point of an oil, which is an adequate indication of whether the oil will dispense, may not provide a reliable guide to whether the oil will flow to the pump suction. Most conventional oils will flow to the pump suction at temperatures below their pour points because the pump suction creates a considerably greater pressure head than is present in the pour point test. However, many multigrade oils will not circulate adequately at temperatures considerably above their pour points. Improved methods of predicting the minimum temperature at which an oil will pump satisfactorily in an automotive engine lubrication system are under development. At present, there is no method that has been shown to work for all oils. The correlation between pour point and flow in instrumented engines is poor, so that, at best, pour point can only be considered to be a rough guide to the minimum temperature at which an oil may be used safely. However, it is the only guide available at the present time.

A similar situation arose several years ago in regard to the viscous drag exhibited by oils under cold cranking conditions. As a result of cooperative studies, an instrument called the cold cranking simulator was developed to measure the low temperature viscosities of engine oils. The instrument is designed to reproduce the elements of viscous drag that affect cranking speed, and has given good correlation with field performance. The $-18°C$ (0°F) viscosities in the SAE viscosity numbers are now determined in the cold crank simulator, and two oils labeled as meeting the same SAE W grade will permit roughly the same cold cranking speed, at least down to the measuring temperature of $-18°C$ (0°F).

Oxidation Stability (Chemical Stability)

High resistance to oxidation is an important requirement of a good engine oil in view of the high temperatures the oil is exposed to and, in the crankcase, the agitation of the oil in the presence of air. Deterioration of an engine oil by oxidation tends to increase viscosity, create deposit forming materials, and promote corrosive attack on some hard alloy bearings. Where the oil supply is relatively small, the rate of deterioration tends to be higher, other factors being equal.

An oil's natural oxidation stability is determined in part by the crude oil from which it is made and the refining processes to which it is subjected. Where engine design or operating conditions require a high degree of oxidation stability, oxidation inhibitors are used. As a general rule, the need for greater oxidation stability increases as oil service temperatures increase. Among the principal factors that make enhanced oxidation stability necessary are high engine specific power output (high horsepower per unit of displacement), small crankcase charge volume, long oil drain intervals, and, in recent model passenger car and light truck engines, modifications and devices to control emissions that result in high operating temperatures. For example, a heavily loaded truck engine requires an oil with excellent oxidation stability because of the high operating temperatures involved, while a large, low speed diesel engine in central station (stationary) service, with a large crankcase oil supply at a moderate temperature, requires an oil with good stability because the oil is expected to remain in service for thousands of hours. Regardless of the necessity for it, good stability is always desirable in view of its helpful influence on engine cleanliness.

Oxidation stability plays only a minor role in the process by which combustion chamber deposits are formed. Under the conditions found in the combustion chambers, any oil will decompose to at least some degree. The contribution of the lubricating oil to such deposits can, however, be controlled partly by base stock refining and selection, see Engine Oil Characteristics, Chapter 2.

Thermal Stability

Thermal stability, or resistance to cracking and decomposition under high temperature conditions, is a fundamental characteristic of the lubricating oil that cannot be improved by means of additives. However, careful selection of additives is important in formulating thermally stable oils, since decomposition of the additives can contribute to the formation of deposits under operating conditions where thermal cracking of the oil occurs.

As noted earlier, thermal stability of the engine oil is of special concern in certain large, highly supercharged, two cycle diesel engines used for marine propulsion. Thermal cracking of the system oil has been experienced in the piston cooling spaces, resulting in deposits that interfere with heat transfer. Oils for these applications must be manufactured from base stocks made from selected crudes by carefully chosen refining processes.

Detergency and Dispersancy

The natural detergent and dispersant properties of most oils is slight and where these characteristics are important, they are obtained through the use of additives.

In nearly all current internal combustion engine applications, oils with enhanced detergency and dispersancy are necessary to control engine deposits and maintain engine performance. The levels of detergency and dispersancy required depend on a number of factors such as engine operating temperatures, type of fuel, continuity of operation, and exposure to low ambient temperatures. In general, conditions that tend to promote oil oxidation, such as supercharging or the use of high sulfur fuels, dictate the use of oils with higher levels of detergency. Conditions that promote condensation of water and unburned or partially burned fuel in the engine, require the use of oils with higher dispersancy.

As noted in the preceding discussion, detergency and dispersancy are not clearly differentiated properties. There is a trend in additive development to improve the dispersancy of the so-called "detergents" and the detergency of the "dispersants." Thus, some oils formulated only with dispersants may give entirely adequate control of high temperature deposits in some services, while other oils formulated only with detergents may give good control of low temperature emulsions and sludges in some services. In general, however, most engine oils contain a mixture of the two types of material with the concentrations and relative proportions depending on the type of engine service an oil is designed for.

One of the main functions of both detergents and dispersants is to suspend potential deposit forming materials in the oil. In this form, these materials are relatively harmless and may be removed from the system by draining the oil. Regular oil drains for this purpose are important, particularly for engines with a relatively small oil supply. If the oil drain intervals are too long, the ability of the additives to suspend the deposit forming materials may be exceeded, deposits will begin to form on engine surfaces, and engine performance will deteriorate. This is the reason manufacturers specify more frequent oil drains for "severe service" such as short trips, sustained high speeds, trailer towing, and so forth.

There are no absolute measures of detergency and dispersancy. It is now customary to describe engine oils, on the basis of performance testing, according to the type of engine service for which they are suitable. While this description includes an evaluation of detergency and dispersancy, the intent is to provide a more comprehensive description that includes all of the factors that make a particular oil suitable for a particular type of engine service.

Alkalinity

Most detergents, and to a lesser extent many dispersants, have some ability to neutralize the acidic end products of fuel combustion and oil oxidation. However, where a considerable ability to neutralize acids is required, as in oils for diesel engines burning high sulfur fuels, highly alkaline (overbased) detergent type materials are used. The concentration of these materials in an oil, and an indication of the oil's ability to neutralize acids, is given by the total base number (TBN), or alkalinity value as it is also called. There is only a general relationship between TBN and the ability of an oil to control wear and corrosion caused by strong acids, since it has been found that some newer additive systems are more effective in this respect than would be predicted by consideration of the TBN alone.

Many of the highly alkaline oils are used as cylinder oils in large engines, which is a once-through use. In these cases, the changes in the oil in service are of little or no concern. In smaller engines, where the same oil serves as both the cylinder and crankcase oil, it is customary to monitor the alkalinity of the oil as a method of determining if it is still capable of performing its neutralization function.

Antiwear   In addition to the corrosive wear caused by acidic products of combustion, metallic wear may occur in areas where, due to loads or operating conditions, effective lubricating films cannot be maintained. The main areas where this occurs are on cylinder walls and rings, particularly of large, high output trunk piston engines, and the valve mechanisms of small, high speed engines. In these cases, it is usually necessary to use oils that are formulated with additives that provide enhanced protection against wear and scuffing under boundary lubrication conditions.

Rust and   All petroleum oils have some ability to prevent rusting and corrosion of en-
Corrosion   gine metals. However, in most cases, this natural ability is not sufficient to:
Protection

1. Protect hard alloy bearings from corrosion caused by oil oxyacids
2. Prevent rusting and corrosion due to condensation of water and combustion products in low temperature or stop-and-go service
3. Control the corrosive wear caused by acidic end products of combustion

Since one or more of these conditions, which can cause troublesome corrosion, are encountered to some extent in nearly all internal combustion engine service, most oils for internal combustion engines are formulated to provide additional protection against corrosion.

Automotive engine oils usually are formulated to provide protection against corrosion and particularly corrosion of hard alloy bearings. In the case of oils intended for gasoline engine service, protection against corrosion and rusting caused by condensation of water and unburned or partially burned fuel components is emphasized. Diesel engine oils are usually formulated to provide protection against corrosion caused by acidic end products of combustion. As stated previously, in these latter oils, the function of corrosion protection is closely related to detergency and alkalinity.

Special preservative oils are available for the protection of engines that are to be laid up seasonally or stored for extended periods. These oils are formulated to provide protection against rust and corrosion caused by atmospheric conditions and, usually, are also formulated to be suitable for short term use in the engines under moderate operating conditions. As a result, an engine can be run safely on the preservative oil prior to being laid up to distribute the oil over the internal engine surfaces. Also, an engine can be run for a reasonable period after being taken out of storage before the preservative oil is drained and replaced by the normally recommended type of oil.

Foam Resistance

All oils will foam to some extent when agitated. If excessive foaming occurs in an internal combustion engine, several problems may result. Overflow and spillage of oil is, of course, one of the most obvious, but foaming can also result in starvation at the oil pump inlet, or slugs of foam being drawn into the pump with the oil. Foam entrained in the oil can cause failure of lubricating films, and noisy, erratic operation of hydraulic valve lifters. To reduce foaming, particularly of oils intended for small, high speed engines where agitation is severe, a defoamant is often included in the formulation. Overfilling the crankcase can cause foaming even with defoamants.

Effect on Gasoline Engine Octane Number Requirement

Although not classed as a lubricating function directly, the characteristics of the lubricating oil have a considerable influence on the octane number requirement (ONR) of gasoline engines. Some oil always finds its way past the rings into the combustion chambers, where it is partially burned to form deposits that adhere to the piston tops and combustion chamber surfaces. As these deposits build up, the ONR of the engine gradually increases. The increase in ONR is a function of both the quantity and nature of the deposits. Heavy, ragged deposits (Fig. 7-5, left) cause the octane number requirement to increase more than smooth, even deposits (Fig. 7-5, right), primarily because the sharp projections on the ragged deposits can be more easily heated to incandescence. In this condition, the deposits may ignite the fuel charge, either before the spark plug fires or before the normal flame front reaches portions of the charge. In either case, knock and power loss result.

It has been found that the contribution of the lubricating oil to the increase in engine ONR may be reduced considerably by careful attention to such factors as crude oil source, refining processes, and the use of blending stocks having a minimum of heavy hydrocarbons. One of the major factors in accomplishing the latter is the use of all distillate blending stocks, rather than the older practice of incorporating some bright stock in the blends. Many multiviscosity automotive engine oils are manufactured to provide maximum benefits in the control of octane number requirement increase.

Fig. 7-5 Combustion Chamber Deposits The heavy, ragged deposits on the left, formed when a conventional oil was used, generally contribute significantly more to the engine octane number requirement than the smooth deposits on the right.

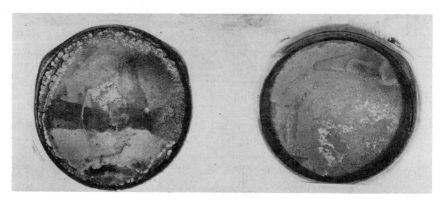

Identification The use of the SAE viscosity classification system has already been discussed in Chap. 2. It is important to remember that this system is concerned only with viscosity. Thus, it may be used to describe the viscosity characteristics of oils intended for widely different services. Two other systems in general use for describing the performance qualities of automotive engine oils are the API "Engine Service Classifications for Engine Oils" and CCMC "Performance Specifications."

API Engine Service Classifications for Engine Oils As originally set up by the American Petroleum Institute (API), this system included three gasoline engine service classifications, ML, MM, and MS, and two diesel engine classifications, DG and DS. A third diesel classification, DM, was added later, and in 1969 the ML classification for gasoline engines was dropped. However, certain inherent difficulties with the system resulted in a joint effort by API, ASTM (American Society for Testing and Materials), and SAE (Society of Automotive Engineers, Inc.) to develop a new system that would be more effective in communicating engine oil performance and engine service classification information between the petroleum and automotive industries. This new system, which was finalized during 1970 and consists of two categories, S (Service Station) and C (Commercial), is described in API Bulletin 1509 and SAE Recommended Practice SAE J183a. At the present time, there are nine engine service classifications, as follows:

S—Service (Service Stations, Garages, New Car Dealers)

SA, formerly for Utility Gasoline and Diesel Engine Service: This service, is typical of older engines operated under such mild conditions that the protection afforded by compounded oils is not required. This classification has no performance requirements, and oils in this category should not be used in any engine unless specifically recommended by the equipment manufacturer.

SB, Minimum Duty Gasoline Engine Service: This service is typical of gasoline engines operated under such mild conditions that only minimum protection afforded by compounding is desired. Oils designed for this service have been used since the 1980s and provide only antiscuff capability, and resistance to oil oxidation and bearing corrosion. They should not be used unless specifically recommended by the equipment manufacturer.

SC, 1964 Gasoline Engine Warranty Maintenance Service: This service is typical of gasoline engines in 1964 through 1966 models of passenger cars and some trucks operating under engine manufacturers' warranties in effect during those model years. Oils designed for this service provide control of high and low temperature deposits, wear, rust, and corrosion in gasoline engines.

SD, 1968 Gasoline Engine Warranty Maintenance Service: This service is typical of gasoline engines in 1968 through 1970 models of passenger cars and some trucks operating under engine manufacturers' warranties in effect during those model years. Also may apply to certain 1971 and/ or later models as specified (or recommended) in the owners' manuals. Oils designed for this service provide more protection against high and low temperature engine deposits, wear, rust, and corrosion in gasoline engines than oils which are satisfactory for API engine service classification SC and may be used when API engine service classification SC is recommended.

SE, 1972 Gasoline Engine Warranty Maintenance Service: This service is typical of gasoline engines in passenger cars and some trucks beginning with 1972 and certain 1971 models operating under engine manufacturers' warranties. Oils designed for this service provide more protection against oil oxidation, high temperature engine deposits, rust, and corrosion in gasoline engines than oils which are satisfactory for API engine service classifications SD or SC and may be used when either of these classifications is recommended.

C — Commercial (Fleets, Contractors, Farmers)

CA, Light Duty Diesel Engine Service: This service is typical of diesel engines operated in mild to moderate duty with high quality fuels and occasionally has included gasoline engines in mild service. Oils designed for this service provide protection from bearing corrosion and from ring belt deposits in some naturally aspirated diesel engines when using fuels of such quality that they impose no unusual requirements for wear and deposit protection. They were widely used in the late 1940s and 1950s, but should not be used in any engine unless specifically recommended by the equipment manufacturer.

CB, Moderate Duty Diesel Engine Service: This service is typical of diesel engines operated in mild to moderate duty, but with lower quality fuels which necessitate more protection from wear and deposits. It occasionally has included gasoline engines in mild service. Oils designed for this service provide necessary protection from bearing corrosion and from high temperature deposits in normally aspirated diesel engines with higher sulfur fuels. Oils designed for this service were introduced in 1949.

CC, Diesel and Gasoline Engine Service: This service is typical of certain naturally aspirated, turbocharged, or supercharged diesel engines operated in moderate to severe duty service and certain heavy duty gasoline engines. Oils designed for this service provide protection from high temperature deposits and bearing corrosion in these diesel engines and also from rust, corrosion, and low temperature deposits in gasoline engines. These oils were introduced in 1961.

CD, Diesel Engine Service: This service is typical of certain naturally aspirated, turbocharged, or supercharged diesel engines in severe duty when using fuels of a wide quality range including high sulfur fuels where highly effective control of wear and deposits are vital. Oils designed for this service were introduced in 1955 and provide protection from bearing corrosion and from high temperature deposits in these diesel engines.

The API Classification System This system is open ended, so that additional classifications can be added when needed. When the letter designations are used by oil marketers or engine manufacturers to indicate the service for which oils are suitable or required, it is the intent that they be preceded by the words, "API Service." To illustrate, a service station oil suitable for servicing new cars under warranty would be referred to as "for API Service SE." If oils are suitable for more than one service, it is appropriate that these oils be so designated, for example, "for API Services CC and SE."

There is a relationship between API Engine Service Classifications and some of the oil quality standards previously outlined. These relationships are shown in Table 7-1.

These relationships result from the fact that in addition to the definitions of the service classifications, there are also minimum performance requirements for each classification that are defined in terms of performance in prescribed engine tests. The prescribed engine tests for any service classification include all of those required for the corresponding related designations, or acceptable alternates. Also, the minimum performance requirements in each test are at least equal to those required by any of the

Table 7-1   Related Service Designations

New API Engine Service Classifications	Related Designations Military and Industry
Service Station Engine Services	
SA	Straight mineral oil
SB	Inhibited oil (non detergent)
SC	1964 Warranty approved Ford ESE-M2C101-A

SD	1968 Warranty approved
	    Ford ESE-M2C101-B,
	    GM 6041-M
	    (Prior to July 1970)

SE	1977 Warranty approved
	    Ford ESE-M2C144-A
	    1979 Warranty approved
	    Ford ESE-M2C153-A,
	    (Sequence III D),
	    GM 6136-M (previously
	    GM 6041-M, Rev.)
	    MIL-L-46152[a]

Commercial and Fleet Engine Services	
CA	MIL-L-2104A,
CB	Supp. 1
CC	MIL-L-2104B, MIL-L-46152[a]
CD	MIL-L-45199B, Series 3 MIL-L-2104C[b]

[a]Oils meeting performance requirements of MIL-L-46152 meet the performance requirements of both API Service SE and CC.

[b]Oils meeting performance requirements of MIL-L-2014C meet the performance requirements of both API Service CD and SC.

corresponding related designations. Thus, for example, the engine test requirements for an oil for API Service SE include all of the engine test requirements for GM 6136-M, and the minimum performance levels in those tests are at least equal to those required to meet the General Motors standard.

This approach of providing a complementary definition of minimum performance requirements for the service classifications will provide the user with better assurance that oils marked as being suitable for a particular service classification will provide an acceptable minimum performance level in service. However, performance differences can still exist among oils designated as for the same service classification, since many oils are, and will be, formulated to exceed the minimum performance levels. These oils will, of course, provide additional benefits in service.

CCMC Performance Specifications The Comite des Constructeurs du Marche Commun (CCMC), which represents the interests of European motor manufacturers, has set up a committee to examine engine oil quality requirements for the eingines they produce. Engine tests have been agreed on, and performance ratings in those tests have been proposed. On the basis of the proposed ratings, the majority of the member organizations of CCMC are incorporating these requirements into their own specifications.

In contrast to the API Engine Service Classifications, which define several oil qualities, the CCMC system defines only one oil quality which is roughly equivalent to API Service SE. The tests included in the CCMC system are the Fiat 600D (alternate Sequence VD), Petter W1 (alternate L-38 plus Sequence IIID), Ford Cortina, Fiat 124AC, Mercedes Benz OM 615/616, Bosch Injector Rig, and Sequence IID.

U.S. Military Specifications In the past, the Department of the Army, U.S. Department of Defense, had responsibility only for issuing specifications for oils for use in military vehicles. However, in order to provide standardization of lubricating oils used in the many commercial vehicles operated by other branches of the government, the Army was also given the responsibility for preparing an oil specification for those vehicles. Since the Army specifications were becoming outdated, two new specifications were prepared and issued in 1971, and previous specifications became obsolete. Brief descriptions of these current specifications, along with the obsolete specifications which are still used to some extent as performance references, follow.

U.S. Military Specification MIL-L-46152 This specification describes oils for both gasoline and diesel engines in commercial vehicles used in U.S. federal and military fleets. In contrast to earlier U.S. Military specifications, it places strong emphasis on gasoline performance in addition to diesel performance. The gasoline engine performance requirements are the same as API Service SE, and the diesel engine performance requirements are the same as the former MIL-L-2104B (API Service CC). It covers oils in the SAE 10W, 30, 10W-30, and 20W-40 viscosities.

U.S. Military Specification MIL-L-2014C This specification describes oils for gasoline and diesel engines in tactical vehicles in U.S. military fleets. The diesel engine performance requirements are the same as the former U.S.

Military Specification MIL-L-45199B (API Service CD). Gasoline engine performance in Sequences IID and VD is required at a level approximately intermediate between API Services SC and SD. The U.S. Military operates very few gasoline engines in tactical service, so the gasoline engine performance requirements were set at a moderate level that would provide adequate performance without risk of compromising the severe diesel requirements desired.

MIL-L-2104A (obsolete December 1, 1964) The performance requirements of this specification were the same as those now required for API Service CA.

Supplement 1 This terminology dates back to the time of U.S. Army Specification 2-104B which preceded MIL-L-2104A. The performance requirements for Supplement 1 were the same as those now required for API Service CB.

MIL-L-45199B (obsolete November 20, 1970) The performance requirements of this specification were the same as those now required for API Service CD.

Manufacturer Specifications Many engine manufacturers issue specifications for oils for their engines. Some of these specifications apply only during the break in or warranty period, while others represent the type of oil that the manufacturer believes should be used for the life of the engine. Some of these specifications are compatible with standard oil qualities available in the marketplace. For example, Ford and General Motors passenger car engine oils are of API Service SE quality, while others represent special formulations designed to meet conditions that are specific to certain engines in certain applications. In the latter category is the oil required for certain railroad diesel engines. Where specifications of this type exist, special oils are usually developed and marketed to meet the specification requirements, assuming the volume required in the marketplace is sufficient to justify the costs of the development.

OIL RECOMMENDATIONS BY FIELD OF ENGINE USE

The following comments provide a guide to the types and viscosities of oils usually recommended for internal combustion engines used in the various major fields of application. It must be remembered that a particular engine may have different requirements from other engines in the same field, and an engine may also have different requirements when it is used in different fields of application.

Passenger Car Most passenger cars have four stroke cycle gasoline engines. A growing number have diesel engines, and a few have rotary engines. Two stroke cycle gasoline engines were phased out many years ago. The usual recommendations for current production, four stroke gasoline engines are oils for API Service SE. These oils provide good protection against low temperature deposits and corrosion, protect against wear, and provide excellent protection against oxidation, thickening, and high temperature deposits under the most severe conditions of high speed operation or trailer towing, even at high ambient temperatures.

Many older model four cycle gasoline engines can be satisfactorily operated on oils for API Service SC or SD, although oils for API Service SE will provide somewhat better overall performance and are more readily available.

The viscosities usually recommended are SAE 20W-20 or 30 for temperatures down to about 32°F (0°C), and SAE 20W-20 or 10W for lower temperatures. For extreme low temperatures, SAE 5W, 5W-20, or 5W-30 oils may be recommended. Multiviscosity SAE 10W-30 or 10W-40 oils are often recommended for year round use, and are now the primary recommendation of several engine manufacturers. European engine manufacturers often recommend higher viscosities, particularly of multiviscosity oils where SAE 20W-40 and 20W-50 oils are frequently perferred.

The recommendations for passenger car diesel engines are generally similar to those for four cycle gasoline engines, that is, oils for API Service SE or SE, CC. Some manufacturers permit the use of multiviscosity oils, while others prefer single viscosity types.

The rotary engines used in passenger cars generally require SAE 10W-30 oils approximately API Service SD or SE quality, and which have shown evidence of satisfactory performance in service. The performance of oil additives in these engines is not necessarily the same as in reciprocating engines, hence the desire of the manufacturers for assurance that the particular oil formulation recommended has given evidence of satisfactory performance.

Two cycle gasoline engines are used to some extent in various areas of the world. These engines are lubricated either by premixing the oil with the fuel, or by injecting oil into the fuel at the carburetor. The oil used is usually of either SAE 30 or 40 viscosity and is formulated specifically for this service. Generally, four cycle engine oils are not satisfactory because the additives in them will form undesirable deposits in the combustion chambers of two cycle engines.

Truck and Bus   A significantly larger proportion of diesel engines are used in trucks and buses than in passenger cars. Both two cycle and four cycle diesel engines are used; the gasoline engines are all four cycle.

The recommendations for gasoline engines in trucks and buses are similar to those of passenger cars, that is, oils for API Service SE or SE, CC. The viscosities used are somewhat heavier, typically SAE 30 or 40 for summer use and SAE 20W-20 or SAE 10W for winter use. SAE 30 oils are used year round in many cases where operation is more or less continuous, vehicles are stored inside, or starting aids can be used. Multiviscosity SAE 10W-30, 15W-40, 20W-40, and 20W-50 oils are being used to an increasing extent to take advantage of the improved starting and fuel economy they provide.

There is considerable variation in the oils recommended for diesel truck and bus engines. In general, oils for API Service CC are recommended for naturally aspirated engines, and oils for API Service CD for supercharged engines. However, the two cycle engines are somewhat sensitive to ash content in the oil; thus, the usual recommendation for them—whether supercharged or naturally aspirated—is an oil for API Service CC with certain restrictions on the ash content. Some of the four cycle engines also have shown sensitivity to the additive system; therefore, various manufacturers have special requirements over and above the basic API Service CC or CD recommendation. Viscosities used are usually SAE 30 for summer use and SAE 20W-20 or 10W for winter use. SAE 30 oils are specified for year round use by some manufacturers.

Where liquefied petroleum gas (LPG) engines are used in trucks or buses, oils for API Service SE are often used for convenience, although somewhat lower quality oils, such as those for API Service SC or SD, may be satisfactory. In some cases, special oils containing no metallo-organic detergents (ashless additives) are recommended.

Farm Machinery The engines used in farm machinery include gasoline, diesel, and LPG. Some two cycle gasoline engines are used for utility purposes such as water pumps and lighting plants. In general, the oil recommendations parallel those of passenger cars and trucks. There is a trend toward the recommendation of oils for API Service CD for naturally aspirated diesel engines as well as supercharged engines. Some manufacturers express a preference for special oils without metallo-organic detergents for LPG engines. Viscosities recommended closely parallel those for truck and bus engines.

The small two cycle gasoline engines used for utility purposes are lubricated by oil mixed into the fuel. For convenience, these engines are sometimes lubricated with SAE 30 or 40 oils of API Service SC, SD, or SE quality, but, where available, special oils designed for use in two cycle gasoline engines are usually preferred. These latter oils are formulated for the conditions encountered in the two cycle engines and will generally give lower port, combustion chamber, and spark plug deposits. Viscosities are usually in the SAE 30 or 40 viscosity range, and there is a trend to prediluting the oils with approximately 10 percent of a petroleum solvent so they will mix more readily with the fuel. Usually, fuel oil mixtures on the order of 16:1 to 24:1 are used, but some of the newer engines are designed for the use of higher ratios.

Contractor Machinery Contractor machinery covers a wide variety of engines, ranging from small, two cycle gasoline engines, through automotive type gasoline and diesel engines, to larger diesel engines used to power giant cranes and earth moving machines. Generally, oil recommendations closely parallel those for similar engines in other automotive equipment.

Aviation Primarily, reciprocating piston engines for aviation use are of the four stroke cycle gasoline type. By far, the majority of these engines are relatively small and are used in personal and civil aircraft.

For many years aircraft engines were operated on high quality, straight mineral oils. In recent years, however, dispersant type oils have been developed which offer benefits in engine cleanliness, and most aircraft engine manufacturers now accept or recommend the use of these oils. Straight mineral oils may still be recommended for the breakin period.

Aircraft engine oils are usually formulated from special base oils made from selected crudes by carefully controlled refining processes. Quality is usually controlled to meet U.S. and U.K. government specifications, although some of the engine manufacturers have their own closely related specifications. Dispersant type oils are usually manufactured by combining

Table 7-2   Aircraft Engine Oil Specifications

Common Grade Designation	U.S. Specifications		U.K. Specifications		Approx. SAE Grade
	St. Mineral MIL-L-6082	Dispersant MIL-L-22851	St. Mineral DERD 2472	Dispersant DERD 2450	
65	1065	None	None	None	30
80	None	Type III(a)	A/O	D65	40
100	1100	None	B/O	D80	50
120	None	Type II(a)	None	D100	—

an approved additive system with proven straight mineral aircraft engine oils.

The viscosities of aircraft engine oils may be designated by SAE viscosity grade, or by a grade number which is the approximate viscosity in SUS at 99°C (210°F). In the U.S. Military designations this latter number is preceded by the digit 1. The specification numbers and viscosity grades covered are shown in Table 7-2. It will be noted that the U.K. grade designations for the ashless dispersant oils are one grade lower in each case. This results from the fact that the grade designation applies to the base oil, and because of the thickening effect of the dispersant a 65 grade base oil is required to make an 80 grade finished oil.

Industrial   Nearly every type of internal combustion engine has some industrial application. The smaller engines are usually of the automotive type, so the oils used are similar to those recommended for automotive applications. The following discussion pertains to the larger engines used as prime movers in power plants, mills, pipeline pumping stations, refineries, and so forth. Both two and four stroke cycle engines are used in these applications. Diesel engines are used with fuels ranging from clean distillates to heavy residuals. Spark ignition engines are usually run on natural gas, producer gas, or LPG. Dual fuel engines, which are operated on a combination of diesel fuel (about 5 percent) as an igniter and natural gas, are also used.

Diesel Engines   Diesel engines range in size from a few hundred horsepower up to engines rated 50,000 hp or more. Generally, they can be divided into three classes:

1. High speed engines
2. Medium speed engines
3. Low speed engines

The lubrication requirements differ considerably for these classes, to some extent because of the types of fuels that the different classes of engines can burn satisfactorily.

High Speed Engines   Generally, these engines are considered to be those that operate at speeds over 1000 rpm. Engine sizes range up to about 250 mm (10 in.) bore, with power outputs up to about 200 hp per cylinder. Multicylinder engines with outputs up to about 4000 hp are available. These engines are all of the trunk piston type, may be either supercharged or naturally aspirated, and may be either two stroke or four stroke cycle. The rings and cylinder walls are lubricated by oil splashed from the crankcase.

High speed engines require high quality fuels, and, therefore, are usually operated on distillate fuels similar to those used in automotive diesel engines. As a result, the sulfur content of the fuel rarely exceeds 1 percent, and often is considerably lower. With fuels of this quality, corrosive wear of cylinders and rings can be controlled satisfactorily by the types of oils developed for automotive diesel service, such as oils for API Service CC or CD. Where the engine is one of the types developed for railroad service, one of the special railroad engine oils may be used. Viscosities recommended are usually either SAE 30 or SAE 40.

Medium Speed Engines These engines operate at speeds ranging between 375 and 1000 rpm. Engine sizes and outputs in this classification range from about 225 mm (8.85 in.) bore with an output of 125 to 135 hp per cylinder, to 600 mm (2.36 in.) or larger bore developing 1500 or more per cylinder. Large medium speed engines with power outputs to 30,000 hp or more from V type configurations of up to 20 cylinders are available. All engines in this class are of the trunk piston type, generally four cycle. Newer engines are usually supercharged.

Smaller engines in this class have the rings and cylinder walls lubricated by oil splashed or thrown from the crankcase to the lower parts of the cylinder walls. Larger engines have separate cylinder lubricators to supply supplemental oil to the cylinder walls.

Medium speed engines are operated on a wide range of fuels. Many of the smaller engines are operated on high quality distillate fuels. Somewhat heavier fuels, such as distillates, may be used in some engines. The disadvantages of this fuel include a higher boiling range and a higher sulfur content. Also, some residual components may be included in the blend. Many of the large engines are designed to operate on residual fuels, although the residual fuels used are somewhat lower in viscosity than the heavy, bunker-type fuels that are often used in low speed engines. Sulfur contents of the residual fuels may range upward to about 2.5 percent, and in some cases even higher.

While some medium speed engines may be operated on automotive diesel engine oils of about API Service CD quality, in general, the oils used in these engines are developed specifically for them and for similar engines used in marine propulsion service.

For convenience, these oils can be described in terms of their total base number (TBN). Smaller engines, and larger engines burning high quality fuels, are usually lubricated with oils of 10 to 20 TBN. Where residual fuels are used or operation is severe with better quality fuels, oils of 30 to 40 TBN are usually used. The 30 to 40 TBN oils are frequently used for the cylinders of engines with separate cylinder lubricators. In some cases with poor quality, high sulfur fuels, oils in the 50 to 70 TBN range may be used. Some care should be taken with this latter approach since the cylinder oil after use drains into the crankcase. Any incompatibility between the cylinder oil and crankcase oil could, therefore, cause difficulties.

Viscosity grades used are usually SAE 30 or 40 for the crankcases, and SAE 40 or 50 for cylinders with separate lubricators. In some cases, where the engines are in intermittent service in exposed locations, SAE 20 oils may be used in the crankcase.

Low Speed Engines   The low speed engine class consists of the large, crosshead type engines, most of which operate at speeds below 300 to 400 rpm. Engines in this class usually range from about 700 mm (27.5 in.) to 1060 mm (41.5 in.) bore, with outputs for the largest engines as high as 4500 hp per cylinder, or 54,000 hp from a 12 cylinder engine, the largest built. Almost all of these engines operate on the two stroke cycle. Separate cylinder lubricators are used.

Low speed engines are usually operated on residual fuels, some with sulfur content of 4.0 percent or more. Since the combustion of fuels of this quality results in the formation of large amounts of strong acids, highly alkaline oils are needed to control corrosive wear and corrosion of rings and cylinder liners. Where the fuel sulfur content is relatively low, oils in the 20 to 40 TBN range may be used; with higher sulfur fuels, oils in the 60 to 80 TBN range are usually used. The viscosity grade is usually either SAE 40 or 50.

The crankcase, or system, oil in these engines lubricates the bearings and crossheads and may lubricate the supercharger bearings. Since a large volume of oil is involved, extremely stable oils designed for long service life are desirable. Little or no contamination of the system oil by combustion products occurs; therefore, high levels of detergency and alkalinity are not required. Many engines are still operated on straight mineral oils, or inhibited straight mineral oils. Other engines are operated on the lower alkalinity oils used for trunk piston engines, usually oils of 10 or less. A number of newer, large engines have oil cooled pistons. The high temperatures that the system oil is exposed to in the piston cooling spaces of these engines may cause thermal cracking of the oil with a buildup of deposits. In turn, these deposits may interfere with heat transfer so that excessive piston temperatures can occur. This had led to the development of special system oils designed to provide better thermal stability and better ability to control the buildup of deposits that result from thermal cracking. Usually, these oils are manufactured from selected crudes by carefully controlled refining processes. The additives used in them are also carefully selected for their ability to resist thermal decomposition at high temperatures.

Usually, system oils are of SAE 30 grade, although SAE 40 oils may occasionally be used.

Gas and LPG Engines   The clean burning fuels used in these engines result in relatively little contamination of the oil by combustion products. However, degradation of the oil caused by oxidation or nitration can occur, particularly in supercharged engines or engines operating under severe service conditions. Some engines are equipped with hard alloy bearings which require protection against corrosion. Valve seat and face wear are problems in some engines.

Oils used in gas engines are all now products developed specifically for the service. Most now contain dispersants to control varnish type deposits resulting from oxidation or nitration. Some manufacturers recommend oils containing only ashless additives (no metallo-organic detergents), others recommend the use of oils containing metalloorganic oxidation and bearing corrosion inhibitors in combination with the dispersant, while others recommend that metallo-organic detergents also be included. The amount of metallo-organic detergent required, as measured by the sulfated ash content, also varies considerably with different engine manufacturers. The main reason for including the metallo-organic detergent is to control valve seat and face wear. The residue from combustion of the detergent appears to act as a solid lubricant for these surfaces. Depending on such factors as metallurgy and operating conditions, different amounts of this type of residue have been found to be effective in different engines. As a result, premium gas engine oils are classified as follows:

1. Oils containing ashless inhibitors
2. Ashless dispersant oils containing about 0.10 percent ash contributed by a metallo-organic oxidation and corrosion inhibitor
3. Oils containing about 0.4–0.9 percent ash contributed by a small amount of metallo-organic detergent in combination with the inhibitor
4. High ash oils containing about 2.0–3.5 percent ash

Each of these four general types is recommended, or specified, by one or more major engine manufacturers.

Viscosities used are usually SAE 40; however, SAE 30 oils are used for some engines operated in exposed locations.

Dual Fuel Engines To some extent, the lubricating oil in a dual fuel engine is subjected to the deteriorating influences found in both diesel and gas engines. The diesel fuel burned as a pilot fuel tends to produce soot and varnish deposits, while the gas fuel, combined with high operating temperatures, tends to cause oxidation and nitration of the oil. While these conditions are not necessarily conflicting, there is some indication that oils specially developed for the service may provide superior engine cleanliness to both diesel and gas engine oils. SAE 30 or 40 oils are used as crankcase oils, with SAE 50 or even higher viscosity oils preferred for cylinder lubrication of engines with separate cylinder lubricators.

Marine Inboard (Small Craft) The engines used in small marine craft are often modified versions of automotive engines, both gasoline and diesel. As a result, the qualities and viscosities of oils recommended closely parallel those used in comparable automotive engines.

Marine Outboard Almost all the outboard motors used for boats are two stroke cycle gasoline engines because of the high power to weight ratio of this design. Oil is mixed with the fuel for lubrication. In spite of a trend to very lean oil to fuel ratios, some oil is always carried into the combustion chambers where, when it is burned, deposits can be formed. Excessive deposit buildup can affect combustion and obstruct the ports causing loss of power. Metallo-organic additives in the oil may aggravate this deposit problem; thus, most oils for outboard engines are not formulated with ashless dispersants. A corrosion inhibitor to provide additional protection against rusting during shutdown periods is usually included. The base oils are selected to minimize carbonaceous deposits in the combustion chambers. Data obtained in a standardized outboard motor test may be submitted to the Boating Industry Association (BIA) for certification that an oil is suitable for Service TC-WC (two cycle water cooled). Oils certified in this manner are generally acceptable to the outboard motor manufacturers.

Oils used in outboard engines are generally at the heavy end of the SAE 30 viscosity range, or the light end of the SAE 40 range. Fuel to oil ratios of 50:1, or higher, are now recommended, particularly for high output engines. Many oils now are prediluted with a small proportion (about 10 percent) of a special petroleum solvent to facilitate mixing with gasoline. At some locations, where there is a high volume of fuel sales for outboard motors, premixed oil and fuel may be offered.

Marine Propulsion

The propulsion engines used in vessels range from small automotive type engines used in fishing boats, harbor craft, and other small craft, to the largest diesel engines now manufactured. In general, the discussion of industrial-diesel engines applies to marine propulsion engines.

Marine Auxiliary

Auxiliary engines used on ships are medium or high speed diesel engines of moderate size. The engines are operated on distillate fuels of fairly good quality. In some cases, convenience dictates the type of oil used in these engines. Where the main propulsion engines are operated on a 10 to 20 TBN oil, for example, the same oil can be used satisfactorily in the auxiliary engines. In other cases, the oils used in the propulsion engines are not suitable for use in the auxiliary engines, and a special oil must be carried for the auxiliaries. This oil will be either one of the 10 to 20 TBN marine diesel engine oils, or an automotive diesel engine oil, depending on the preference of the engine manufacturer.

Railroad

Diesel engines used in railroad locomotives, because of space limitations, often have high specific ratings in order to get high power output from a relatively small engine. Combined with restrictions on cooling water capacity, this results in both crankcase and cylinder temperatures being high. Long idling periods and rapid speed and load changes also contribute to severe operating conditions for the lubricating oil.

Both two and four stroke cycle engines are used in this service, and cylinder lubrication is supplied by oil throw from the crankcase.

U.S. railroad engine manufacturers recommend oils developed specifically for their engines, usually either the Superior Class I oils or, for the latest high output models, the "third generation," or Superior Class II oils. These oils are of SAE 40 viscosity. European manufacturers recommend automotive diesel engine oils, usually of API Service CC or CC, SE quality in the SAE 30 viscosity.

Specialty Applications

Large numbers of small gasoline engines are used in such applications as snowmobiles, chain saws, lawn mowers, garden tractors, home lighting plants, and trail bikes. Most of these engines are air cooled and two cycle. Four cycle engines are used where weight is not a major concern; or in larger engines, where the higher fuel consumption of the two cycle engines might be a concern. The two cycle engines are lubricated by oil mixed with the fuel. Both the oils developed for outboard motors and oils developed specially for air cooled two cycle engines are used. Special oils are used most frequently in snowmobiles and chain saw engines. Mixing is critical with snowmobile oils because of the low temperatures at which the engines are operated, so prediluted oils are frequently used. Operating temperatures in snowmobiles also may be high because the engine is shrouded to provide heat for the operator, and this may cause piston varnish problems. Mixing is also a problem with chain saws, as is high temperature operation with piston varnish and port deposits.

Bibliography

Mobil Technical Book
Diesel Engine Lubrication in Stationary Service

Mobil Technical Bulletins
Engine Oil Specifications and Tests — Significance and Limitations
Diesel Engine Operation
Rotary Engines — Wankel

8

Stationary Gas Turbines

There are four commercially important types of prime movers which provide mechanical power for industry: the steam turbine, the diesel engine, the gas or gasoline piston engine, and the gas turbine — a relative late-comer to the stationary power field. All four types achieve their result by converting heat into mechanical energy; any machine which does this (uses chemical reactions to supply thermal energy) is properly called a heat engine.

Of the four basic prime movers, the gas turbine is the most direct in converting heat into usable mechanical energy. The steam turbine introduces an intermediate fluid (steam), so the products of combustion (from the chemical reaction) do not act directly on the mechanism creating motion. Piston engines initially convert heat into linear motion which must be transmitted through a crankshaft to produce usable shaft power. But the gas turbine, in its simplest form, converts thermal energy into shaft power with no intermediate heat or mechanical redirection.

The theory of gas turbines was well understood by about 1900, and although development work started shortly after that, it was not until 1935 that engineers overcame low compressor efficiency and temperature limitations imposed by available materials and succeeded in building a practical gas turbine.

Just before World War II, the Swiss built and operated successful gas turbine plants for industrial use. In the United States, the first large-scale gas turbine application was in a process for cracking petroleum products.

Near the end of World War II, both Germany and England succeeded in developing turbojet-propelled aircraft. Turbojet and related designs for aircraft propulsion are an extension of the industrial gas turbine, and ironically enough, aircraft turbines have now been adapted for stationary power applications.

Although gas turbines cannot yet equal the fuel economy of the other prime movers mentioned, they have gained a place in stationary applications due to several advantages: light weight, small size, ease of installation, quick starting, multifuel capabilities, no cooling water requirements (of some designs), and availability of large amounts of exhaust heat. Use of this exhaust heat in a steam turbine is raising the efficiency of the combined cycle to over 40 percent.

PRINCIPLES OF GAS TURBINES

The Simple Cycle Open System

The simple cycle, open system gas turbine consists of a compressor, combustor, and turbine (Fig. 8-1). The compressor draws in atmospheric air, raises its pressure and temperature, and forces it into the combustor. In this chamber, fuel is added, which burns in contact with the compressed air, boosting its temperature and heat energy level. The hot, compressed mixture travels to the turbine where it expands and develops mechanical energy, i.e., torque applied to a shaft. A part of this energy is needed to drive the

Fig. 8-1   Gas Turbine of Simple Cycle Open System Single Shaft Design

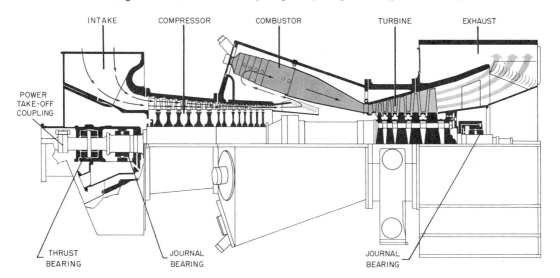

compressor; the rest is available to drive a useful load such as a generator, compressor, etc. Fig. 8-2 illustrates how temperature and pressure rise and fall within this type of gas turbine system. It is the gaseous working fluid that gives the gas turbine its name, not the fuel, which may be liquid or gaseous (or even solid).

The flow of energy through the cycle is revealing (Fig. 8-3). Entering air of itself contributes no energy. Actually, work applied to the compressor shaft is transferred to the compressed air to add some energy to it as it enters the combustor. Fuel induced into the combustor represents net energy input, balanced by work output (useful shaft work) plus exhaust energy vented to the atmosphere. The turbine work required to drive the compressor simply circulates within the cycle — from shaft to compressed air to combustor energy and back to turbine work. Surprisingly, of the total shaft work

Fig. 8-2 Graph Showing Approximately How Gas Temperatures and Pressures Change in Flowing through Gas Turbine of Type Shown in Fig. 8-1

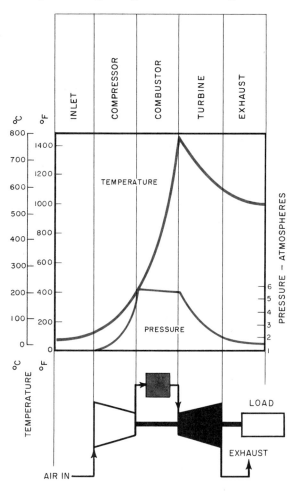

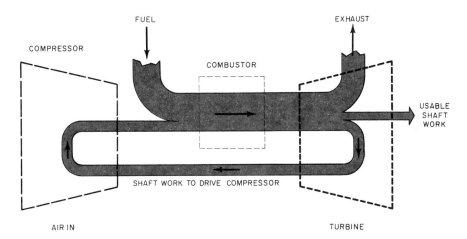

Fig. 8-3   Diagram Showing Flow of Energy in Simple Cycle, Open System Gas Turbine

developed by the turbine, roughly two thirds must go to drive the compressor, leaving only one third for useful shaft work.

A common variation of the simple cycle, open system design is the two shaft type (Fig. 8-4). The turbine is divided into two stages — a high pressure stage which drives the compressor and a low pressure stage which drives the load. The low pressure, free or power, turbine is not mechanically connected to the high pressure turbine. Its speed can be controlled independently to suit the speed of the driven unit, while the speed of the high pressure turbine can vary as needed to develop the required power.

The thermal efficiency of the simple, open cycle gas turbine is relatively low, reaching a maximum of about 25 percent. Efforts to improve thermal efficiency include designing for as high a turbine inlet temperature as practicable and as high a compressor pressure ratio as practicable up to an optimum value. Improvement can also be made through recovery of exhaust heat, which is discussed later.

Fig. 8-4   Simple Cycle, Open System Two Shaft Gas Turbine

Fig. 8-5   Regenerative Cycle Open System Showing Use of Exhaust to Heat Entering Air and Improve Thermal Efficiency

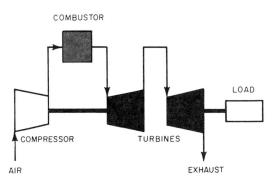

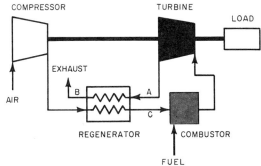

Regenerative Cycle, Open System

The regenerative cycle was designed to make use of exhaust heat (Fig. 8-5). The regenerator receives hot, turbine exhaust gas at A, rejects it at B. Compressed air passes through the unit in a counterflow direction and picks up heat from the turbine exhaust; then the heated air leaves the regenerator at C and passes into the combustor. Hotter air in the combustor needs less fuel to reach maximum temperature before flowing into the turbine; thus, thermal efficiency is improved. This conforms to a basic rule for heat engines, which states that the effect of raising the mean effective temperature at which heat is added while lowering the mean effective temperature at which heat is rejected increases a cycle's thermal efficiency.

Increased efficiency, generally exceeding 30 percent, is a factor favoring the regenerative cycle. Drawbacks to the regenerative cycle are increased costs of equipment and installation.

Intercooling, Reheating

The compressor tends to be an inefficient machine; the ratio of energy at its outlet to that at its inlet is low. A detailed discussion of compressors appears later. Elementary theory shows, however, that compression in which the temperature of the gas does not rise during compression requires less work than compression with increasing temperature. To partially achieve the former, the compression system is sometimes divided into two stages and the working fluid is withdrawn and cooled between stages, as shown in Fig. 8-6.

Intercooling gives more net work per pound of working fluid, allowing a smaller turbine plant for the same output or a greater output from the same size turbine. Also, where combustor temperature is maintained, the portion of usable shaft work developed increases because less shaft work is required at the compressor.

As the compression process may be divided, so may the expansion process, with reheating between stages instead of intercooling (Fig. 8-7). Reheating, ideally to the same top temperature in each stage, offers a straight forward way of boosting output with little increase in plant size, although a combustor is needed for each stage and the turbine is correspondingly more complex. With higher temperatures at the stack exhaust, the regenerative

Fig. 8-6 Intercooling between Compressor Stages to Reduce Work of Compression and Size of Machine

Fig. 8-7 Reheating between Turbine Stages to Increase Power Output

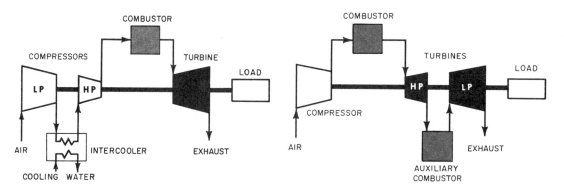

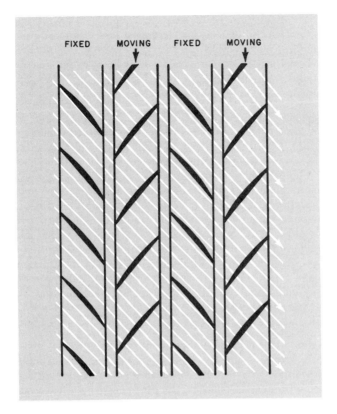

Fig. 8-8   Axial Flow Compressor Blading

cycle is almost always used. Without a regenerator, there is little gain in thermal efficiency.

Essential Gas Turbine Components

Compressor   The key to successful gas turbine operation is an efficient compressor. Early turbine plants were troubled with units that could not handle large-volume air flows efficiently. Today, almost all compressors in industrial gas turbine applications are axial flow or centrifugal, or a combination of the two.

In the axial flow compressor, blades on the rotor have air foil shapes to provide optimum air flow transmission. Moving blades draw in entering air, speed it up, and force it into following stationary vanes. These are shaped to form diffusers that convert the kinetic energy of moving air to static pressure (Fig. 8-8). A row of moving blades and the following row of fixed blades are considered a stage; air flows axially through these stages. The number of axial flow compressor stages in a modern industrial gas turbine varies from 12 to 16. Pressure ratios range from 5.5:1 to 7.5:1. With intake air at 14.7 psia (1.03 kP/cm^2) and 60°F (15.6°C), and pressure ratio of 6.45:1, air will be compressed, in an uncooled compressor, to about 94.7 psia (6.66 kP/cm^2) and 426°F (219°C). Blade heights become smaller as air travels through the axial flow compressor because the air's specific volume diminishes at the increasingly higher pressures.

Air displacement of axial flow compressors varies from 8 to 20 cfm (13.6 – 34 m³/h) of working fluid per horsepower. Large-volume, multistage machines may displace up to 300,000 cfm (510,000 m³/h), with temperature rises of 400°F (204°C) and more. Compressor efficiencies range from 85 to 92 percent, although performance hinges on minimum fouling of blades and diffusers.

Combustor For efficient combustion, the combustor, or combustion chamber, must assure low pressure losses, low heat losses, minimum carbon formation, positive ignition under all atmospheric conditions, flame stability with uniform outlet temperatures, and high combustion efficiencies.

The combustion section may comprise one or two large cylindrical combustors; several smaller tubular combustors; an annular combustor with several fuel nozzles compactly surrounding the turbine; or an annular combustor in which several tubular liners, or cans, are arranged in an annular space (Fig. 8-1). For all these, the design is such that less than a third of the total volume of air entering the combustor is permitted to mix with the fuel; the remainder is used for cooling. The ratio of total air to fuel varies among engine types from 40 to 80 parts of air, by weight, to one of fuel. For a 60:1 ratio, only about 15 parts of air are used for burning; the rest bypasses the fuel nozzles and is used downstream to cool combustor surfaces and to mix with and cool the hot gases before they enter the turbine.

Turbine This component is an energy converter, transforming high temperature, high pressure gases into shaft output. Gas turbines follow impulse and reaction designs—terms which describe the blading arrangement. The distinction depends upon how the pressure drop at each stage is divided. If the entire drop takes place across the fixed blades and none across the moving blades, it is an impulse stage (Fig. 8-9A). If the drop takes place in the moving blades, as well as the fixed blades, it is a reaction stage (Fig. 8-9B).

Fig. 8-9 Impulse (A) and Reaction (B) Turbine Blading

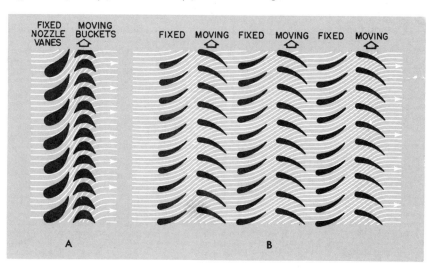

The combination of a row of fixed blades ahead of a row of moving blades constitutes a stage. Since the amount of power that can be developed by a single stage of reasonable size, shape, and speed is limited, many turbines are multistaged. For best performance, the gas turbine must work with high inlet temperatures. Manufacturers continually strive to build turbines that can accommodate higher temperatures, either through improved materials or through new designs. For the most part, design improvements focus on cooling the parts subjected to highest temperatures and stresses, particularly the first stage or two of the turbine. One cooling technique is to take air bled from the compressor outlet and channel it over parts holding the blades. Another method is to extend the shank linking blade bucket to the rotor, thus removing rotor parts from the hottest concentration of gases.

Twenty years ago, inlet temperatures of 1200°F (649°C) were considered extreme; today, as a result of continuing research, the ability of turbine materials to withstand high temperatures permits operation in the range of 1600 to 1700°F (871–927°C) (and even higher in recent aircraft engines). Most industrial turbines now operate at temperatures between 1450 and 1550°F (788 and 843°C) without blade cooling.

By the time these gases reach the turbine exhaust, their temperature has dropped at least 600°F (333°C), and possibly by as much as 900°F (500°C) depending upon inlet temperature and the number of stages.

Jet Engines for Industrial Use

In recent years, aircraft jet engine designs have been adapted for industrial service. Instead of providing propulsion power directly, the hot compressed gases from the engine are fed to a power or free turbine which converts the heat energy into rotative power (Fig. 8-10). Used in this manner, the jet engine acts as a gas generator, providing the working fluid to power a turbine. Actually, as many as ten jet engines have been used as gas generators for one large tubine.

At first glance, this might appear to be a more complicated way of providing shaft output, but advantages of the scheme are numerous. Compact design, light weight, high thermal efficiency, and rapid replacement are key benefits. A 10,500 hp jet gas turbine may weigh 80 tons less than a conventional industrial gas turbine of equal power and efficiency. Higher pressure ratios are possible, in the range of 10:1 to 12:1 or more, so a regenerator isn't required; yet thermal efficiency matches or exceeds comparable industrial (conventional) gas turbines. Higher turbine inlet temperatures (1500°F, 816°C, and above) are also possible, which in turn leads to higher thermal efficiencies. Compactness of the design makes possible plants that occupy only a third or less of the space of the conventional plant.

The compactness and light weight of the jet engine is partially explained by its relatively high speed—usually in the range 8000 to 18,000 rpm. Large industrial gas turbines, on the other hand, usually run at speeds of 3000 to 9000 rpm. Since jet engines are mass produced, they offer several advantages. Replacement parts are readily available, and even entire engines can easily be replaced. This makes possible interchangeability and factory overhaul instead of maintenance in the field. The light weight reduces foundation needs and makes installation easier. Another advantage is cold starts to full load operation in less than two minutes, in most applications, are possible.

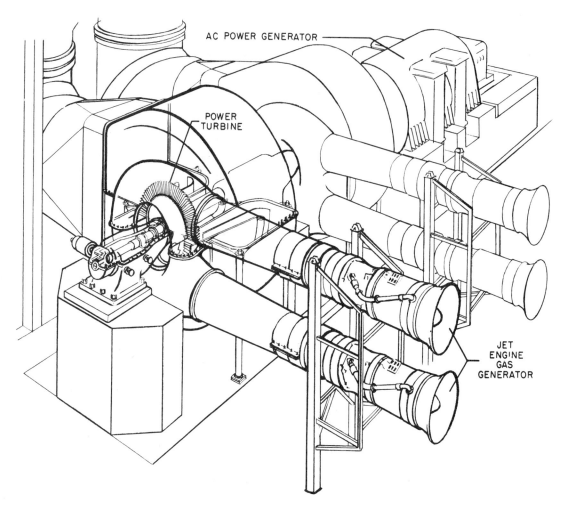

Fig. 8-10 Two Jet Engines Provide Hot Gas to Each of Two Power Turbines Coupled to a Single Generator

However, bearings of jet engines (practically always of the rolling element type) run hotter than those of typical industrial gas turbines. Most of the bearings are "buried" within the engines (as shown schematically in Fig. 8-11) and surrounded by hot gases. Bearing housings are sealed and pressurized with hot air bled from appropriate compressor stages. The resulting problem is twofold:

1. Removing the heat rejected at the bearing
2. Finding lubricants that can stand up to bearing temperatures of 400°F (204°C) and more

The answers are higher lubricant flow rates and the use of synthetic oils—high temperature lubricants which are discussed in Chapter 4.

Although jet gas turbines are more selective in the type of fuel they can burn and were originally built for continuous runs of not more than a few hours, they have performed well burning natural gas in runs of several thousand hours. They are competing favorably with conventional gas turbines in a variety of applications, including pipeline transmission and peaking generation.

Aircraft type gas turbines will never, however, completely replace the rugged, heavy duty, industrial gas turbines. The latter offer several advantages:

1. They can use plain bearings with long life capability not limited by eventual fatigue failure, as in the case of rolling element bearings.
2. They can operate on a wide range of lower cost fuel.
3. Their bearings are usually located in relatively cool areas (Fig. 8-1), so that they run at relatively low temperatures.

This last factor not only means that the lubricant requirement is less strict, it also means that special high temperature construction materials need not be used.

Small Gas Turbine Features

All industrial turbines are not large machines delivering thousands of horsepower. A great many are machines of less than 1000 hp rating, and a surprisingly large number are under 100 hp in size.

Many different makes of small gas turbines are available. Most follow design practices of larger industrial turbines, using plain (journal) bearings located located in relatively cool areas. Others, however, including those designed for frequent start-stops, especially at low temperatures, use rolling element bearings, which are also located in relatively cool areas. Some makes are available with either plain or rolling element bearings, depending on the type of service anticipated.

Some relatively small gas turbines use axial flow compressors and turbines. Others, in order to achieve compactness, use centrifugal compressors (usually one but sometimes two stages) or, as in Fig. 8-12, an axial stage followed by a centrifugal stage. Some use a radial in-flow turbine rather than an

Fig. 8-11   Aircraft Type Gas Generator of Dual-Axial, or Twin-Spool, Design

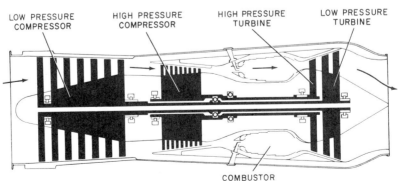

LOW PRESSURE COMPRESSOR   HIGH PRESSURE COMPRESSOR   HIGH PRESSURE TURBINE   LOW PRESSURE TURBINE

COMBUSTOR

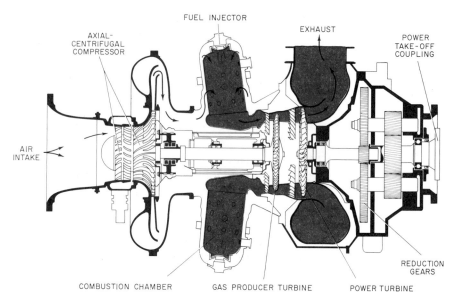

Fig. 8-12 Small (360-hp max.) Two Shaft Gas Turbine for Mechanical or Generator Drive Shaft bearings of the gas producer and the power section are of plain, or sleeve, type. Both plain and rolling element bearings are used in the reduction gear box.

axial one. Small turbines usually run at speeds ranging from about 18,500 to 60,000 rpm.

Several small gas turbines are being developed for vehicle use, using regenerators, high turbine inlet temperatures, and other means to increase thermal efficiency and reduce fuel consumption. Most of these machines are of two shaft type, with gas generator speeds ranging from 22,000 to 65,000 rpm and power turbine speeds ranging from 12,000 to 46,000 rpm. When mass produced versions of these turbines are available, it is likely that they will find many applications for stationary purposes as well as vehicle drives.

GAS TURBINE APPLICATIONS

Gas turbines, in sizes from 50 hp to nearly 50,000 hp, have wide application in industry today. Four broad areas of usage stand out—electric power generation, pipeline transmission, process operations, and total energy applications. These four areas indicate the gas turbine's versatility and suggest its wide ranging potential as an efficient power source. The gas turbine's success in these applications will doubtless make them more desirable in other areas as well.

Electric Power Generation Utilities use gas turbines for peaking generation and for emergency service. Most units exceed 10,000 hp, and both conventional turbines and jet engines adapted for industrial service have been used. System loads in a utility vary daily and seasonally. For efficient operation, base loads are usually supplied by steam turbines, with peak loads handled by gas turbines. In emergency shutdowns of base load equipment, gas turbines can take over when necessary. In peaking applications, a gas turbine may operate only a few hundred hours a year; this arrangement is the most efficient for year round generation. As gas turbine operation has become more economical and convenient, the units are being used for longer periods because they are capable of giving reliable service for several thousand hours a year.

Outside the utility, gas turbines are used to drive generators in industrial plants and buildings on an emergency standby basis. Sometimes they are used as peaking units to supplement on site, steam plant power. Along with diesel and gas engines, gas turbines are finding application in a new concept called continuous duty/standby, in which they are normally coupled to a nonessential load like the compressors in an air conditioning system. In emergency blackout situations, the nonessential load is dropped and a generator supplying essential electrical loads such as elevators, exit lighting, and critical equipment is picked up. Hospitals are finding this form of emergency power extremely reliable.

Pipeline Transmission

In the transmission of natural gas, centrifugal compressors are often used, with gas turbines, fueled by natural gas, driving them. This is an ideal arrangement because this preferred turbine fuel is readily available and the large loads are well suited for gas turbine drives. Units range in size from 1000 to 13,000 hp, with capacities in the 5000 to 10,000 hp range being the most prevalent.

Process Operations

In this area, more than in any other, the full potential of the gas turbine has yet to be realized. In fact, if properly designed, the gas turbine can become an integral part of the process, as essential as any other key element in the system. In process applications, the turbine is designed not so much to satisfy unit efficiency as to satisfy system efficiency. For example, in addition to generating shaft power, a turbine with an oversized compressor can produce pressurized air for use in a blast furnace. Instead of attempting to utilize exhaust gas to raise turbine efficiency, the gas can be channeled to a process requiring heat, for example, a dryer. Such uses may be accomplished with supplementary firing using the hot exhaust gas (which contains up to 17 percent of oxygen) as preheated combustion air for kilns or boilers. Hot, pressurized gases from certain processes can be ducted to the turbine's combustor, reducing the amount of fuel needed; the turbine itself may drive process equipment.

Combined Cycle Operation

Increasing use is being made of a recent concept of combining a gas turbine and a steam turbine as shown in Fig. 8-13. Combined cycles to generate power achieve thermal efficiency as much as 40 percent. The high temperature energy in the gas turbine's exhaust gas is extracted in a waste heat boiler. The steam is fed to a steam turbine and the exhaust steam is used either for process heating or returned to a condenser as boiler feed water.

Another method of combined cycle use passes the exhaust gases through two exhaust heat boilers in series. The first boiler produces superheated steam at 400 psia and the second, saturated steam at 104 psia. The high pressure steam is fed to the inlet of a mixed pressure steam turbine and the low pressure steam is fed at an intermediate stage of the same steam turbine. The combined steam flow exits the turbine to the main condenser. The gas turbine, steam turbine, and compressor are mounted on a single shaft.

Combined cycle plants are offered ranging from a net plant base load output of 12 to 594 MW and a heat rate of approximately 8000 Btu/kWh.

Total Energy In installations of this type, all building energy is provided at the site (Fig. 8-14). Typically, natural gas supplies a gas turbine which drives a generator to carry the electrical load. Waste heat from the turbine exhaust supplies a waste heat boiler, which generates steam for absorption air conditioning, space heating, and domestic hot water. Some installations use diesels or reciprocating gas engines instead of gas turbines. Plant capacity depends upon the size and type of building—shopping center, school, offices, apartments, etc. Power requirements today range from 300 hp to about 1500 hp. Compact turbines, either single shaft for constant speed applications or two shaft for variable speed applications, are selected. The units are small and light (frequently under 1500 lb) and can accept full load from start up in a minute or less. Well designed total energy systems are capable of thermal efficiencies of 65 to 85 percent.

LUBRICATION OF GAS TURBINES

The principal purposes of lubrication are to reduce wear, reduce friction, remove heat, and prevent rust. The elements of the gas turbines which need lubrication include plain and rolling element bearings, gears, and contact type seals.

As noted previously, aircraft type gas turbines are equipped with rolling element (ball or roller) bearings, whereas industrial gas turbines most often use plain (sliding) bearings. Plain bearings for radial loads may be:

Fig. 8-13 Gas Turbine, Steam Turbine Combined Cycle Plant

Fig. 8-14 Use of Exhaust Heat from Gas Turbine to Provide Steam for Absorption Refrigeration in "Total Energy" Plant

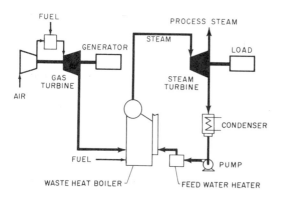

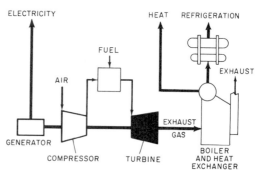

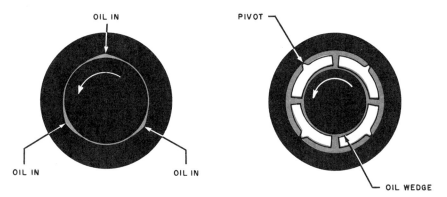

Fig. 8-15 Anti-Whip Bearings — Three-Lobe Type (left) and Tilting-Pad Type (right)
Multiplicity of fluid films suppress tendency for self-induced vibration of lightly loaded high-speed shafts.

1. Conventional heavy steel-backed babbitt-lined cylindrical type
2. Precision-insert type (as in reciprocating engine practice)
3. "Antiwhip" type in order to suppress selfinduced vibration, which is sometimes a problem in high speed machinery.

Two of the last type are shown schematically in Fig. 8-15.

Gears for speed reduction and for accessory drive are usually of spur, helical, herringbone, or bevel types. Some accessory drives are of worn type.

Large Industrial Gas Turbines

Lubrication systems for the large, heavy duty, industrial gas turbines are generally similar to those for steam turbines or other high speed, rotating machines.

A typical lubrication system may be described as a pressure circulation system, complete with reservoir, pumps, cooler, filters, protective devices, and connecting tubing, fittings, and valves. Its purpose is to provide an ample supply of cooled, cleaned lubricating oil to the gas turbine, accessory and reduction gearing, driven equipment, and hydraulic control system.

An example of the arrangement of these parts is shown in Fig. 8-16. A gear type, positive displacement, main pump, driven by the power turbine, supplies pressurized oil to each of the main compressor and turbine bearings and to accessory gearboxes. Return oil from the bearings and gearboxes drains by means of gravity to a sump tank. Foot valves in the tank prevent oil in the supply system from draining back to the sump during shutdown periods.

The pressurized oil passes through an oil cooler and filter (specified 5 μm by some manufacturers) before reaching the bearings. A pressure relief valve installed in the supply line maintains pressure at a constant level.

A motor driven standby pump automatically supplies pressurized oil to the bearings before starting and for a few minutes after shutdown to prevent "heat soak" damage to the bearings. The standby pump also takes over au-

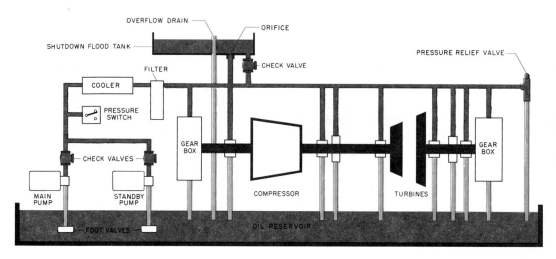

Fig. 8-16   Lubrication System for Large Industrial Gas Turbine

tomatically should the main pump fail during operation. Some lubrication systems, as shown in Fig. 8-16, also include a flood tank to build up a reserve supply of oil during operation. This supply can flood the bearings by means of gravity for 15 to 20 minutes after system failure due to emergency power loss.

Protective devices are incorporated into the lubrication system to guard equipment against low oil supply, low oil pressure, and high oil temperature. The devices sound a warning or shut down the unit if any of these conditions occurs. In some installations, in the event of emergency power loss a pump driven by a battery powered motor is available for supplying pressurized oil on a short term basis. Also, in some installations, an additional cooler and filter may be included. Oil flow can be directed to one filter and cooler while the other filter and cooler are cleaned during operation of the gas turbine, avoiding shutdown.

Bearing oil header pressures are usually controlled at 15 to 25 psi (1.1–1.8 kP/cm²). Hydraulic control pressures are often in the 50 to 100 psi (3.5–7 kP/cm²) range, but much higher pressures are used in some systems.

Oil Operating Conditions   Bearing and gear loads are generally moderate and shaft speeds are high, permitting use of circulating type oils of "light" or "medium" viscosity.* Oil temperatures entering bearings and gears are usually in the 130 to 160°F (54–71°C) range, and temperature rise in these elements is 25 to 50 Fahrenheit degrees (14–28 Celsius degrees). Bulk oil temperatures in the reservoir will usually be 155 to 200°F (68–93°C). These temperatures are well within the thermal limitations of high quality petroleum lubricating oil of suitable type. Starting temperatures may be low, so that oils of high viscosity index (VI), which do not thicken excessively at low temperatures, are recommended. One manufacturer specifies a maximum viscosity of 800 SUS (172.3 cSt) for reliable circulation of oil at starting.

*Approximately 150, 225, and 300 SUS @ 100°F (32, 48.5, and 64.7 cSt @ 38°C), respectively.

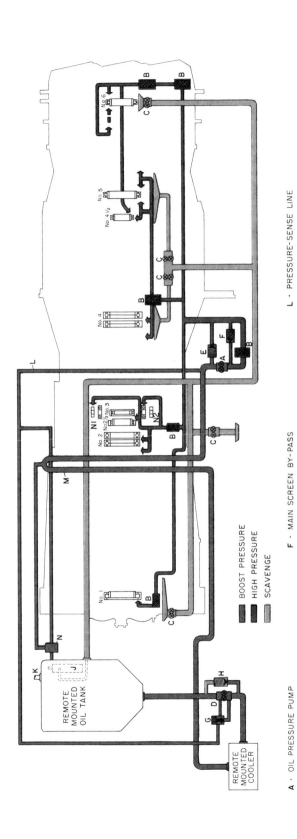

A · OIL PRESSURE PUMP
B · OIL STRAINER
C · SCAVENGE PUMP
D · OIL BOOST PUMP
E · MAIN PRESSURE REGULATING VALVE

F · MAIN SCREEN BY-PASS
G · BOOST PUMP REGULATING VALVE
H · BOOST PUMP RELIEF VALVE
J · OIL DEAERATOR
K · BREATHER CONNECTION

L · PRESSURE-SENSE LINE
M · ANTI-SIPHON LOOP
N · PRESSURE-OPERATED VALVE
N1 · ACCESSORY-DRIVE BEARINGS
N2 · ACCESSORY-DRIVE BEARINGS

Fig. 8-17 Lubrication System for Aircraft Gas Turbine in Stationary Service

The lubricating oil is recirculated in contact with air. Portions of it are broken into small drops or mist so that intimate mixing with air occurs. Under these conditions, oil tends to oxidize, and the tendency is increased by high operating temperatures, by excessive agitation or splashing of oil, and by the presence of certain contaminants that act as catalysts. The rate of oil oxidation also depends on the ability of the oil to resist this chemical change.

Oxidation is accompanied by the formation of both soluble and insoluble products and by a gradual thickening of the oil. The insoluble products may be deposited as gum or sludge in oil passages, and their accumulation will restrict the supply of oil to bearings, gears, or hydraulic control devices. For long time performance in reuse systems, the lubricating oil must have high chemical stability in order to resist oxidation and the formation of sludge.

Water may find its way into bearing housings as a result of condensation of moisture from the atmosphere during idle periods or from other sources. The churning of oil with moisture in circulation systems tends to create emulsions. To resist emulsification, the oil should have the ability to separate quickly from water. With such an oil, any water that enters the bearing housings or circulation systems will collect at low points from which it may be drained.

Adapted Aircraft Gas Turbines

The lubrication of aircraft gas turbines adapted for stationary service differs substantially from that of large, heavy duty, industrial gas turbines. The lubrication systems are quite different, being much more compactly designed of necessity. The bearings, which are rolling element types, and gears, in some instances, are "buried" within the turbines and require more intricate passageways and other arrangements for supplying and removing oil. The bearings and contact seals run very much hotter than the plain bearings and clearance seals of the industrial turbines, usually requiring the use of synthetic lubricating oil and the placing of greater emphasis on the cooling function of the oil. Because of the need for compactness, bearing and gear loading is much higher in aircraft than in stationary practice, requiring extra load carrying ability in the lubricant.

The lubrication system for a dual, axial compressor, gas generator is shown schematically in Fig. 8-17. Oil from the tank is gravity fed to boost pump D, which supplies oil at constant pressure to oil pressure pump A regardless of changes in pressure drop through the oil cooler and piping. During operation, pressure at the suction of A acts through pressure-sense line L (1) on boost pump regulating valve G, regulating the pressure rise of oil boost pump D, and (2) on pressure operated valve N, keeping it closed. During shutdown, pressure in line L falls, permitting valve N to open and vent antisiphon loop M. Since loop M extends above oil level, this prevents leakage of oil from the tank through pumps D and A into the auxiliary gear case. Oil pressure pump A forces oil—at about 45 psi (3.2 kP/cm^2)—through strainers B to main bearing and accessory drive locations. Here the oil is jetted through calibrated orifices to parts requiring lubrication and cooling, including bearings, contact seals, and gears. The oil flows by gravity from these parts to sumps from which it is removed by individual scavenge pumps C and returned to the tank through a deaerator J. All bearing and accessory drive housings are vented by means of a breather system, not shown, which includes a rotary breather pump.

Power turbines used with aircraft type gas generators have their own separate lubrication systems.

Oil Operating Conditions Thermal conditions in land based, aircraft gas turbines are not only more severe than in heavy duty, industrial turbines, as noted previously, but also are more severe than in their airborne counterparts. In aircraft use, the turbines operate at full power only during takeoff and climb—a small fraction of the time—while in stationary service they may operate for long periods at full load. The lubricating oil comes in contact with hot surfaces in the 400 to 600°F (204−316°C) range and mixes with hot gases which leak into bearing housings. Bulk oil temperatures in the tank may be in the 160 to 250°F (71−121°C) range. This is a very severe condition and requires a lubricant having excellent thermal and oxidative stability to resist degradation and deposit formation. Measures which may be taken to moderate this severe thermal condition include derating engines, increasing oil reservoir sizes, and improving oil cooling facilities. In any case, it is common to idle gas turbines for several minutes before shutdown to cool them and prevent exposing the oil to high "soak back" temperatures.

Small Gas Turbines Small gas turbines, as indicated in a previous section, vary considerably in design. Lubrication systems, however, are usually relatively simple versions of the type of circulation system described for large industrial turbines. Essential elements include a sump, strainer, shaft driven main pump, pressure-regulating valve, cooler, full-flow filter, manifold to bearings, gears, etc., and gravity return to sump. Some systems employ scavenge pumps to return oil to the main sump or tank.

Most small turbines are designed to use petroleum lubricating oils. Turbine type oils similar to those described for large industrial gas turbines may be used, but oils of lower viscosity may be specified for some very high speed machines. Some are designed to use automotive engine oils or automatic transmission fluids. In all cases, the gas turbine manufacturer specifies a suitable lubricant, and his recommendation should be followed unless specific approval is obtained for the use of some other product.

9
Steam Turbines

In a steam turbine, hot steam at a pressure above atmospheric is expanded in the nozzles where part of its heat energy is converted into kinetic energy. This kinetic energy is then converted into mechanical energy in the turbine runner either by the impulse principle or the reaction principle. If the nozzles are fixed and the jets directed toward movable blades, the jets' *impulse* force pushes the blades forward. If the nozzles are free to move, the *reaction* of the jets pushes against the nozzles, causing them to move in the opposite direction.

In small, purely impulse turbines, steam is expanded to exhaust pressure in a single set of stationary nozzles. As a result of this single expansion, the steam issues from the nozzles in jets of extremely high velocity. In order to obtain maximum power from the force of the jets' impact on a single row of moving blades, the blades must move at about half the velocity of the jets. Thus, single stage impulse turbines operate at very high rotative speeds. To reduce rotative speed while maintaining efficiency, the high velocity can be absorbed in more than one step, which is called *velocity compounding*.

Such a turbine (Fig. 9-1) has one velocity compounded stage followed by four pressure compounded stages. In the velocity compounded stage, steam

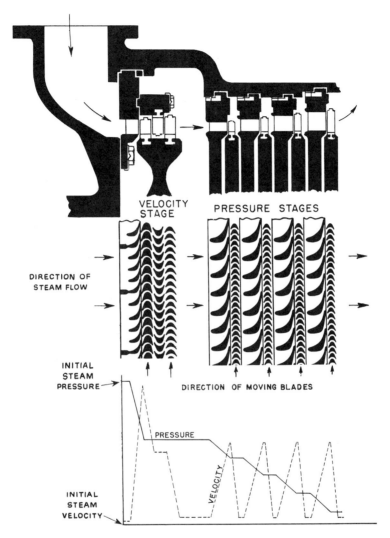

Fig. 9-1 Turbine with Impulse Blading Velocity compounding is accomplished in the first two stages by two rows of moving blades between which is placed a row of stationary blades that reverses the direction of steam flow as it passes from the first to the second row of moving blades. Other ways of accomplishing velocity compounding involve redirecting the steam jets so that they strike the same row of blades several times with progressively decreasing velocity.

is first expanded in the stationary nozzles to high velocity. This velocity is reduced in two steps through the first two rows of moving blades. The steam is then expanded again in a set of stationary nozzles and delivered to the first pressure stage, then through additional sets of nozzles to the succeeding pressure stages. In each case, velocity is increased and pressure is decreased in the stationary nozzles. In the moving blades, velocity decreases but the pressure remains constant. The graph at the bottom (Fig. 9-1) shows the changes in pressure and velocity as the steam flows through the various rows of nozzles and moving blades.

Another method of reducing rotor speed while maintaining efficiency is to decrease the velocity of the jets by dividing the drop in steam pressure into a number of stages. This is called *pressure compounding*. Refer again to Fig. 9-1, the steam pressure is reduced by somewhat less than one-half in the velocity compounded stage. The steam then passes to the pressure compounded stages where it is reduced in four steps to the final exhaust pressure. Each pressure stage consists of a row of stationary nozzles and a row of movable blades, and the whole assembly is equivalent to mounting four single stage impulse turbines on a common shaft. The arrangement of Fig. 9-1 using velocity compounding in the first stage followed by pressure compounding in the remaining stages is an approach used in many large turbines.

In reaction turbines, steam is directed into the blades on the rotor by stationary nozzles formed by blades that are designed to expand the steam enough to give it a velocity somewhat greater than that of the moving blades. (As a result of this expansion to provide steam velocity, this arrangement may be referred to as a 50 percent reaction turbine.) The moving blades form the walls of moving nozzles that are designed to permit further expansion of the steam and to partially reverse the direction of steam flow, which produces the reaction on the blades. The distinguishing characteristic of the reaction turbine is that a pressure drop occurs across both the moving and stationary nozzles or blades, see Fig. 9-2. Normally reaction turbines employ a considerable number of rows of moving and stationary nozzles through which the steam flows as its initial pressure is reduced to exhaust pressure. The pressure drop across each row of nozzles is, therefore, relatively small and steam velocities are correspondingly moderate, permitting medium rotative speeds.

Reaction stages are usually preceded by an initial velocity compounded impulse stage, as in Fig. 9-2, in which a relatively large pressure drop takes place. This results in a shorter, less costly turbine.

In the radial flow reaction, or Ljungstrom, turbine, instead of the steam flowing axially through alternating rows of fixed and moving blades, it flows radially through several rows of reaction blades. Alternate rows of blades move in opposite directions. They are fastened to two independent shafts which operate in opposite directions, each shaft driving a load.

After expansion in the turbine, the steam usually exhausts to a condenser, where it is condensed to provide a source of clean water for boiler feed. This simple cycle, see Fig. 9-3—water to steam to power generation, and steam to water—forms the basis on which most steam power plants operate.

TURBINE TYPES

Steam turbines are made in a number of different arrangements to suit the needs of various power plant or industrial installations. Turbines up to 40 to 60 MW capacity are generally single cylinder machines. Larger units are usually of compound type, that is, the steam is partially expanded in one cyl-

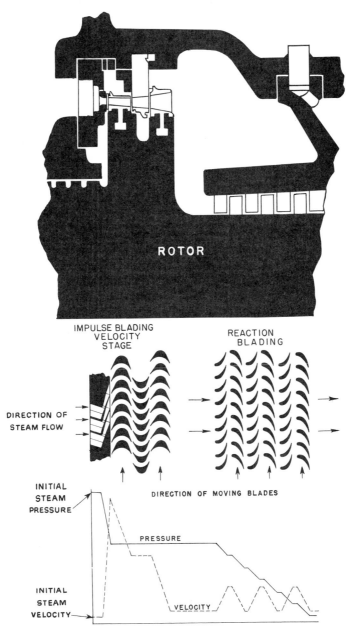

Fig. 9-2 Reaction Turbine with One Velocity Compounded Impulse Stage   The first stage of this turbine is similar to the first, velocity compounded, stage of Fig. 9-1. However, in the reaction blading of this turbine, both pressure and velocity decrease as the steam flows through the blades. The graph at the bottom shows the changes in pressure and velocity through the various stages.

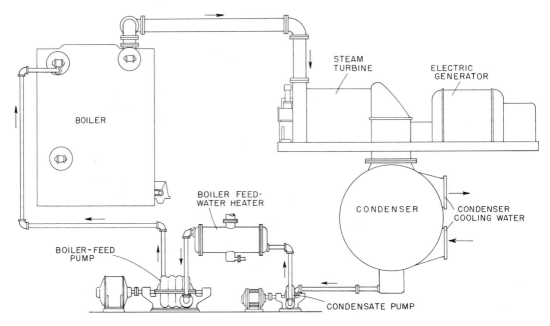

Fig. 9-3 Simple Power Plant Cycle This diagram shows that the working fluid, steam and water, travels a closed loop in the typical power plant cycle.

inder then passed to one or more additional cylinders where expansion is completed.

Single Cylinder Turbines Single cylinder turbines are of either the *condensing* or *back-pressure* (or noncondensing) type. These basic types and some of their subclassifications are shown in Fig. 9-4.

When the steam from a turbine exhausts to a condenser, the condenser serves two purposes:

1. By maintaining a vacuum at the turbine exhaust, it increases the pressure range through which the steam expands. In this way, it materially increases the efficiency of power generation.
2. It causes the steam to condense, thus providing clean water for the boilers to reconvert into steam.

Industrial plants frequently require steam at low to moderate pressures for process use. One of the more economical ways of generating this steam is with a *combined cycle* plant, or as it is also called, cogeneration. The steam is generated at high pressure and, after expansion through a turbine to the pressure desired for process use, delivered to the process applications. This permits power to be generated by the turbine without appreciably affecting the value of the steam for process use. It may be done with a back pressure turbine designed to exhaust all the steam against the pressure required for process use, or it may be done with an *automatic extraction* turbine in which part of the steam is withdrawn for process use at an intermediate stage (or stages) of the turbine and the remainder of the steam exhausted to a condenser. Such a turbine requires special governors and valves to maintain constant pressure of the extracted steam and constant turbine speed under varying turbine load and extraction demand.

Steam can also be extracted without control from various stages of a turbine to heat boiler feed water (which is called *regenerative* heating). Such units are called *uncontrolled extraction* turbines, since the pressure at the extraction points varies with the load on the turbine.

To obtain higher efficiency, some turbines (called *reheat* turbines) are arranged so that the steam after expanding part way is withdrawn, returned to the boiler, and reheated to approximately its initial temperature. It is then returned to the turbine for expansion through the final turbine stages to exhaust pressure.

High pressure noncondensing turbines have been added to many moderate pressure installations to increase capacity and improve efficiency. In such installations, high pressure boilers are installed to supply steam to the noncondensing turbines which are designed to exhaust at the pressure of the original boilers and supply steam to the original turbines. The high pressure turbines are called *superposed* or *topping* units.

Where low pressure steam is available from process work, it can be used to generate power by admitting it to an intermediate stage of a turbine de-

Fig. 9-4 Single Cylinder Turbine Types Typical types of single cylinder turbines are illustrated. As shown, condensing turbines, as compared to back-pressure turbines, must increase more in size toward the exhaust end to handle the larger volume of low pressure steam.

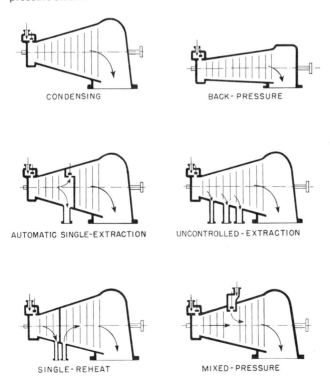

CONDENSING

BACK-PRESSURE

AUTOMATIC SINGLE-EXTRACTION

UNCONTROLLED-EXTRACTION

SINGLE-REHEAT

MIXED-PRESSURE

signed for the purpose and expanding it to condenser pressure. Such a machine is a *mixed-pressure* turbine, and is another form of cogeneration.

Compound Turbines

Compound turbines have at least two cylinders or casings — a high pressure one and a low pressure one. In order to handle large volumes of low pressure steam, the low pressure cylinder is frequently of the double flow type. Very large turbines may have an intermediate pressure cylinder, and two, three, or even four double flow, low pressure cylinders. The cylinders may be in line using a single shaft, which is called *tandem compound,* or in parallel groups with two or more shafts, which is called *cross compound.* Reheat between the high and intermediate pressure stages may be employed in large turbines. Steam may be returned to the boiler twice for reheating. Some of these arrangements are shown diagrammatically in Fig. 9-5.

TURBINE CONTROL SYSTEMS

All steam turbines are provided with at least two independent governors that operate to control the flow of steam. One of these operates to shut off the steam supply if the turbine speed should exceed a predetermined maximum. It is often referred to as an emergency trip. On overspeed, it closes the main steam valve, cutting off steam from the boiler to the turbine. The other, or main, governor may operate to maintain practically constant speed, or it may be designed for variable speed operation under the control of some outside influence. Extraction, mixed pressure, and back-pressure turbines are provided with governors that control steam flow in response to a combination of speed and one or more pressures. The governors of such units are extremely complex, whereas the direct acting speed governors on small, mechanical drive turbines are relatively simple.

Speed Governors

Speed governor systems may be divided into three principal parts:

1. Governor or *speed sensitive element*
2. *Linkage* or force amplifying mechanism that transmits motion from the governor to the steam control valves
3. *Steam control valves*

The most widely used speed sensitive element is the centrifugal, or flyball, governor, see Fig. 9-6. Weights which are pivoted on opposite sides of a spindle and are revolving with it, are moved outward by centrifugal force against a spring when the turbine speed increases, and inward by spring action as the turbine slows down. This action may operate the steam admission valve directly through a mechanical linkage, as shown, or it may operate the pilot valve of a hydraulic system, which admits and releases oil to opposite sides of a power piston, or on one side of a springloaded piston. Movement of the power piston opens or closes steam valves in order to control turbine speed.

Moderate and large sized, high speed turbines are provided with a double relay hydraulic system to further boost the force of the centrifugal governor and to increase the speed with which the system responds to speed changes.

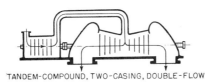

TANDEM-COMPOUND, TWO-CASING, DOUBLE-FLOW

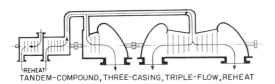

TANDEM-COMPOUND, THREE-CASING, TRIPLE-FLOW, REHEAT

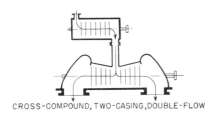

CROSS-COMPOUND, TWO-CASING, DOUBLE-FLOW

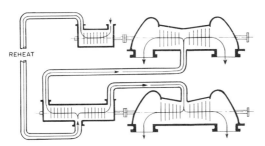

CROSS-COMPOUND, FOUR-CASING, QUADRUPLE-FLOW, REHEAT

Fig. 9-5 Some Arrangements of Compound Turbines While many arrangements are used, these diagrams illustrate some of the more common ones.

A second type of speed sensitive element is the oil impeller. Oil from a shaft driven pump flows through a control valve to the space surrounding the governor impeller. The impeller, mounted on the turbine shaft, consists of a hollow cylindrical body with a series of tubes extended radially inward. As the oil flows inward through the tubes it is opposed by centrifugal force and a pressure is built up that varies with the square of the turbine speed. This pressure is applied to springloaded bellows which positions a pilot valve. The pilot valve, in turn, controls the flow to a hydraulic circuit which operates the steam control valves.

Many newer turbines are equipped with electrical or electronic speed sensitive devices. Signals from these devices, along with signals derived

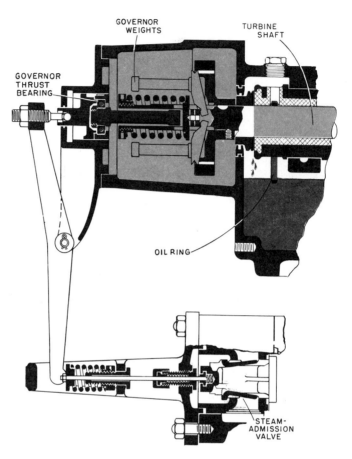

Fig. 9-6 Mechanical Speed Governor A simple arrangement such as this using a flyball governor is suitable for many small turbines.

from load, initial steam pressure, and other variables, are fed to an electronic computer where they are compared and the appropriate signals sent to hydraulic servo valves to adjust the steam control valves.

As indicated, the linkage between the speed sensitive element and the steam control valves may be anything from a simple lever to an extensive hydraulic system controlled by an electronic computer.

In small turbines, the flow of steam is controlled by a single valve, usually of the balanced type to reduce the operating force required. In large units, a valve for each of several groups of nozzles controls steam flow. The opening or closing of a valve cuts in or out a group of nozzles. The number of open valves, and thus the number of nozzle groups in use, is varied according to the load. The valves may be operated by a barlift arrangement, by cams, or by individual hydraulic cylinders.

Additional control valves, called *intercept* valves, are required on reheat turbines. These are placed close to the intermediate, or reheat, cylinder and are closed by a governor system if the turbine starts to speed up as a result of a sudden large load reduction. This is designed to prevent the large volume of high energy steam in the piping from the high pressure turbine exhaust, the reheat section of the boiler, and the piping to the intermediate pressure turbine from continuing to flow and possibly causing overspeeding and emergency tripping of the turbine. In older turbines, the intercept valves were controlled by a separate governor system, but some newer machines have the intercept valves operated from the main hydraulic control system. As an additional safety measure, intercept valves are preceded by stop valves which are actuated by the main emergency overspeed governor.

Pressure Governors Automatic extraction and back-pressure turbines are provided with governors arranged to maintain constant extraction or exhaust pressure irrespective of load (within the capacity of the turbine). The pressure sensitive element consists of a pressure transducer, and its response to pressure changes is communicated through the control system to the valves that control steam extraction, and to the speed governor which controls admission of steam to the turbine. On automatic extraction turbines, the action of the pressure and speed responsive elements is coordinated so that turbine speed is maintained. This may not be the case for back-pressure turbines.

LUBRICATION REQUIREMENTS

The lubrication requirements of steam turbines can be considered in terms of the parts that must be lubricated, the type of application system, the factors affecting lubrication, and the lubricant characteristics required to satisfy these requirements.

Lubricated Parts The main lubricated parts of steam turbines are the bearings, both journal and thrust. Depending on the type of installation, a hydraulic control system, oil shaft seals, gears, flexible couplings, and turning gear may also require lubrication.

Journal Bearings The rotor of a single cylinder steam turbine, or of each casing of a compound turbine, is supported by two hydrodynamic journal bearings. These journal bearings are located at the ends of the rotor, outside of the cylinder. Because of the extremely small clearances between the shaft and shaft seals and between the blading and the cylinder, the bearings must be accurately aligned and must run without any appreciable wear in order to maintain the shaft in its original position and avoid damage to shaft seals or blading.

Primarily, the loads imposed on the bearings are due to the weight of the rotor assembly. The bearings are conservatively proportioned so that pressures on them are moderate. Horizontally split shells lined with tin base babbitt are usually used. The bearings are enclosed in housings and supported on spherical seats or flexible plates to reduce any angular misalignment.

The passages and grooves in turbine bearings are sized to permit the flow of considerably more oil than is required for lubrication alone. The additional oil flow is required to remove frictional heat and the heat conducted to the bearing along the shaft from the hot parts of the turbine. The flow of oil must be sufficient to cool the bearing enough to maintain it at a proper operating temperature. In most turbines, the temperature of the oil leaving the bearings is on the order of 160°F (71°C), but in special cases it may be as high as 180°F (82°C).

Where a turbine is used to drive a generator, the generator bearings are similar in design to the turbine bearings, and are normally supplied from the same system.

Large turbines are now frequently provided with "oil lifts" in the journal bearings to reduce the possibility of damage to the bearings during starting and stopping, and to reduce the starting load on the turning gear. Oil under high pressure from a positive displacement pump is delivered to recesses in the bottoms of the bearings. The high pressure oil lifts the shaft and floats it on a film of oil until the shaft speed is high enough to create a normal hydrodynamic film.

A phenomenon that occurs in relatively lightly loaded, high speed journal bearings, such as turbine bearings, is what is known as "oil whip" or "oil film whirl." The center of the journal of a hydrodynamic bearing ordinarily assumes a stable, eccentric position in the bearing that is determined by load, speed and oil viscosity. Under light load and high speed, the stable position closely approaches the center of the bearing. There is a tendency, however, for the journal center to move in a more or less circular path about the stable position in a self-excited vibratory motion having a frequency of something less than one-half shaft speed. In certain cases, such as some of the relatively light weight, high pressure rotors of compound turbines that require large diameter journals to transmit the torque, this whirling has been troublesome and has required the use of bearings designed especially to suppress oil whip.

Several types of bearings designed to suppress oil whip are available. Among the common types are the pressure or pressure pad bearing (Fig. 9-7), the three lobe bearing (Fig. 9-8), and the tilting pad antiwhip bearing (Fig. 9-9). The pressure pad bearing suppresses oil whip because oil carried into the wide groove increases in pressure when it reaches the dam at the end. This increase in pressure forces the journal downward into a more eccentric position that is more resistant to oil whip. The other types illustrated depend on the multiple oil films formed to preload the journal and minimize the tendency to whip.

Thrust Bearings Theoretically, in impulse turbines, the drop in steam pressure occurs almost entirely in the stationary nozzles. The steam pressures on opposite sides of the moving blades are, therefore, approximately equal and there is little tendency for the steam to exert a thrust in the axial direction. In actual turbines, this ideal is not fully realized and there is always a thrust tending to displace the rotor.

In reaction turbines, a considerable drop in steam pressure occurs across each row of moving blades. Since the pressure at the entering side of each of the many rows of moving blades is higher than the pressure at the leaving side, the steam exerts a considerable axial thrust toward the exhaust end. Also, where rotors are stepped up in diameter, the unbalanced steam pressure acting on annular areas thus created adds to the thrust. Usually the total thrust is balanced by means of dummy, or balancing, pistons on which

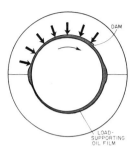

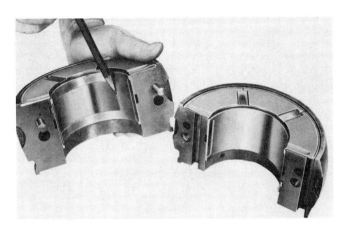

Fig. 9-7 Pressure Bearing The wide groove in the upper half ends in a sharp dam at the point indicated. As shown in the insert this causes a downward pressure which forces the journal into a more eccentric position that is more resistant to oil whirl.

the steam exerts a pressure in the opposite direction to the thrust. In double flow elements of compound turbines, steam flows from the center to both ends so that thrust is well balanced.

Regardless of the type of turbine, thrust bearings are always provided on each shaft in order to take axial thrust and, thus, hold the rotor in correct

Fig. 9-8 Three Lobe Bearing The shape of the bearing is formed by three arcs of radius somewhat greater than the radius of the journal. This has the effect of creating a separate hydrodynamic film in each lobe, and the pressures in these films tend to keep the journal in a stable position.

Fig. 9-9 Tilting Pad Antiwhip Bearing As in the three lobe bearing, the multiple oil films formed tend to keep the journal in a stable position.

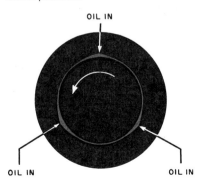

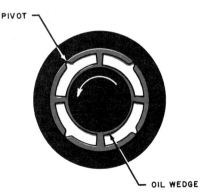

axial position with respect to the stationary parts. Although thrust caused by the flow of steam is usually toward the low pressure end, means are always provided to prevent axial movement of the rotor in either direction.

The thrust bearings of small turbines may be babbitt faced ends on the journal bearings, or rolling element bearings of a type designed to carry thrust loads. Medium and large turbines are always equipped with tilting pad (Fig. 9-10), or tapered land (Fig. 9-11) type thrust bearings.

Hydraulic Control Systems   As discussed earlier, medium and large turbines have hydraulic control systems to transmit the motion of the speed or pressure sensitive elements to the steam control valves. Two general approaches are used for these systems.

In mechanical hydraulic control systems the operating pressure is comparatively low and oil from the bearing lubrication system may be used safely as the hydraulic fluid. Separate pumps are provided to supply the hydraulic requirement. An emergency tripping device is provided to shut down the turbine if there is any failure in the hydraulic system.

Larger turbines now being installed are equipped with electrohydraulic control systems. In order to provide the rapid response needed for control of these units, the hydraulic systems operate at relatively high pressures, typically in the range of 1500 to 2000 psi (10.3 to 13.8 MPa). The systems consist of an independent reservoir and two separate and independent pumping systems. The large fluid flow required for rapid response to sudden changes in load is usually provided by gas charged accumulators.

The critical nature of the servo valves used in these systems requires that careful attention be paid to the filtration of the fluid, and strict limits on particulate contamination usually are observed. The need for precise control also necessitates that both heaters and coolers be used to maintain the temperature of the fluid and, thus, its viscosity in a narrow range. Since

Fig. 9-10 Combined Journal and Tilting Pad Thrust Bearing   A rigid collar on the shaft is held centered between the stationary thrust ring and a second stationary thrust ring (not shown) by two rows of tilting pads.

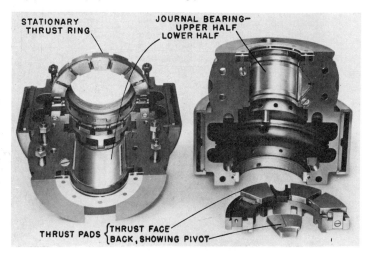

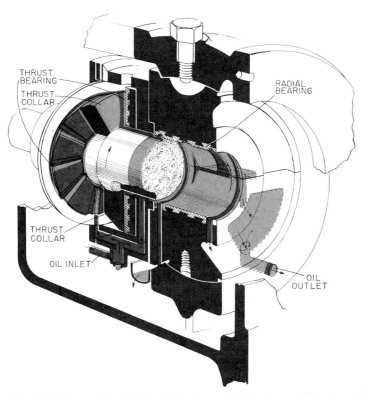

THRUST
BEARING

THRUST
COLLAR

RADIAL
BEARING

THRUST
COLLAR

OIL INLET

OIL
OUTLET

Fig. 9-11 Tapered Land Thrust Bearing and Plain Journal Bearing   The thrust bearing consists of a collar on the shaft, two stationary bearing rings, one on each side of the collar. The babbitted thrust faces of the bearing rings are cut into sectors by radial grooves. About 80 percent of each sector is beveled to the leading radial groove, to permit the formation of wedge oil films. The unbeveled portions of the sectors absorb the thrust load when speed is too low to form hydrodynamic films.

a leak or a break in a hydraulic line could result in a fire if the high presssure fluid sprayed onto hot steam piping or valves, fire-resistant hydraulic fluids are widely used in these systems.

Oil Shaft Seals for Hydrogen Cooled Generators   Because it is a more effective coolant than air, hydrogen is commonly used to cool medium and large generators. Shaft mounted blowers circulate the gas through rotor and stator passages, then through liquid cooled hydrogen coolers. Gas pressures up to 60 psi (413 kPa) are used. A further development, which has permitted increases of generator ratings over hydrogen cooling alone, is the direct liquid cooling of stator windings. Some liquid systems use transformer oil while others use water. Even with water cooled stators, the interior of the generator is still filled with hydrogen.

The main connection between the type of cooling and turbine lubrication is that when hydrogen is used for cooling, some of the oil is exposed to the hydrogen. Oil shaft seals; see Fig. 9-12, are used to prevent the escape of the hydrogen. Turbine oil for these seals may be supplied from an essentially

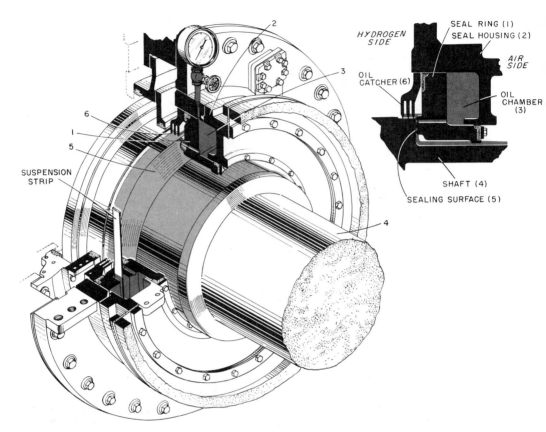

Fig. 9-12 Shaft Seal for Hydrogen Cooled Generator Oil under pressure in the annular chamber formed between the seal ring and housing (see inset) forces the seal ring against the shaft shoulder. From the annular chamber, oil flows through the passage shown to the sealing face formed by a shoulder on the generator shaft.

separate system having its own reservoir, pumps, and so forth, or may be supplied directly from the main turbine lubricating system. In either case, before entering the reservoir, oil returning from the seals must be passed through a detraining tank to remove any traces of hydrogen. Otherwise hydrogen could accumulate in the reservoir and form an explosive mixture with air. In addition, the main turbine oil reservoir of all units driving hydrogen cooled generators must be equipped with a vapor extractor to remove any traces of hydrogen that may be carried back by the sealing oil or the oil from the generator bearings.

Gear Drives Efficient turbine speed is often higher than the operating speed of the machine being driven. This may be the case, for example, where a turbine drives a direct current generator, paper machine, centrifugal pump, or many other industrial machines. It is also the case where a turbine is used for ship propulsion. In these applications, reduction gears are used to connect the turbine to the driven unit.

Reduction gears used with moderate and large sized turbines are usually precision cut, double helical type. Double reduction gear sets are required with marine propulsion turbines, and epicyclic reduction gears are sometimes used instead of conventional gear sets. Usually, the gears are enclosed in a separate oiltight casing and are connected to the turbine and driven machine through flexible couplings. Small machines may have the gear housing integral with the turbine housing and the pinion on the turbine shaft.

Reduction gears may have a circulation system that is entirely separate from the turbine system, or may be supplied from the turbine system. In the latter case, a separate pump (or pumps), is provided for the gears. Some small geared turbines have the turbine bearings ring oiled, and the gears splash oiled.

In marine propulsion applications, the bearing through which the propeller thrust is transmitted to the ship's hull is frequently an integral part of the reduction gear set. It may be placed just behind the low speed bull gear, or at the forward end of the bull gear shaft. It is usually of the tilting pad type. With a fixed pitch propeller, the thrust bearing must be designed to operate with either direction of rotation, since turbine rotation (and propeller rotation) is reversed for astern movement of the ship.

Flexible Couplings   Most modern turbogenerator sets are rigidly coupled. Many other turbines, especially geared units, are connected to their loads by flexible couplings. Gear type couplings are used most commonly.

Where gear type couplings are enclosed in a suitable housing, they may be lubricated from the turbine circulation system. A coupling of this type acts as a centrifugal separator, and means are provided for a flow through of oil so that any separated solids will be flushed out.

Smaller couplings may be lubricated by a bath of oil carried inside the case, or with grease.

Turning Gear   When starting and stopping large turbines, it is necessary to turn the rotor slowly to avoid uneven heating or cooling which could cause distortion or bowing of the shaft. This is done with a barring mechanism or turning gear. The turning gear usually consists of either an electric or hydraulic motor that is temporarily coupled to the turbine shaft through reduction gears. Rotor speed, while the turning gear is operating, is usually below 100 rpm. In order to provide adequate oil flow to the bearings during this low speed operation, a separate auxiliary oil pump usually is provided. The oil coolers are used at maximum capacity to increase oil viscosity and to help maintain oil films in the bearings. If oil lifts are provided in the turbine bearings, they are also operated while the turning gear is operating.

Lubricant Application   One of the essential factors in reliable steam turbine operation is the provision of a lubricating system that will assure an ample supply of lubricant to all of the parts requiring it. The size of the turbine generally determines whether the system is simple or extensive and complex. Small turbines, such as those used to drive auxiliary equipment, are usually equipped with ring oiled bearings; other moving parts being lubricated by hand. Moderate sized units, driving through reduction gears may have ring oiled bearings for the turbine, and a circulation system to supply oil to the gears and bearings in the reduction gear set. Most moderate sized units and all large turbines are equipped with circulation systems (see Chap. 6) that supply oil to all parts of the unit requiring lubrication. As pointed out herein, separate circulation systems may be provided for the seal oil for hydrogen cooled turbogenerators, and for the hydraulic control systems.

Factors Affecting Lubrication Steam turbines in themselves do not represent particularly severe service for petroleum base lubricating oils. However, due to the costs involved in shutting down most turbines to change the oil, and the relatively large volume of oil contained in large turbine systems, turbine operators expect extremely long service life from the turbine oil. In order to achieve long service life, oils must be carefully formulated for the specific conditions encountered in steam turbine lubrication systems.

Circulation and Heating in the Presence of Air The temperature of the oil in steam turbine systems is raised both by the frictional heat generated in the lubricated parts, and by heat conducted along the shaft from the rotor. As the oil flows through the system, it is broken into droplets or mist which permits greater exposure to air. System designs are usually conservative so that maximum oil temperatures are not excessive, but in long term service some oxidation of the oil does occur. This oxidation may be further catalyzed by finely divided metal particles resulting from wear or contamination and by water.

Slight oxidation of the oil is harmless. The small amounts of oxidation products formed initially can be carried in solution in the oil without noticeable effect. As oxidation continues, some of the soluble products may become insoluble, or insoluble materials may be formed. The oil also gradually increases in viscosity.

Insoluble oxidation products may be carried with the oil as it circulates. Some may then settle out as gum, varnish, or sludge on governor parts, in bearing passages, on coolers, on strainers, and in oil reservoirs. Their accumulation may interfere with the supply of oil to bearings and with governor operation.

Some oxidation products that are soluble in warm oil become insoluble when the oil is cooled. This can result in insulating deposits forming on cooling coils or other cool surfaces. The resultant reduction in effectiveness of cooling may cause higher oil temperatures and contribute to more rapid oxidation.

Some increase in viscosity within a range that has proved satisfactory in service is not necessarily harmful. However, excessive viscosity increase may reduce oil flow to bearings, increase pumping losses, and increase fluid friction and heating in bearings. All of these effects tend to increase the operating temperature of the oil and contribute to more rapid oxidation of it.

Contamination Water is the contaminant that is most prevalent in turbine lubrication systems. Common sources of water contamination are:

1. Steam from leaking shaft seals or the shaft seals of turbine driven pumps
2. Condensation from humid air in the oil reservoir and bearing pedestals
3. Water leaks in oil coolers

The lubrication systems of turbines that operate intermittently are more likely to become contaminated with water than are the systems of turbines which operate continuously.

When a turbine oil is agitated with water some emulsion will form. If the oil is new and clean, the emulsion will separate readily. The water will then settle in the reservoir where it can be drawn off or removed by the purification equipment. Oxidation of the oil, or contamination of it with certain types of solid materials such as rust and fly ash, may increase the tendency of the oil to emulsify, and also to stabilize emulsions after they are formed. Persistent emulsions can join with insoluble oxidation products, dirt, and so forth to form sludges. The character of these sludges may vary; but, if they accumulate in oil pipes, passages, and oil coolers, they can interfere with oil circulation and cause high oil and bearing temperatures.

Water in lubricating oil, combined with air that is always present, can cause the formation of common red rust and also black rust that is similar in appearance to pipe scale. Rusting may occur both on parts covered by oil and on parts above oil level. In either case, in addition to damage to the metal surfaces, rusting is harmful for a number of reasons. Particles of rust in the oil tend to stabilize emulsions and foam and act as catalysts that increase the rate of oil oxidation. Rust is abrasive and when carried by the oil to the bearings may scratch the journals and cause excessive wear. Carried into the small clearances of governor mechanisms, it can cause sluggish operation or, in extreme cases, sticking.

Many types of solid materials can contaminate turbine oil systems. Pulverized coal, fly ash, dirt, rust, pipe scale, and metal particles are typical examples. These solid materials may enter the system in the following ways:

1. During erection and may remain even after the initial cleaning and flushing
2. Through openings in the bearing pedestals or reservoir covers
3. Carried in with water
4. Generated within the system

Solid contaminants contribute to deposit or sludge formation, and some tend to reduce the ability of the oil to separate from water. Some are abrasive and may be the direct cause of scoring or excessive wear. Fine metal particles may act as catalysts to promote oil oxidation. Purification of the circulating oil is discussed in Chap. 16.

In a sense, air is also a contaminant. Air entrained in the oil can produce what is called "bubbly" oil. Bubbly oil which may have a higher viscosity than clear oil is compressible. The latter effect can cause sponginess in hydraulic controls, and may reduce the load carrying ability of oil films. Entrained air also increases the exposure of the oil to air, therefore increases the rate of oxidation. Excessive amounts of air in the system may lead to foaming in the reservoir or bearing housings. This could result in overflow with oil loss and unsightly and hazardous spills. In some cases, foam escaping from bearings adjacent to a generator may be drawn into the windings or onto the collector rings where it may cause breakdown of the insulation, short circuits, or arcing.

Lubricating Oil Characteristics   The first requirement of a steam turbine oil is that it have the proper viscosity at operating temperature to provide effective lubricating films, and adequate load carrying ability to protect against wear in heavily loaded mechanisms, and under boundary lubrication conditions. Other characteristics are concerned with providing long service life, protecting system metals, and maintaining the oil in a condition to perform its lubricating function.

Viscosity   In direct connected steam turbines, the main lubrication requirement is for the journal and thrust bearings. Higher viscosity oils provide a greater margin of safety in these bearings, but at the same time, increase pumping losses and friction losses due to shearing of the lubricant films. In high speed machines, particularly, the latter can become an important cause of power loss and heating of the lubricant. General practice is to use oils as low in viscosity as possible within a range that has proved suitable service. Most larger units are designed to operate on oils of ISO Viscosity Grade 32 (28.8 – 35.2 cSt @ 40°C). Some applications require somewhat higher viscosity oils, ISO Viscosity Grade 46 (41.4 – 50.6 cSt @ 40°C). Both of these grades of oil also provide excellent performance in hydraulic control systems.

For geared turbines with a common circulation system supplying both the bearings and gears, higher viscosity oils are required to provide satisfactory lubrication of the gears. Oils of ISO Viscosity Grade 68 (61.2 – 74.8 cSt @ 40°C) are used in many of these systems. Oils of ISO Viscosity Grade 100 (90 – 110 cSt @ 40°C) are also used in some machines, particularly in marine propulsion turbines. With some geared turbines, the oil is passed through a cooler immediately before being delivered to the gears. The resultant cooler oil has a higher viscosity, thus it provides better protection for the gears.

Reservoir temperatures of small, ring oiled turbine bearings vary widely from one design to another depending principally on size and whether water cooling is used. The oils used for these range in viscosity from ISO Viscosity Grade 32 for the cooler units, up to as high as ISO Viscosity Grade 320 (288 – 352 cSt @ 40°C) for the latter units.

Load Carrying Ability   As noted earlier, steam turbine lubrication systems are conservatively designed. Bearing loads are moderate. Under these conditions, mineral oil lubricants of the correct viscosity normally provide adequate load carrying ability. However, in turbines not equipped with oil lifts, boundary lubrication conditions occur in the bearings during starting and stopping. Under boundary lubrication conditions, some wear will occur unless lubricants with enhanced film strength are used.

To some extent, the increased load carrying ability of the oil films needed during start up is provided by the higher viscosity of the cool oil. However, the additive systems in turbine oils are frequently selected to provide some improvement in film strength to ensure an additional margin of safety.

In some cases, particularly in marine propulsion installations, the need for size and weight reduction has led to the use of heavily loaded gear reducers. Performance of these gears on conventional turbine lubricants is not satisfactory, and lubricants with extreme pressure properties are used. These lubricants are formulated to have the properties of good turbine oils with the addition of the extreme pressure properties, thus, they may be used throughout the turbine lubrication system.

Oxidation Stability   The most important characteristic of turbine oils from the standpoint of long service life is their ability to resist oxidation under the conditions encountered in the turbine lubrication system. Resistance to oxidation is important from the standpoint of retention of viscosity; resistance to the formation of sludges, deposits, and corrosive oil oxyacids; and retention of water separating ability, foam resistance, and ability to release entrained air.

Turbine oils usually are manufactured from base oils refined from selected crudes. These oils are selected both for their natural oxidation stability, and their response to oxidation inhibitors. Processing is carefully controlled and is often the most extensive applied to any base oils. Inhibitors are selected to be effective under the temperature conditions encountered in steam turbines and to provide the optimum improvement in the particular base oils used.

Oxidation life of steam turbine oils is frequently specified in terms of ASTM D943-76 Standard Test Method for Oxidation Characteristics of Inhibited Steam-Turbine Oils; see Chap. 2. Manufacturer specifications typically call for a test life to a Neutralization Number of 2.0 of 1000 h minimum. Commercial products of ISO VG 32 typically run from 2000 h to 5000 h or more. Higher viscosity products may run somewhat less. In the IP 280 (CIGRE) test, limits on Total Oxidation Products of 1.0 mass percent and sludge of 40 percent of the T.O.P. are typical. Regardless, commercial inhibited oils available today will generally give many years of service in well-maintained systems.

Protection against Rusting   Straight mineral oils have some ability to protect against rusting of ferrous surfaces, but this ability is inadequate for the conditions encountered in steam turbine lubrication systems. Effective rust inhibitors are, therefore, required in steam turbine oils.

In action, rust inhibitors "plate out" on metal surfaces, forming a film that resists displacement and penetration of water. This results in gradual depletion of the rust inhibitor. Normal makeup of new oil usually maintains an adequate level of rust inhibitor in the system. In certain cases, where excessive water contamination has been experienced, special treatment to replenish the rust inhibitor may be necessary.

Rust inhibitors must be carefully selected to provide adequate protection without affecting other properties of the oil, especially water separating ability.

Water Separating Ability   New, clean, highly refined mineral oils will generally resist emulsification when water is mixed with them and any emulsion that is formed will break quickly. Certain additives, such as some rust inhibitors, some contaminants, and oxidation products can both increase the tendency of an oil to emulsify and make any emulsion that is formed more stable. As a result, careful selection of additives is necessary if a turbine oil is to have good initial water separating ability, and excellent oxidation stability is necessary if this water separating ability is to be maintained in service.

Foam Resistance   Turbine circulation systems are, in general, designed and constructed to minimize or eliminate conditions that have been found to cause foaming. However, turbine oils may contain defoamants to reduce the tendency to foam, and the stability of any foam that does form. Since oxidation can increase the foaming tendency and also the stability of the foam, good oxidation stability is an important factor in maintaining foam resistance in service.

Entrained Air Release Rapid release of entrained air is particularly important in systems supplying turbines with hydraulic governors. Excessive amounts of entrained air in these systems can cause sponginess, producing delayed or erratic response.

The rate at which entrained air is released from a mineral oil is an intrinsic characteristic of the oil itself. It is dependent on factors such as the source of crude, and type and degree of refining. Currently there are no known additives that will improve the ability of an oil to release entrained air, but there are many additives that will degrade this ability. Thus, formulating a steam turbine oil with good air release properties is a process of selecting base oils with good air release properties, and then selecting additives that will perform the functions desired of them without seriously degrading the air release properties of the base oil.

Fire Resistance As discussed under "Hydraulic Control Systems", fire resistant fluids are usually used in electro-hydraulic governor control systems. The fluids used most commonly are based on either phosphate esters or blends of phosphate esters and chlorinated hydrocarbons. As noted also, these systems are extremely critical of solid contaminants in the fluid so considerable attention must be paid to filtration of the fluid.

The wide use of steam at high temperatures, up to as high as 1200°F (649°C), has increased the possibility of fires from oil leaks in turbine lubrication systems. This has led to some attempts to use fire resistant fluids in turbine lubricating systems. Fluids used are the same as, or similar to, those used in hydraulic governor systems. However, these fluids have a deteriorating effect on conventional generator insulating materials so cannot be used in older machines.

Bibliography

Mobil Technical Book
Steam Turbines and Their Lubrication

10

Hydraulic Turbines

The primary application of hydraulic turbines is to drive electric generators in central power stations. Since the turbine shaft is usually rigidly coupled to the generator shaft with one set of bearings supporting both, for lubrication purposes, the two units can be considered as one machine.

Many of the locations suitable for large hydroelectric installations are in remote areas, often mountainous and with difficult access. Under these conditions, capital costs for plant construction and transmission lines are high. Therefore, in order for the power generated to be competitive in cost, plants must be designed for minimum maintenance and long service life. Generally, this has been accomplished, and the large majority of hydroelectric units installed during the last 40 to 50 years are still in service.

Although many of the large hydroelectric installations are in remote locations, there are many installations in comparatively accessible locations. In recent years, the development of some of these locations has been aided by the introduction of the pumped storage concept and the use of bulb turbines.

Thermal and nuclear power plants, as well as open flume hydroelectric turbines, operate most efficiently at relatively constant loads. Thus, during off-peak periods, if the plant can be operated at or near full load, the power in excess of the load requirement is comparatively low in cost. Where a suitable location is available, this power can be used to pump water up to a storage reservoir. During periods of peak demand, the water is then available to drive hydroelectric generators to supplement the power available from the base load stations. In effect, pumped storage is a method of storing low cost, off-peak power for use during peak demand periods.

Some pumped storage plants are equipped with both turbines and pumps. The generator is built as a combination motor/generator. During pumping operations, the turbine may be disconnected from the main shaft, or the turbine casing may be blown dry with compressed air so that the turbine operates without load. Similar arrangements are used with the pump. Most new installations are equipped with reversible pump/turbines, and reversible motor/generators. There may be some sacrifice in efficiency with the combination pump/turbine, for example:

1. As a pump it may be less efficient than a unit designed as a pump alone.
2. As a turbine it may be less efficient than a unit designed as a turbine alone.

However, the lower cost of the combination machine generally offsets this loss of efficiency.

Bulb turbines (also called tubular or Tube®* turbines) are low head machines for what are referred to as "flow-of-stream" river applications. They can be used for relatively small, low cost installations that can be readily blended into the surrounding country side. These machines have permitted the development of hydroelectric power in locations where installation of the older types of turbines would not be practical or economic.

TURBINE TYPES

Several types of hydraulic turbines are used. They can be considered as impulse or reaction (pressure) types. Reaction turbines include the inward flow (Francis), diagonal flow (Deriaz) and propeller types. The choice of which type of unit to use in a particular application is a function of the pressure head and the quantity of flow available.

Impulse Turbines

In an impulse turbine, usually called a Pelton turbine, jets of water are directed by nozzles against shaped buckets on the rim of a wheel. The impulse force of the jets pushes the buckets on the rim of a wheel. The impulse force of the jets pushes the buckets and causes the wheel to revolve. The buckets move in the same direction as the jets of water. In order for a turbine of this type to operate efficiently, the velocity of the jets must be high, thus, a high pressure head of water is required. Pelton turbines are usually designed for pressure heads in the range of about 500 to 3900 ft. (150–1200 m). Single units with outputs up to 200 MW have been built.

Pelton turbines are built with either a horizontal or vertical shaft. Horizontal shaft machines are built with either one or two nozzles per runner.

*A registered trademark of Allis-Chalmers Manufacturing Company.

Usually, they are used for small to moderate sized installations. A single runner may be connected directly to a generator, or two runners may be used, both on the same side of the generator or one on each side.

Vertical shaft Pelton turbines with four to six nozzles are now being used for larger installations. A cutaway of an installation with four nozzles is shown in Fig. 10-1. In this machine, the nozzle tips are actuated by hydraulic pistons located inside the nozzle bodies. The deflectors are actuated by a

Fig. 10-1 Pelton Turbine Installation

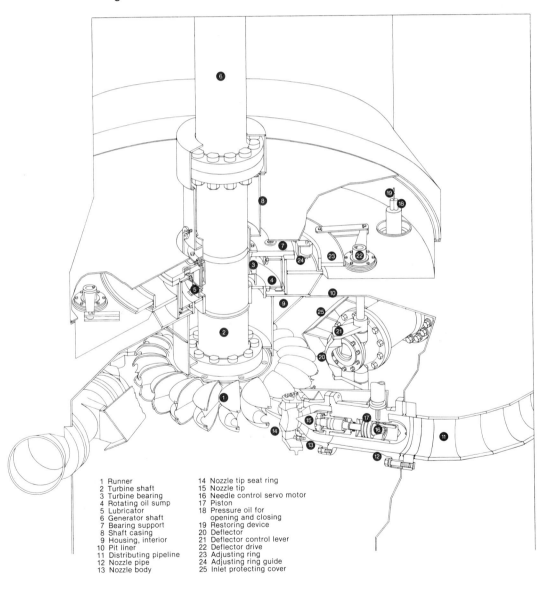

1 Runner	14 Nozzle tip seat ring
2 Turbine shaft	15 Nozzle tip
3 Turbine bearing	16 Needle control servo motor
4 Rotating oil sump	17 Piston
5 Lubricator	18 Pressure oil for
6 Generator shaft	opening and closing
7 Bearing support	19 Restoring device
8 Shaft casing	20 Deflector
9 Housing, interior	21 Deflector control lever
10 Pit liner	22 Deflector drive
11 Distributing pipeline	23 Adjusting ring
12 Nozzle pipe	24 Adjusting ring guide
13 Nozzle body	25 Inlet protecting cover

ring and lever arrangement, which in turn is actuated by hydraulic servo pistons. In many older machines, servo pistons that operate the nozzle tips are located outside the nozzles, and a mechanical linkage is used to operate the nozzle tips.

Pelton turbines are used in a number of pumped storage applications. Both horizontal and vertical shaft machines are used for this purpose. Where a horizontal shaft machine is used, the turbine is often mounted at one side of the generator/motor with the pump at the other side. Other arrangements can be used. In vertical shaft machines, the turbine normally is mounted above the pump so that the pump will operate with a positive pressure at the suction. The pump usually is coupled to the shaft through a clutch so it can be disconnected and stopped during turbine operation. In some installations, the water is blown out of the pump and it is left coupled to the turbine during turbine operation. The water is blown out of the turbine for pump operation. In some cases, the turbine is used to start the pump and motor and bring them up to speed. The water flow to the turbine is then shut off and the water is blown out of the casing.

Reaction Turbines   In reaction turbines, the flow of water impinges on a set of curved blades which, in effect, are the nozzles. The reaction of the water on the nozzles (blades) causes them to move in the opposite direction.

In a Francis turbine (Fig. 10-2), the water flows radially inward from a volute casing and is turned through 90 degrees in the blades before flowing to the tail race outlet. In a Deriaz turbine (Fig. 10-3, 10-7) the direction of

Fig. 10-2   Francis Turbine

Fig. 10-3 Deriaz Turbine

water flow is partially turned in the volute casing so that the flow is diagonally through the blades. The shaped boss then turns the water through the remaining angle to direct it into the tail race outlet. In propeller turbines, the direction of flow is controlled by a volute casing or a flume so that the water flows axially through the turbine.

Francis and Deriaz turbines are intermediate to high head machines. The various types of propeller turbines (fixed blade, Kaplan, and bulb) are low head machines.

Francis Turbines Francis turbines are probably the most widely applied hydraulic turbine. They are now also widely used as reversible pump turbines. Originally used for intermediate head installations, the range of use of the Francis turbine has been extended well up into the head range that was formerly the exclusive province of impulse turbines. Francis turbines are being used at heads ranging from about 65 to 1650 ft. (20 to 500 m), and even higher head units are under development. Outputs in the 200 to 400 MW per unit range are common, and units rated at up to 750 MW are under construction.

Small Francis turbines are sometimes built with horizontal shafts, but larger Francis turbines are nearly always vertical shaft machines. Water is brought in from the penstock through a spiral (volute) casing (Fig. 10-4) from which it is directed into the turbine guide vanes by fixed vanes in the stay ring or speed ring (Fig. 10-5). The movable guide vanes control the flow of water into the turbine to maintain the turbine speed constant as the load var-

Fig. 10-4 Spiral Casing for Francis Turbine

ies. They can be operated by a single regulating ring actuated by one or two hydraulic servo motors (Fig. 10-5), or by individual Servo Motors (Fig. 10-6). With a regulating ring, the individual vanes are connected to the ring through shear pins so that if one vane is jammed by a foreign object during closing, the remainder of the vanes can still be closed to shut off the turbine.

Diagonal Flow Turbines The efficiency of Francis turbines drops off quite rapidly if the flow is less than the design value, or if the pressure head varies significantly. Where either of these conditions exists, a diagonal flow or Deriaz turbine can sometimes be used.

In the Deriaz turbine, both the guide vanes and the runner blades are adjustable (Fig. 10-7). The runner blades are adjusted by a hydraulic Servo Motor located either in the turbine shaft, or in the runner boss. The guide vanes are usually controlled by a regulating ring, similar to those used with Francis turbines. Movement of the guide vanes and runner blades is synchronized by the control system to maintain runner speed with changes in load, and keep the efficiency high as changes in the pressure head occur.

Deriaz turbines are presently built for pressure heads in the range of about 60 to 425 ft. (18 to 130 m). Designs suitable for heads up to about 650 ft. (200 m) are available. The usual construction is with a vertical shaft. Single units with ratings up to about 150 MW are in service. The larger units are reversible pump/turbines.

Fixed Blade Propeller Turbines Because of the low heads at which propeller turbines operate, comparatively small changes in the head water or tail water level can make significant changes in the total head acting on the turbine.

With a fixed blade turbine this may cause marked changes in the efficiency of the turbine. For that reason, fixed blade propeller turbines (usually referred to simply as propeller turbines) are used only in locations where the head is fairly constant. Relatively few installations of this type of turbine exist. Outputs range up to about 40 MW per unit.

Propeller turbines are vertical shaft machines. The blades of the runner (Fig. 10-8) are cast integrally with the hub or welded to it.

Kaplan Turbines The Kaplan turbine (Fig. 10-9) is the largest class of low head hydraulic turbine. The runner blades are adjustable, operated by either a hydraulic Servo Motor inside the runner hub, or by a Servo Motor in the shaft with a mechanical linkage to the blade adjustment mechanism in the hub. The former arrangement is used for most new machines. As with Deriaz turbines, the controls for the guide vane adjustment and runner blade adjustment are synchronized to provide an optimum setting for each load and head condition. A typical Kaplan turbine installation is shown in cross section in Fig. 10-10.

Kaplan turbines are built for heads from about 16 to 250 ft. (5 to 75 m). Single units with power outputs on the order of 100 MW are in service, and units with outputs in the 175 MW range are under construction.

Fig. 10-5 **Assembly of Francis Turbine**

Fig. 10-6   Guide Vane Adjustment with Individual Servo Motors   This shows operating mechanism for a Kaplan turbine, but is equally applicable to Francis.

Fig. 10-7   Cross Section of Deriaz Turbine

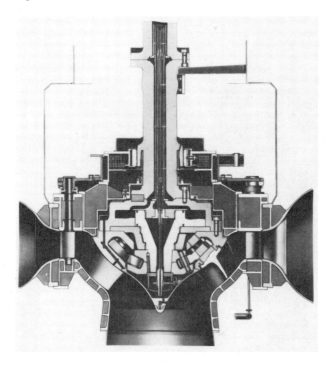

Fig. 10-8 Fixed Blade Propeller Turbine

Fig. 10-9 Kaplan Turbine

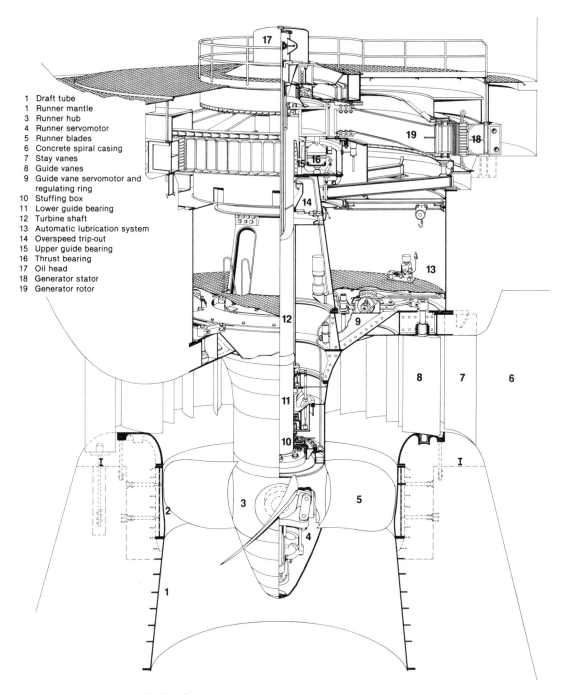

1   Draft tube
1   Runner mantle
3   Runner hub
4   Runner servomotor
5   Runner blades
6   Concrete spiral casing
7   Stay vanes
8   Guide vanes
9   Guide vane servomotor and
    regulating ring
10 Stuffing box
11 Lower guide bearing
12 Turbine shaft
13 Automatic lubrication system
14 Overspeed trip-out
15 Upper guide bearing
16 Thrust bearing
17 Oil head
18 Generator stator
19 Generator rotor

Fig. 10-10   Cross Section of Kaplan Turbine

Bulb Turbines   The bulb turbine is actually a special application of the Kaplan turbine. As shown in Fig. 10-11, the conventional Kaplan turbine is mounted vertically with a spiral casing carrying the water into the stay ring. From the turbine, a draft tube carries the water out to the tail race, creating a suction head on the turbine. In contrast, the bulb turbine has the turbine mounted in a bulb shaped section of the flume with the water flowing essentially straight through the turbine. The arrangement is much more compact, and is less costly to construct.

The shaft of bulb turbines is either horizontal or angled slightly downward toward the turbine. With a horizontal shaft, the generator is usually mounted in a bulb-shaped casing inside the flume, and is driven directly from the turbine (Fig. 10-12). A mechanical drive to a generator mounted on the surface may also be used. With an angled shaft, the shaft may be extended out of the flume to drive a surface mounted generator directly, or the generator may be mounted inside the flume.

Bulb turbines for heads up to 70 ft. (21 m) are in service. Unit outputs are usually less than about 10 MW, but units up to 50 MW are in service.

LUBRICATED PARTS

The main parts of hydraulic turbines requiring lubrication are the turbine and generator bearings, the guide vane bearings, the contol valve, governor and control system, and compressors.

Turbine and Generator Bearings   Horizontal shaft machines require journal bearings to support the rotating parts, including the generator armature. With the exception of Pelton turbines, thrust bearings are also required to absorb the thrust of the water acting on the runner. Vertical shaft machines require guide bearings to keep the shaft centered and aligned, and thrust bearings to carry the weight of the rotating parts and absorb the thrust of the water acting on the runner.

Vertical shaft machines have a guide bearing above the turbine, and one or two guide bearings at the upper, or generator, end of the shaft. In some cases with long shafts, an additional guide bearing may be installed about midway between the turbine guide bearing and the generator guide bearing. For reasons of accessibility, normally the thrust bearing is installed at the upper end of the shaft, either above the armature or just below it. Bearing arrangements are sometimes referred to as *two bearing* where a combination guide and thrust bearing is located above the armature and a guide bearing below. They are referred to as *umbrella* type where a combination thrust and guide bearing (Fig. 10-13) is located below the armature, or *semiumbrella* type where separate guide and thrust bearings are located below the armature. Most current medium and large machines are of either the umbrella or semiumbrella type.

Journal and Guide Bearings   The bearings of horizontal shaft machines are of the fluid film type, with babbitt lined, split shells. Oil lifts to assist starting

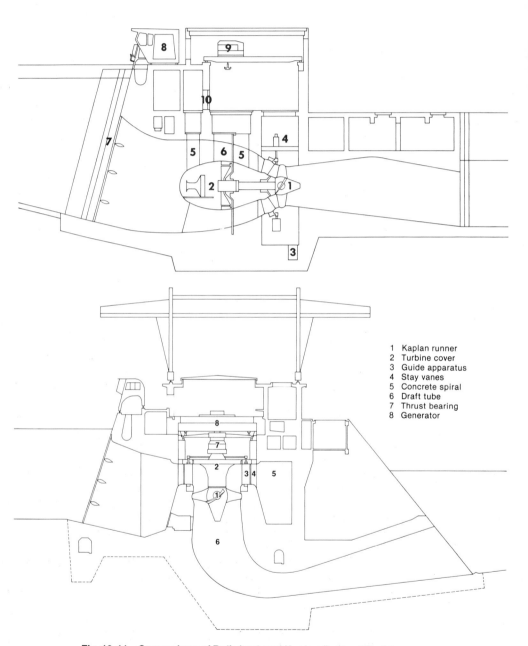

1 Kaplan runner
2 Generator casing
3 Guide wheel closing weight
4 Servomotor
5 Entry shafts

6 Generator dismantling shaft
7 Inlet rack
8 Rack cleaning machine
9 Power house crane
10 Switchboards

1 Kaplan runner
2 Turbine cover
3 Guide apparatus
4 Stay vanes
5 Concrete spiral
6 Draft tube
7 Thrust bearing
8 Generator

Fig. 10-11   Comparison of Bulb (top) and Kaplan (bottom) Turbines

Fig. 10-12 Bulb Turbine Installation

may be used in the journal bearings when the rotating parts are extremely heavy.

Two general types of turbine guide bearings are used in vertical shaft machines. Many older turbines are equipped with water lubricated rubber or composition bearings. This construction minimizes sealing requirements at the top of the turbine casing, but is not too satisfactory where the water

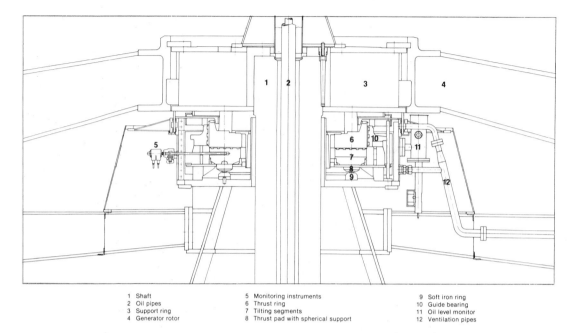

1	Shaft	5	Monitoring instruments	9	Soft iron ring
2	Oil pipes	6	Thrust ring	10	Guide bearing
3	Support ring	7	Tilting segments	11	Oil level monitor
4	Generator rotor	8	Thrust pad with spherical support	12	Ventilation pipes

Fig. 10-13   Umbrella Type Bearing Construction

carries silt and other solids. As a result, most new machines are built with a stuffing box at the top of the turbine, and a babbitt lined guide bearing (Fig. 10-10). Various types of seals are used in the stuffing boxes, including carbon ring packing and gland packing.

The upper guide bearings are either babbitt lined, split cylindrical shell type bearings (Fig. 10-14), or segment bearings (Fig. 10-15). The segment type bearings are becoming increasingly popular because they permit easy adjustment of shaft alignment and bearing clearance. The segments may be crowned, that is, machined to a slightly larger radius than that of the journal plus the thickness of the oil film in order to permit easier formation of oil films. The construction shown in Figs. 10-14 and 10-15 with the journal formed by an overhanging collar on the shaft, and an oil dam extending up under the collar, is now common. It eliminates the need for an oil seal below the bearing. Bearings running directly on a machined journal on the shaft, with either an oil seal below the bearing or an oil reservoir fastened to the shaft and rotating with it are also used.

Thrust Bearings   Thrust bearings of horizontal shaft machines are of the tilting pad or fixed pad type. Where a reversible pump/turbine is used, tilting pad bearings designed for operation in either direction must be used. Thrust bearings for vertical shaft hydroelectric units are among the most highly developed forms of these bearings now in use. As pointed out earlier, the thrust bearing must support the weight of the rotating parts plus the hydraulic thrust of the water acting on the turbine runner. Single bearings capable of supporting loads in excess of 2000 tons (1800 Mg) are in service.

Tilting pad thrust bearings are used on all larger machines. Some older machines are equipped with tapered land bearings. Kingsbury and Michell type bearings in which the pads tilt on a pivot, or on a rocking edge on the bottom of the pad, were used on many older machines. Most newer designs use flexible supports under the pads to permit the slight amount of tilt needed to form wedge oil films. Bearings with flexible pad supports can be run in either direction so are particularly suited to reversible pump/turbine units. Springs (Fig. 10-16), elastomeric pads, interconnected oil pressure cylinders, and flexible metallic supports are all used. Spherical supports are also used. The bearing segments may be flat, or crowned slightly to aid in the formation of oil films. In many of the larger machines, provision is made to pump up the bearings with high pressure oil to assist starting (Fig. 10-16).

Methods of Lubricant Application   The bearings of hydroelectric sets are either self-lubricated or supplied by a central circulation system. Circulation systems may be either unit systems where a separate system is used for each unit in a station, or station systems where all the units in a station are supplied from one system. In many cases, one or more of the bearings may be of the self-lubricating type, with a unit system supplying the other bearings.

In self-lubricated bearings, the oil is contained in a tank surrounding the shaft (Figs. 10-13, 10-14 and 10-15). Oil is lifted by grooves in the bearings, or by a ring pump on the shaft. Cooling coils can be located in the tank, or with a cylindrical shell bearing, a cooling jacket may be located around the bearing shell. External cooling coils may also be used, but are generally suitable only for relatively high speed machines where sufficient pumping force is generated to circulate the oil through the external circuit.

Fig. 10-14 Structural View of Cylindrical Bearing for Vertical Shaft Turbine

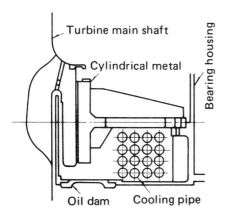

Fig. 10-15 Cylindrical Shell and Segment Type Guide Bearings

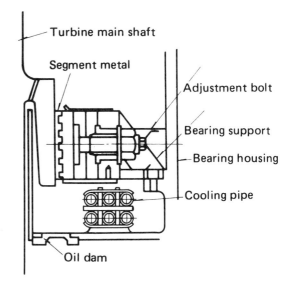

Fig. 10-16 **Spring Supported Tilting Pad Thrust Bearing** Stationary portion showing oil lift slots.

Governor and Control Systems

Older hydroelectric units were equipped with mechanical hydraulic control systems with a mechanical speed sensitive device and a hydraulic system to actuate the guide vanes, and the runner blades if a Deriaz or Kaplan turbine were used. Newer machines are often equipped with electrical speed sensitive devices, and electronic systems are also coming into use.

The hydraulic systems are now usually independent of the bearing lubrication system. Pumps are driven by electric motors. Air charged accumulators are used to maintain system pressure and supply the large fluid flow necessary to adjust rapidly to meet sudden changes in load. They also provide a source of fluid under pressure to shut the turbine down in the event of a failure in the system. Emergency shutdown may also be assisted by the use of closing weights on the guide vane operating mechanism, or by designing the vanes so that water pressure will close them if the hydraulic system fails.

Guide Vanes

The guide vanes, or wicket gates, are manufactured with an integral stem at each end which serves as the bearing journal. One bearing is used at the bottom and one or two bearings at the top. A thrust bearing may also be required. These bearings, as well as the bearings of the operating mechanism, are grease lubricated. Centralized application systems are now usually used to supply these bearings.

Control Valves

In some turbines, the guide vanes are arranged to close tightly and act as the shutoff valve for the turbine. In most installations, however, separate closing devices on the water inlet are used. In the case of pump turbines, closing devices on both the inlet and outlet are used.

Closing devices include sluice valves, rotary valves (Fig. 10-17), and butterfly valves. All are designed for hydraulic operation. Closing weights may be used for emergency shutdown. Bearings are grease lubricated.

Fig. 10-17 Rotary Closing Valve

Compressors   In most hydroelectric plants, compressed air is required to maintain the pressure in the hydraulic accumulators. Compressed air is also used to blow out the draft tube and turbine casing when maintenance is to be performed. Compressed air also blows out the pump or turbine when the changeover from pump to turbine operation, or vice versa, is made in pump/turbine installations. Compressed air is also used in some impulse turbine installations to keep the tail water out of the turbine when the tail water level is high.

LUBRICANT RECOMMENDATIONS

The need for extreme reliability and long service life of hydroelectric plants generally dictates that premium, long life lubricants be used. Inhibited, premium oils are usually used for oil applications. Viscosities usually are of ISO Viscosity Grade 32, 46, 68 or 100 depending on bearing design, speeds, and operating temperatures. Oils with excellent water separating characteristics are desirable. While start up temperatures are rarely below freezing, the oils used must have adequate fluidity for proper circulation at those temperatures. Where oil lifts are not used for starting, oils with enhanced film strength may be desirable to provide additional protection during starting and stopping.

Hydraulic system requirements can generally be met with the same types of oils used for bearing lubrication. Good air separation properties are desirable to ensure that air picked up in the accumulators separates readily in the reservoir.

Greases used in grease lubricated bearings require good water resistance and rust protection. They should be suitable for use in centralized lubrication systems, and have good pumpability at the lowest water temperatures. Both lithium and calcium soap greases are used. NLGI No. 2 consistency greases are usually used, with No. 1 consistency greases used in some extremely cold locations.

Compressors used in hydroelectric plants can be lubricated as outlined in Chap. 14.

11

Nuclear Power Plants

POWER REACTORS

Basically, all nuclear power reactors are similar. They differ in function, makeup, and operating characteristics from research or testing reactors or from production reactors, such as those at Hanford, Washington, which are used to manufacture plutonium. The power reactor, whose main function is to furnish energy, consists broadly of a core containing nuclear fuel, a moderator (although this is eliminated in fast reactors), a cooling or heat transfer system, a control system, and shielding. In practice, although they are basically similar, it is possible to design an almost endless number of different reactor types by using various combinations of fuel, coolant, and moderator. Table 11-1 lists the main elements involved in the design of a reactor. Under each design element is listed the basic variations possible for fuel, fertile material, moderator, coolant, neutron energy level, and geometry. Multiplying all these variables, we find that there are at least 1200 combinations from which to choose.

It would seem that such variety could lead to confusion, but in actuality certain combinations are ruled out because of the availability of some of the

Table 11-1 Reactor Component Variables

Fuel	Fertile Material	Moderator	Coolant	Neutron Energy	Geometry
Natural uranium	Th-232	Light water	Gas (CO_2, He)	Fast	Heterogeneous
U-235	U-238	Heavy water	Light water	Intermediate	Homogeneous
U-233		Graphite	Heavy water	Thermal	
Pu-239		Terphenyl	Liquid metal		
		None	Hydrocarbons		

components or the economics; for example, many areas must use natural uranium, for they lack the capability of enrichment. This rules out certain types of reactors, such as the fast flux. Also, natural uranium puts a limitation on the type of moderator, critical size, and power level, and although heavy water is a good moderator, especially for naturally fueled reactors or low enrichment, its cost has mitigated against its widespread use. Helium as a coolant for gas cooled reactors has found limited use because it is found in sufficient quantities only on the North American continent.

For these reasons, various countries throughout the world have pursued a particular course of design that depended on the availability of materials for construction and fuel; for example, most European nations have based their first generation designs on the use of natural uranium because of a lack of enrichment facilities. On the other hand, the United States, with its extensive system built for defense purposes, has concentrated its reactor designs on enriched fuel. The almost exclusive use of natural uranium in Europe, however, is changing with the introduction of enrichment facilities.

In addition, the environmental conditions existing in certain areas dictate the use of nuclear energy as well as the particular reactor design. England, because of a critical shortage of fossil fuel, acted quickly to provide nuclear power and was the first country to develop electric energy from the fissioning process. The urgency of the situation made it necessary to use the simplest and most reliable reactor available at the early stages. Therefore, England developed the gas cooled, graphite moderated, natural uranium reactor to a high degree of usefulness and efficiency.

In the early 1950s, fossil fuels were relatively plentiful and cheap in the United States (and to a large extent in Russia) and, therefore, it was less urgent to develop nuclear power, which could not then compete with fossil fueled generation. As a consequence, the long term view was taken, and various reactor concepts were developed on which to base the most feasible and economical design for use in the future. The choice today in the United States is a light water cooled and moderated, enriched fuel reactor, although much attention is still being directed to the fast breeder and high temperature, gas cooled convertor type.

Basic Reactor Systems

Among the hundreds of combinations of fuel, coolant, moderator, etc., which have theoretical possibilities as reactor systems, eight basic types have been studied in the research stages and have resulted in the demonstration or commercial power reactors.

1. Pressurized water reactor
2. Boiling water reactor
3. Sodium graphite reactor
4. Fast breeder reactor
5. Homogeneous reactor
6. Organic cooled and moderated reactor
7. Gas cooled reactor
8. High temperature, gas cooled reactor

Pressurized Water Reactor Fission heat is removed from the reactor core by water pressurized at approximately 2000 psi to prevent boiling (Fig. 11-1). Steam is generated from a secondary coolant in the heat exchanger.

Major Characteristics

Light water is the cheapest coolant and moderator.
Water is a well documented heat transfer medium and the cooling system is relatively simple.
High pressure requires a costly reactor vessel and leakproof primary coolant system.
High pressure, high temperature water at rapid flow rates increases corrosion and erosion problems.
Steam is produced at relatively low temperatures and pressures (compared with fossil fueled boilers) and may require superheating to achieve high plant efficiencies.
Containment requirements are extensive because of possible high energy release in the event of primary coolant system failure.

Boiling Water Fission heat is removed from the reactor by conversion of water to steam in the core (Fig. 11-2). It may be a single or dual cycle system.

Major Characteristics

Light water is the coolant, moderator, and heat exchange medium, as in a pressurized water reactor.
Reactor vessel pressure is less than the primary circuit of the pressurized reactor.

Fig. 11-1 Pressurized Water Reactor System The key applies to Figs. 11-1 through 11-7.

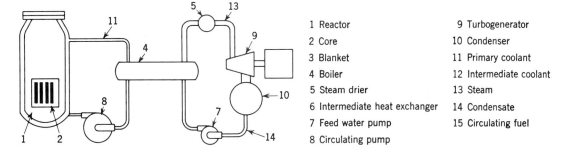

1 Reactor	9 Turbogenerator
2 Core	10 Condenser
3 Blanket	11 Primary coolant
4 Boiler	12 Intermediate coolant
5 Steam drier	13 Steam
6 Intermediate heat exchanger	14 Condensate
7 Feed water pump	15 Circulating fuel
8 Circulating pump	

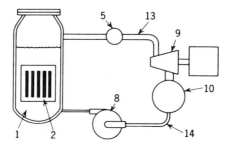

Fig. 11-2 Direct Cycle Boiling Water Reactor

Steam pressures and temperatures are similar to those of pressurized water.
Heat exchanges, pumps, and auxiliary equipment requirements are reduced or eliminated.
Has an inherent safety characteristic in that power surge causes a void formation, thus reducing the core power level.

Sodium Graphite Molten sodium metal (Fig. 11-3) transfers high temperature heat from graphite moderated core to an intermediate exchanger. Intermediate sodium-potassium coolant transfers heat to the final water cooled, steam generation equipment.

Major Characteristics

The high boiling point of liquid metal eliminates pressure on the reactor and primary system.
Permits high reactor temperatures.
Steam generated at relatively high temperatures and pressures.
Corrosion problems are minimized.
Low coolant pressures reduce containment requirements.
Violent chemical reaction with water and high radioactivity of alkali metal requires a triple cycle coolant system with dual heat exchange equipment to minimize hazards.
Relatively complex core.

Fast Breeder Heat from fission by fast neutrons is transferred by sodium coolant (Fig. 11-4) through an intermediate sodium cycle to steam boilers as in the sodium graphite type. No moderator is used. Neutrons escaping from the core into a blanket breed fissionable Pu-239 from fertile U-238.

Fig. 11-3 Soldium Graphite Reactor System

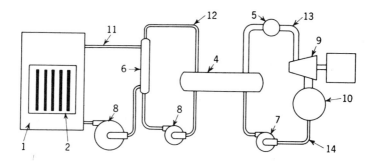

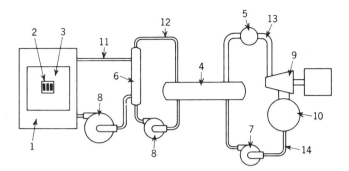

Fig. 11-4   Fast Breeder Reactor System

Major Characteristics

Reactor is designed to produce more fissionable material than is consumed.
Wide choice of structural materials as a result of low absorption of high
    energy neutrons.
Low neutron absorption by fission products permits high fuel burn up.
A small core with a minimum area intensifies heat transfer problems.
Core physics, including short neutron lifetime, makes control difficult.

Aqueous Homogeneous   Heat formed in the core, which is a critical mass of
solution or slurry of fuel and moderator, is carried by fuel solution to the
heat exchangers to form steam (Fig. 11-5). Slow neutrons from the core
breed fissionable U-233 from Th-232 in the blanket.

Major Characteristics

The system has a high degree of inherent stability; mechanical control rods
    are unnecessary.
Fuel element problems are eliminated. Continuous processing of irradiated
    fuel is possible to remove fission products and permit maximum burn up.
Fuel solution is highly radioactive and corrosive.
Core and blanket, including primary system, must be kept at high pressure
    to prevent boiling.
Precautions must be taken to avoid accumulation of critical mass outside
    the reactor vessel.

Fig. 11-5   Homogeneous Reactor System

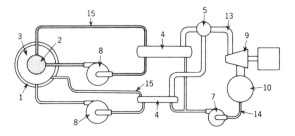

Containment requirements are high, for radioactive material is circulated through the primary coolant and blanket loops.

Organic Moderated   Heat is removed from the core by organic coolant at low or moderate pressure. Steam is generated in the boiler or heat exchanger. (Fig. 11-6).

Major Characteristics

High pressure in reactor and primary circuit is avoided, although higher temperatures can be achieved than in a pressurized water reactor.

Organic coolant becomes only slightly radioactive and causes little corrosion.

Heat transfer characteristics good but lower than water.

Hydrocarbon coolant may deteriorate and cause fouling or scale formation on the fuel elements.

Gas Cooled   Heat removed from the core by gas at moderate pressures is circulated through steam generating heat exchangers that produce low and high pressure steam. It utilizes carbon dioxide gas, graphite moderator, and natural uranium fuel.

Major Characteristics

Utilizes natural uranium fuel and relatively available materials and construction.

Permits low pressure coolant and relatively high reactor temperatures.

Containment requirements are moderate and corrosion problems minimal at low temperatures.

Reactor size is relatively large because of natural fuel and graphite moderator. Power density (kilowatt output per liter of core volume) is extremely low.

Poor heat transfer characteristics of gases require high pumping requirements.

Steam pressures and temperatures are low.

Carbon dioxide gas is relatively cheap, safe, and easy to handle.

High Temperature, Gas Cooled   Heat from the reactor core is carried by inert helium to the heat exchanger for generation of steam or directly to a gas turbine; the gas returns to the reactor in a closed cycle (Fig. 11-7).

Fig. 11-6   Organic Cooled and Moderated Reactor

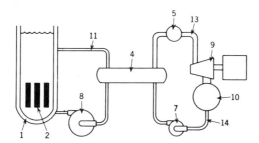

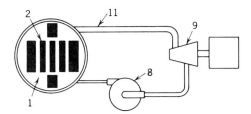

Fig. 11-7 **High Temperature Gas Cooled Reactor**

Major Characteristics

Good efficiency can be achieved in a dual cycle with a minimum gas temperature of 1400°F (760°C).

High fuel burn up is possible and conversion of fertile material permits lower fuel costs.

Minimum corrosion of fuel elements will be caused by inert gas.

High temperature coolant minimizes the disadvantages of poor heat transfer characteristic of the gases.

Possible contamination of turbine in a direct cycle caused by fuel element failure.

Fuel element design for long life is complicated by high temperatures.

The supply of helium worldwide is limited.

RADIATION EFFECTS ON PETROLEUM PRODUCTS

In general, radiation damage may be defined as any adverse change in the physical and chemical properties of a material as a result of exposure to radiation. Radiation damage is, of course, a relative term, for the changes in a material that may be adverse to one process or system may be advantageous to another. This is true of organic materials in particular; for example, the evolution of a gaseous hydrocarbon from a liquid organic material may result in an explosion hazard, an increase in liquid viscosity, or, on the other hand, a new method for synthesizing a hydrocarbon. Similarly, radiation of an organic fluid may result in unwanted growth in molecular size with consequent thickening or solidification of the liquid, or may present a feasible method of polymerization to form an elastomer or plastic. In the study of radiation damage, we are concerned mainly with the adverse or undesirable changes in the lubricants of a power reactor.

Broadly speaking, there are two mechanisms of radiolysis that must be considered in a study of the damage to organic fluids. One is the primary electronic excitation and ionization of organic molecules caused by beta particles, gamma rays, and fast neutrons. The other is the capture of thermal neutrons and some fast neutrons by nuclei that would cause changes in the nuclei and the generation of secondary radiation that would result in further damage.

Two methods are utilized to measure radiation energy. One measures the quantity of energy to which the material is exposed and is called the roentgen (R); the other is the amount of energy that the material absorbs and is called the rad. For gamma radiation, the exposure unit (the roentgen) is defined as the quantity of electromagnetic radiation that imparts 83.8 ergs of energy to 1 g of air.

The radiation dosage of a material is defined as an absorption of 100 ergs of energy from any type of radiation by 1 g of material. Actually, absorbed energy will vary with the type of radiation and the effect will depend on the material exposed. For gamma radiation, however, one rad absorbed is approximately equivalent to 1.2 R of radiation dosage. The rad is useful for comparing the equivalent energy of mixed radiation fluxes but does not distinguish between types.

From a radiation damage standpoint, 1 rad of neutron flux causes 10 times more biological damage to tissues than an equivalent absorbed energy of gamma rays. For petroleum products, however, the dosage, as measured by such effects as viscosity increase, is almost equivalent for the two types. This is discussed in more detail later in this capter.

The general levels of radiation dosage are as follows:

Dosage (roentgens)	Effect
200 – 800	Lethal to humans
< 5 million	Negligible to petroleum products
5 – 10 million	Damaging to petroleum products
> 10 million	Survived by only most resistant organic structures

Based on experimental work to date, the damage to petroleum products may be summarized in the list below. It must be appreciated, however, that the intensity of these effects or, in fact, the incidence of one or more of them will depend on the amount of absorbed energy, the exact composition of the specific petroleum material, and other environmental conditions such as temperature, pressure, and gaseous composition of the atmosphere.

The effects are as follows:

1. Liquid products (fuels and oils) darken and acquire an acrid oxidized odor.
2. Hydrogen content decreases and density increases.
3. Gases such as hydrogen and light hydrocarbons evolve.
4. Physical properties change, higher and lower molecular weight materials are formed, and olefin content increases.
5. Viscosity and viscosity index increase.
6. Polymerization to solid state can occur.

Mechanism of Radiation Damage

Organic compounds and covalent materials do not normally exist in an ionized state and therefore are highly susceptible to electronic excitation and ionization as the result of deposited energy. Covalent compounds, including the common gases, liquids, and organic materials consist of molecules that are formed by a group of atoms held together by shared electron bonding, which yields strong exchange forces. The molecules are bound together by relatively weak van der Waal forces.

Conversely, ionic compounds, such as inorganic materials, which include salts and oxides, are already ionized (metals may also be considered as being in an ionized state) and are not susceptible to ionization but are susceptible to further electronic excitation. Ionic compounds consist of highly electropositive and electronegative ions held together in a crystal lattice by electrostatic forces in accordance with Coulomb's law. There is no actual union of ions in the crystal to form molecules, although all crystals may be considered as being composed of large molecules of a size limited only by the capacity of the crystal to grow.

Therefore, the effect of radiation energy on nonionic compounds is to form ions, radicals, and excited species and thereby make the compounds more reactive with themselves or with the atmospheric environment. On the other hand, the effect of radiation on ionic compounds is to change the properties of the compound related to crystal structure. How each type of radiation affects organic materials is discussed below.

Fast Neutrons Most materials have a low capture cross section for fast neutrons and, therefore, these particles can deposit energy only by elastic and inelastic scattering. Indeed, one important use of organic materials is to moderate or deenergize fast neutrons to make them into thermal neutrons that can be captured in the fission process.

The main bulk of fast neutron deposited energy causes considerable damage to organic materials by exciting the molecule and causing ion formation. Some fast neutrons deposit energy by momentum losses which cause dislocation of atoms, commonly called Wigner type damage.

An appreciable fraction of the neutron's energy is transferred to each struck nucleus. The nucleus is ejected as a recoil ion and the neutron is scattered or reflected to continue striking other nuclei until it is degraded in energy to the thermal energy state. The recoil ion interacts with orbital electrons, thus producing molecular excitation and ionization and becoming neutralized. As recoil ions, they are above the thermal energy state.

Gamma Radiation Gamma rays react with matter either by photoelectric effect, Compton effect, or pair production. The precedence of a particular mechanism depends on the atomic number of the irradiated medium and the energy of the gamma radiation. For materials of low atomic number (<27), such as hydrogen, and carbon, the photoelectric effect predominates at gamma energies below 0.1 MeV; the Compton effect from 0.1 to 10 MeV; and pair production above 10 MeV. For materials of high atomic number (>82), such as lead, the photoelectric effect from 1.0 to 4.0 MeV; and pair production above 4 MeV.

In accordance with current theory, the methods of energy transfer affect material as follows:

1. *Photoelectric effect.* Gamma energy interacts with the entire atom and transfers an entire photon of energy in one encounter. This energy is absorbed by a single electron which is usually in a K or L shell (Fig. 11.8). The electron ejects from the atom with an energy equal to the incident photon minus the binding energy of the electron of the atom.

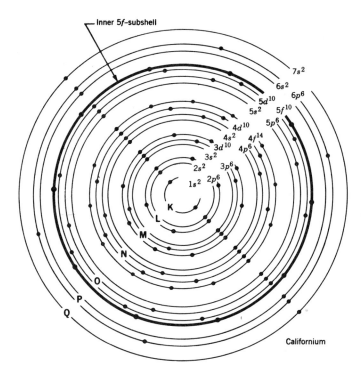

Fig. 11-8 Electron-Shell Structure

2. *Compton effect.* A gamma ray photon interacts with a single electron. It loses a portion of its energy to the electron, which recoils at an angle to the incident gamma photon.
3. *Pair production.* The gamma energy is completely converted in the region of the nuclei's Coulomb field; the result is the formation of a positron-electron pair.

The ejected high energy electrons or electron-ion pairs in turn react with orbital electrons to cause electron excitation and ionization.

Thermal Neutrons Thermal neutrons are captured by nuclei, which in turn become radioactive and give off alpha particles, gamma radiation, beta particles, or protons. These reactions are symbolized as (n,α), (n,γ), (n,β) or (n,p),* respectively. Although thermal neutrons cause very little damage to organic materials, that which occurs is due to the secondary radiation that results predominately from the 2.2 MeV rays given off by the capture of thermal neutrons by hydrogen nuclei. For chloro-organics the (n,β) reaction is important because of high thermal neutron capture cross section of chlorine. Nitro-organics suffer radiation damage when thermal neutrons absorbed in the N^{14} (n,p) C^{14} reaction cause ejected protons which ionize the media.

*These symbols indicate that a neutron (n) is absorbed and alpha, gamma, beta, or proton radiation results.

Calculation of Energy Absorption and Radiation Damage

Energy deposition or absorption leading to radiation damage is described by using the units of the rad. This was accomplished by the International Commission on Radiological Units in 1953, which recommended that the absorbed dose of any ionizing radiation be defined as the amount of energy imparted to matter by ionizing particles per unit mass of irradiated material. The unit of absorbed dose is the rad which is equivalent to 100 ergs of absorbed energy per gram of material. It is often more convenient, however, to measure the absorption of ionizing energy from a radiating source by its effect on a quantity of air rather than the material being irradiated. Therefore a relationship between the rad is convenient; although not strictly accurate, 1 rad may be considered as equivalent to 1 eV. Because, as previously stated, the major energy transferred from source to medium results in electronic excitation and ionization, and organic materials may suffer such damage, methods for approximate calculation of the estimated damage from this energy absorption have been developed.

For example, the average energy required to sever a chemical bond is approximately 2 eV and the typical energy to cause ionization of a hydrocarbon is 10 eV. Therefore, it is reasonable that the MeV level energies available from average fast neutrons and gamma photons radiated from a reactor core are more than sufficient to cause (even by secondary processes) ionization and dissociation in a chemical compound.

The rate of reaction, which will be taken as the extent of radiation damage, is measured in the amount of material reacting for a given energy input. The term G has been assigned to the rate of reaction and is defined as the number of molecules reacting for each 100 eV of energy absorbed.

To particularize these general theories to nuclear power plant applications, several broad generalizations can be made regarding the effect of radiation on organic fluids. Two of the most striking effects of organic liquid irradiation are gas evolution and viscosity changes. The gas evolution is not dependent on the dosage rate but is rather a linear function of the applied energy dose. Similarly, viscosity is a function of dosage with a marked threshold of reaction. The viscosity gradually increases until a definite dose has been absorbed when there is an exponential increase in viscosity with small incremental additions of radiation energy.

Another general observation is that aromatic compounds behave differently than other organic compounds, especially the aliphatics. The theory frequently offered to explain this is that in an aliphatic substance the electrons paired to form the carbon-to-carbon or carbon-to-hydrogen bond have a different absorptive capacity for radiative energy than the bonds in an aromatic C−C or C−H linkage. It is theorized that energy absorbed in the aromatic ring structure is distributed over the entire structure rather than in any particular C−C or C−H bond.

Although the discussion to this point has been theoretical, with emphasis on the particular type of radiation, in practice the radiation from a reactor core is mixed. Therefore, to estimate the radiation stability we must calculate the contribution from each type of radiation. To do this a table of viscosity changes based on increasing radiation exposure has been developed experimentally for various standard organic liquids.* To use this information we furnish the following example:

*Experiments by Bolt and Carroll − California Research Corporation.

Hexadecane is exposed to a radiation flux consisting of

10^{10} thermal neutrons (cm²) (sec)

10^9 fast neutrons/(cm²) (sec)

10^{11} gamma photons/(cm²) (sec)

How many hours of exposure will produce an increase in viscosity of 55 percent at 210°F (99°C)?

First determine the dosage in rad. Based on data developed in experiments by Bolt and Carroll of California Research Corporation, we find that hexadecane will absorb 1.0×10^9 rad when exposed to any one of the following fluxes:

Thermal neutrons 2.82×10^{19} th n/cm²

Fast neutrons 1.95×10^{17} f n/cm²

Gamma radiation 1.95×10^{18} γ/cm²

Now the contribution of each component of that assumed radiation flux may be calculated as follows:

$$\text{Thermal neutron contribution} = \frac{10^{10} \text{ th n/cm}^2 - \text{sec} \times 1.0 \times 10^9 \text{ rad}}{2.82 \times 10^{19} \text{ th n/cm}^2}$$
$$= 0.4 \text{ rad/sec}$$

$$\text{Fast neutron contribution} = \frac{10^9 \text{ f n/cm}^2 - \text{sec} \times 1.0 \times 10^9 \text{ rad}}{1.95 \times 10^{17} \text{ f n/cm}^2}$$
$$= 4.8 \text{ rad/sec}$$

$$\text{Gamma radiation contribution} = \frac{10^{11} \text{ th n/cm}^2 - \text{sec} \times 1.0 \times 10^9 \text{ rad}}{1.95 \times 10^{18} \text{ th n/cm}^2}$$
$$= 0.4 \text{ rad/sec}$$

From these same experiments, a dose of 5.0×10^8 rad causes an increase at 210°F of 55% in the viscosity of hexadecane. Therefore, under the mixed flux being estimated hexadecane is stable to the radiation for the following period:

$$\frac{5.0 \times 10^8 \text{ rad}}{56.2 \text{ rad/sec}} = 8.9 \times 10^6 \text{ sec} = 2470 \text{ h}$$

Chemical Changes in Irradiated Materials

The physical and chemical properties of hydrocarbon fluids which make them important as lubricants change during irradiation to varying degrees based on their chemical composition and the presence of additives. These changes may be traced to alteration of the chemical structure of the materials. As discussed, nuclear radiation either directly, or by secondary radiation, deposits high level energy in the irradiated organic substance and causes ionization and molecular excitation. The ions and excited molecules rapidly react to form free radicals which further combine or condense (Fig. 11-9).

The changes in chemical structure may be measured by various classic methods; for example, it is possible to determine the approximate number of free radicals formed by the use of scavengers such as iodine. In addition, gas

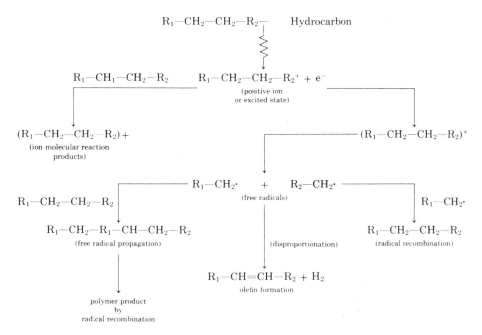

Fig. 11-9   Radiolysis Processes in Hydrocarbons

is evolved, either hydrogen or light petroleum fractions. Investigations have shown that both carbon-hydrogen and carbon-carbon bonds can be broken by radiolysis. The dissociated or ionized molecules can condense, rearrange, form olefins, or other products, depending on the environment. At temperatures below 400°F (204°C), temperature effects do not seem to be significant.

Because most petroleum lubricants are combinations of saturated and unsaturated aliphatics and aromatic compounds, the reactions of these principle hydrocarbon classes have been studied under the influence of ionizing radiation. These studies (Table 11-2) indicate, as would be suspected, that

Table 11-2   Summary of Radiation Changes in Organic Compounds

G-Values[a]	Saturated Hydrocarbons	Unsaturated Hydrocarbons	Aromatic Hydrocarbons
Polymerization	None	10–10,000	None
Cross linking	~1	6–14	<1
H_2 evolution	2–6	1	0.04–0.4
CH_4 evolution	0.06–1	0.1–0.4	0.001–0.08
Destruction of irradiated material	4–9	6–2000	<1

[a]G = molecules produced, destroyed, or reacted per 100 eV of energy absorbed.

Source: From *Nuclear Engineering Handbook*, by H. Etherington, © 1958 McGraw-Hill Book Company, Inc., used by permission of McGraw-Hill Book Co., Inc.

Table 11-3 Radiation Sources for Research

Type	Power	Types of Radiation	Energy of Radiation	Application[b]	Remarks
Heavy-water tank reactor	1–5 MW	Neutrons (fast and thermal), secondary particles, and radiations	Thermal neutron flux: $2-8 \times 10^{13}$ n/cm^2 sec Gamma flux: 10^8 R/hr	High flux, general research reactor; quite large; isotope production; large equipment physics experiments; medical research; loop experiments	Uses forced cooling; D$_2$O is expensive and requires special handling increasing cost over a comparable light-water-moderated reactor; reactor can be made critical with natural uranium, although enriched uranium is normally used; very high flux is obtained with a small critical mass; reactor can be pressurized; core can be made large, which allows more flexibility for experiments; core may be averaged to give a wide range of neutron spectrum from thermal to fast; somewhat safer than light-water-moderated reactor because of longer neutron lifetime
Aqueous-homogeneous reactor	10W 5–500 W 50 kW	Neutrons (fast and thermal), secondary particles, and radiations	Thermal neutron flux: 2×10^8 n/cm^2 sec Thermal neutron flux: 10^{10} n/cm^2 sec Thermal neutron flux: 10^{12} n/cm^2 sec	Basic physics research; medical and biological research; radioisotope production; testing by the danger coefficient method; exponential experiments; study of effect of neutron and gamma radiation on chemical and biological materials; production of short-lived radioisotopes; educational training facility	Fuel can be added and maintained easily; no fuel element fabrication cost; corrosion and chemical problems increase with power level; limited space for high flux irradiation; large negative temperature coefficient is desirable safety feature; small critical mass required; difficult handling problems of radiolytic gases; can purify radiolytic gases to provide a relatively pure source of gamma emitters

Reactor type	Power	Radiation	Flux	Uses	Remarks
Graphite (pile) reactor	3–30 MW	Neutrons (fast and thermal), secondary particles, and radiations	Thermal neutron flux: $1\text{–}5 \times 10^{12}$ n/cm^2 sec Gamma flux: $10^5\text{–}10^6$ R/hr	Isotope production; large equipment physics experiments; medical research; loop experiments	Can be made critical with natural uranium; very large core with space for a number of sample irradiations; neutron flux per unit of power low; reactor is bulky and expensive; air cooling sufficient for all except very high powered ones; thermal to fast flux is high (approximately 100 to 1)
Swimming pool reactor	10–100 kW	Neutrons (fast and thermal), secondary particles, and	Thermal neutron flux: $10^{11}\text{–}10^{12}$ n/cm^2 sec Gamma flux: $10^6\text{–}10^7$ R/hr	Flexible source of radiation; used for shielding studies; irradiation; training and instruction purposes	Can operate at low power (up to 200 kW rather cheaply; relatively safe; requires demineralized water as shield, coolant, reflector, and moderator; large pool enables large objects to be irradiated; low-power reactors use natural circulation; higher power reactors (1 MW) require forced circulation cooling; 5 MW reactor requires special core encasement; ratio of thermal to fast neutrons is approximately 1 to 1
	1–5 MW		Thermal neutron flux: $10^{13}\text{–}5 \times 10^{13}$ n/cm^2 sec Gamma flux: $10^8\text{–}10^9$ R/hr		
Subcritical assemblies[a]	3 W	Neutrons (fast and thermal), secondary particles, and radiations	Thermal neutron flux: 10^8 n/cm^2 sec	Introductory reactor theory training and studies	Minimum facility required with corresponding low cost; shielding problems small; requires special neutron source of Van de Graaff for operation
Zero power reactor[a]	0.1–10 W	Neutrons (fast and thermal), secondary particles, and radiations	Thermal neutron flux: 10^7 n/cm^2 sec Gamma flux: 80 R/hr/w	Absorption properties of reactor materials, critical reactor source; physics and reactor theory training	Useful for reactor studies and training
Temperature controlled reactor[a]	1 kW	Neutrons (fast and thermal), secondary particles, and radiations	Thermal neutron flux: 10^{10} n/cm^2 sec	Production of short-lived isotopes; physics experiments; radiation testing; source for critical experiments	Minimum cost critical reactor capable of being installed in any location; minimum experience necessary for operation

Table 11-3 Radiation Sources for Research (Continued)

Type	Power	Types of Radiation	Energy of Radiation	Application[b]	Remarks
Argonaut reactor[a]	10 kW	Neutrons (fast and thermal), secondary particles, and radiations	Thermal neutron flux: 10^{11} n/cm^2 sec	Production of short-lived isotopes for medical use and research; reactor training	Low cost; good training reactor; minimum facility cost
Light-water high-flux tank reactor[a]	5–200 MW	Neutrons (fast and thermal), secondary particles, and radiations	Thermal neutron flux: 5×10^{13}–10^{15} n/cm^2 sec Gamma flux: 10^8–10^9 R/hr	Radiation effects on structural materials; in-pile (loop) studies of fuel elements and other components	Higher power at a cheaper cost than for a comparable swimming pool reactor; uses simple forced cooling; pressurization can be used; ratio of thermal to fast neutrons is approximately 1 to 1; 20 MW and above are generally classified as material text reactors; their high flux levels permit studies of radiation damage for structural materials and for fuel element studies in a reasonable time
Van de Graaff machine	0.25 kW 0.5 kW 3.0 kW 50 μA beam 5 μA current	Electrons, positive ions, secondary particles, and gamma radiation	1 MeV 2 MeV 3 MeV 5 MeV 10 MeV	Polymerization reactions; food preservation and sterilization operations; basic research requiring relatively small quantities of other forms of radiations needed in teaching physics, biology, chemistry, etc.; radiography	Versatile machines; can produce neutrons as secondary particles and gamma radiation as secondary radiation
Linear accelerator	4 kW 10 kW 18 kW 60 kW 400 kW	Electrons, secondary particles, and gamma radiation	6–10 MeV 24 MeV 10 MeV 10 MeV	Irradiation of plastics, polymerization, catalytic and other chemical reactions, basic and applied research; higher energy electrons, monoenergetic, sharply defined beams	Nearly monoenergetic beams

Source	Power rating	Radiation	Energy	Application[b]	Remarks
Resonant transformer	5 kW 10 kW 25 kW	Electrons	1 MeV 2 MeV 4 MeV	Irradiation of plastics; food sterilization; chemical reactions, such as polymerization	A relatively low-cost source of lower energy electrons with significant industrial applications
Cockcroft-Walton generator	250 W 500 W	Positive ions	0.25 to 0.5 MeV 1 to 1.75 MeV	Nuclear reactions, chemical and biological irradiations and studies	A high intensity source (0.5 to 3 mA) of relatively low energy positive ions
Circular accelerators such as cyclotrons and betatrons	Very low	Eletrons, positive ions, secondary particles, and gamma radiation	Up to 30 beV	Basic particle physics studies	Principally used to produce high energy particles at low current density; very expensive for general irradiation studies
Cobalt-60		Gamma	1.17 and 1.33 MeV	Food and drug sterilization; chemical reactions; medical research; hydrocarbon research; physics studies; chemical research; radiation damage studies	Useful for environmental irradiation studies; source can be contained in lead cask (small irradiation volume), water well (requires demineralized water), or in a hot cell
Fuel Elements (MTR)		Gamma	10^5 to 10^6 R/hr, equivalent to 5000 C of Cobalt-60	Food sterilization; hydrocarbon research	Wide range and energies; difficult to reproduce conditions of test; high handling expenses; limited availability
Other radioactive isotopes		Neutrons	Variable	Tracer studies (medicine, agriculture, anthropology, etc.); basic physics; density measurements reactor startup, chemical reactions (such as polymerization); radiography; food sterilization; ionization	A summary of applications and a bibliography is included in the report "Isotopes—an eight-year summary of U.S. distribution and utilization," available (price $2) from Superintendent of Documents, U.S. Government Printing Office, Washington 25, D.C. Cost information on isotopes is available from ORNL and from private companies making isotopes available

[a] Light-water tank reactor.
[b] Because all reactors produce neutrons and gamma radiation to a greater or lesser extent, the application of a particular reactor to given use is not sharply defined. The application column therefore lists only those areas to which the particular type of reactor is most suitable. With the exception of those programs that require very high neutron fluxes such as radiation damage studies most of the reactors are applicable to such areas as production of radioisotopes, fundamental investigations of radiation effects on chemical reactions or in genetic mutations, analytical use such as chemical activation analysis and neutron defraction studies, medical therapy, as in neutron capture techniques, and training in reactor engineering and physics.

Table 11-4   Characteristics of Stock Oils Used for Radiation Resistance Comparisons

Description	Sp gr 60/60°F	Color ASTM	Viscosity ∂ 100°F CS	Viscosity ∂ 100°F SUS	Viscosity ∂ 210°F CS	Viscosity ∂ 210°F SUS	Viscosity Index	% Wt Sulfur
A Mineral colza oil	0.838	½	4.21	39.81	. . .	. . .	. . .	0.1
B Empire pale oil	0.932	3	49.4	229.3	5.44	43.76	12	1.91
C 150-sec solvent naphthenic	0.884	2	34.6	161.9	5.04	42.47	67	0.44
D Light neutral turbine	0.872	1	32.2	151.1	5.3	43.31	107	1.14
E 150-sec light neutral low sulfur	0.863	1½	32.6	152.9	5.34	43.45	107	0.21
F Heavy neutral turbine	0.888	2	102	472.8	10.7	61.82	96	1.34
G Solvent bright stock	0.905	4½	549	2544.0	32.7	154.4	97	1.78
H "Aromatic extract"	0.913		62.3	289.1	7.02	49.18	69	2 approx

N-d-m Analysis

Viscosity Gravity Constant	% C^A	% C^S	% C^P	Molecular Weight	Refractive Index na 20°C
A 0.833	8.0	32.0	60.0	250 approx	1.4661
B 0.908	24.8	24.7	50.5	370	1.4886
C 0.860	11.7	26.3	62.1	330	1.5179
D 0.819	5.5	32.3	62.3	400	1.4800
E 0.815	6.2	25.4	68.4	410	1.4778
F 0.822	8.1	23.2	68.7	500	1.4878
G 0.818	7.9	22.9	68.2	670	1.4976
H 0.862	19.4	25 approx	55 approx	390	1.5085

unsaturated hydrocarbons are most reactive and aromatics the least affect-ed. Saturated compounds fall somewhere between the two extremes. The results are expressed as G values; that is, the number of molecules reacting or produced for each 100 eV of ionizing radiation. For example, at least 6 to 15 molecules of unsaturated hydrocarbons react for each 100 eV. In certain instances, high G values (up to 10 000) result for free radical polymerization,

Table 11-5   The Effect of Irradiation (with Air Access at 85° F) for Base Oils of Different Types

Oil	10^8 Rads Dose Color inc	10^8 Rads Dose Sp gr inc	10^8 Rads Dose % Viscosity inc 100°F	10^8 Rads Dose % Viscosity inc 210°F	5×10^8 Rads Dose Color inc	5×10^8 Rads Dose Sp gr inc	5×10^8 Rads Dose % Viscosity inc 100°F	5×10^8 Rads Dose % Viscosity inc 210°F
A	. . .	. . .	. . .	. . .	2½	0.008	60	. . .
B	1½	Nil	13.1	9.5	2	0.002	72	46
C	1½	Nil	13.9	8.3	3½	0.004	89	58
D	2½	Nil	13.3	8.3	4	0.004	84	50
E	1	0.002	9.8	9.9	2½	0.009	106	64
F	2	0.001	14.7	10.3	6	0.002	94	60
G	½	Nil	17.7	1.0	3½	0.003	133	89
H	. . .	. . .	. . .	. . .	4	. . .	83	47

whereas low G values (G = 6) are found for random crosslinking and very low values (G < 1) for methane formation.

Aromatic materials, with a G value for destruction of 1, are highest in radiation resistance. The principal reaction is crosslinking, from which very small amounts of gas evolve.

Additional studies were made on a range of petroleum oils representative of typical paraffinic, naphthenic, and aromatic materials varying in sulfur content. The physical and chemical characteristics of the base oils are listed in Table 11-4 and the effect of irradiation on these oils is given in Table 11-5. The viscosity was plotted against the aromatic and sulfur content (Fig. 11-10). These data show that as the aromatic content increases the viscosity increase is reduced in almost a linear relationship. The effect of sulfur content is similar but more marked. Further, it was noted that radiation damage appears greatest for oils with the highest molecular weights or the highest initial viscosity.

Because it was found that naturally occurring compounds improved the radiation stability of petroleum oils and that these compounds were usually removed by refining procedures, the effect of using synthetic aromatic and sulfur compounds as additives was studied. The results are shown in Table 11-6. From these studies it appears that a disulfide, or an alkyl selenide, provides good radiation damage protection. The disulfides prevent polymerization by a mechanism termed free radical chain stoppers. The disulfides have an advantage also of being good EP and antiwear agents but do not prevent oxidation or olefin formation.

Fig. 11-10   Radiation Stability versus Sulfur and Aromatic Content

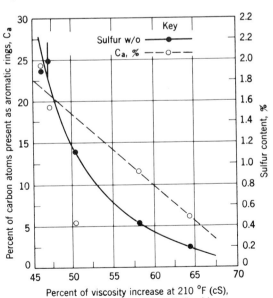

Table 11-6 The Effect of Aromatic Sulfur Compounds on the Radiation Stability of a Petroleum Oil (with Air Access, at 85°F)

Fluid	% Sulfur Added	5 × 10⁸ Rads % Viscosity increase, cSt 100°F	210°F	10⁹ Rads % Viscosity increase, cSt 100°F	210°F	Optical Densities* of IR Peaks 10.34 (olefin)	5.84 (oxidation)
150-sec low sulfur light neutral (oil E, Table 28)	...	106	64	400	194	0.16	0.12
E + cyclic compound	0.7	117	66	256	135	...	...
E + cyclic compound	2.3	103	56	289	141	...	...
E + monosulfide	0.5	83	51	257	142	0.17	0.12
E + monosulfide	1.7	67	35	169	84	...	...
E + disulfide A	0.9	62	38	137	78	0.18	0.13
E + disulfide B	0.8	59	37	...	...	0.17	0.12
E + selenide	...	74	41	216	119	...	...

*Determined by differential infra-red analysis between unirradiated sample and sample irradiated to 10⁹ rads.

It is well known, however, that aromatic compounds possess good thermal and radiation stability and in the latter case protect less stable aliphatic molecules by the transfer of energy. These compounds are usually characterized by complex molecules which resonate between a number of possible electronic structures and, therefore, possess fairly stable excited energy states. In other words, when a paraffinic hydrocarbon absorbs energy, it is raised to an unstable state in which the energy is greater than the electronic forces that constitute the chemical bonds. The result is a bond fracture with residual free radicals. In an aromatic with an equivalent absorbed energy, the higher level is not sufficient to sever the greater electronic binding forces, and the energy is eventually liberated as heat or light. The radiation stability of aromatic petroleum extracts are, in decreasing order, polyglycols, paraffinic hydrocarbons, diesters, and silicones. The effect of aromatic compounds was studied both as antirad additives to mineral oils and as pure synthetic fluids. The results are given in Figs. 11-11 and 11-12. These data show the following relationships:

1. The aromatics with bridging methylene groups between the aromatic molecules are less efficient as protective agents than antirad additives with direct links between aromatic rings.
2. Long chain alkyl groups attached to the aromatic rings make them less protective agents, probably because of a difference in stability of the compound and lowering of the aromatic ring content.
3. Small amounts of a free radical inhibitor in addition to the aromatic additive substantially reduce the viscosity increase.
4. The protection afforded is not simply a direct function of aromatic content; in fact it would appear that 40 percent of added aromatic material

is a practical maximum. Beyond 40 percent it is preferable to use a pure aromatic of suitable physical characteristics.

The study of aromatics showed that radiation stability is a function of the bond dissociation energy of the weakest carbon bond. These are plotted in Fig. 11-13.

Initially, in a study of the suitability of petroleum products for reactor lubrication, tests were conducted on conventional oils and hydraulic fluids by Carroll and Calish of the California Research Corporation. The samples were irradiated by using spent fuel elements from the Materials Testing Reactor in Arco, Idaho. This is essentially a gamma source. The petroleum products tested were four industrial oils, four base oils (pale, neutral, and white oils and bright stock), two gear lubricants, and an automatic transmission fluid. The test results (Figs. 11-14 to 11-17) were determined by changes in viscosity and VI after static irradiation up to 9×10^8 R.

Mobil investigations on turbine oils at similar doses are described in Fig. 11-18.

Most of these investigations show that high quality, mineral oil based lube oils can withstand irradiation up to 100 Mrad and that synthetic fluid lubricants are available with radiation stability above 100 Mrad.

Fig. 11-11   Radiation Protection of Synthetic Aromatic Additives

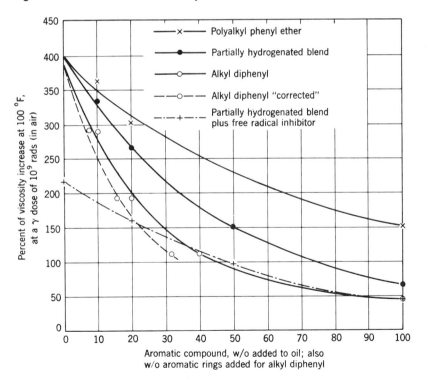

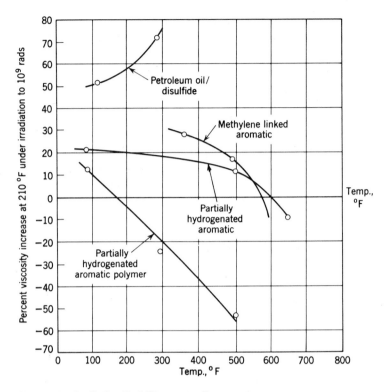

Fig. 11-12   Radiation Stability versus Temperature

A study of the changes in properties and performance of conventional lube oils after irradiation shows the following:

1. Conventional antioxidant additives of the phenolic or amine type confer little radiation stability to base oils and are preferentially destroyed between 10^8 and 5×10^8 rad.
2. Didodecyl selenide, which is known to be an effective antioxidant, also has radiation protective properties. The oxidation stability is effective after an irradiation of 10^9 rad.
3. Diester base oils, phosphate esters (wear additives), and holgenated EP agents produce acids at a low radiation dose.
4. Polymers such as polybutenes and polymethacrylates cleave readily and thus lose their effect as VI improvers.
5. Silicone antifoam agents are destroyed at low radiation dose.
6. In most cases, the presence of air, compared with an inert atmosphere, increases radiation damage by a factor of 1.6 to 2.3 times, as indicated by viscosity increase.

In summary, high quality, conventional lubricating oils are suitable for doses up to 10^8 rad. Further radiation resistance can be formulated into a good quality petroleum oil by the use of antirads such as radical scavengers (antirad A) or aromatic structures (antirad C). These formulated oils will pro-

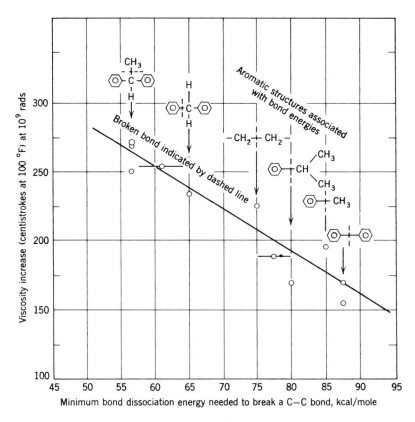

Fig. 11-13   Radiation Stability versus Bond Dissociation Energy

Fig. 11-14   Irradiation of Conventional Base Oils

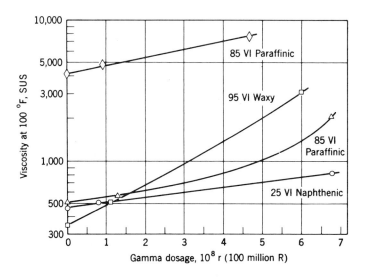

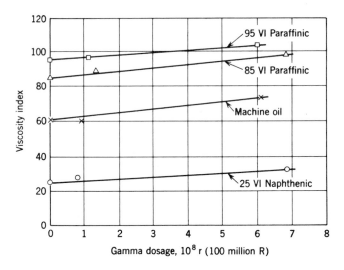

Fig. 11-15   Effect of Irradiation on VI

Fig. 11-16   Irradiation of Conventional Industrial Oils

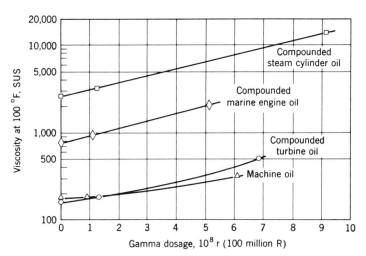

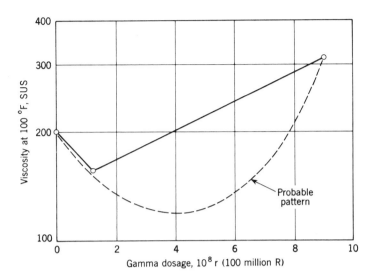

Fig. 11-17 **Irradiation of VI Improved ATF**

Fig. 11-18 **Effect of Radiation on Turbine Oils**

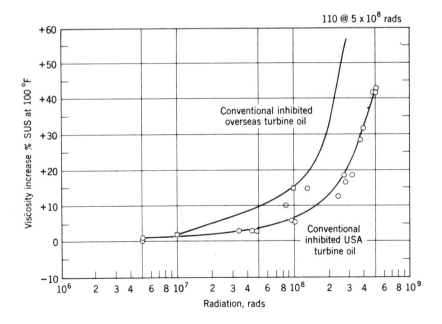

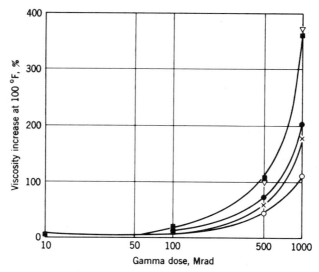

▽ Base oil
■ Base oil + conventional oxidation inhibitors
● Base oil + conventional oxidation inhibitors + antirad *A*
× Base oil + antirad *C*
○ Base oil + antirad *A* and *C*

Fig. 11-19 Effect of Antirads on Viscosity

tect up to 10^9 rad, as shown in Figs. 11-19 and 11-20. Above these doses and at high temperatures, synthetic type lubricants that use partly hydrogenated aromatics blended with aromatic polymers are required. The effect of temperature at high radiation dose (10^9 rad) has been studied under a nitrogen atmosphere for these synthetic fluids (Fig. 11-12).

Fig. 11-20 Effect of Antirads on Oxidation

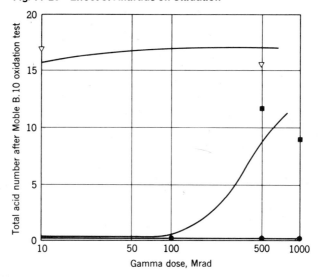

Table 11-7   Wear Characteristics of Synthetic Lubricants (in Comparison with Mineral Oils, Shell Four-ball machine)

| Fluid | Scar Diameter (cm) | | Viscosity at 100°F, cSt |
	100-kg Load for 10 Min	45-kg Load for 10 Min	
Methylene/ethylene linked aromatic	. . .	Seized after 35 sec	25.1
Ethylene linked aromatic	0.37	0.32	195
Partly hydrogenated blend A	0.33	0.22	34.5
A+ polymer	0.34	. . .	251
A+ disulfide (0.8 sulfur)	0.16	0.07	31.1
150-sec light neutral turbine oil	. . .	0.23	32.6
Five-ring polyphenyl ether	0.26	0.08	363

If, however, materials are to be satisfactory as lubricants, they must not only have good thermal and irradiation stability but their wear characteristics must be satisfactory. Such wear tests have been made by a Shell Four-Ball machine at two different loads and are given in Table 11-7. Table 11-8 lists a number of new generation synthetic and mineral oil blends and pure synthetic lubricants and potential applications.

Grease Irradiation   The damage caused by high energy radiation on greases is a dual effect. First, the radiation attacks the thickening structure and causes separation and fluidity. Following this, continued irradiation causes polymerization of the base oil to the original thickened condition and finally solidification. The precise pattern of change is dependent on the type of thickener, the gel structure, and the radiation stability of the thickener and base oil.

In general, greases have been evaluated for radiation damage by determining the worked penetration, following irradiation, and comparing it with the original value. These irradiation evaluations are usually of the static type; but, as mentioned previously, dynamic testing during irradiation has yielded markedly different results. Typical greases that have organic soap components of alkali or alkaline earth metals, although resistant to high amounts of radiation, break down at total doses of approximately 10^8 rad. Micrographs of the soap structure showed a drastic change in the normal fiber structure of the gelling agent. On the other hand, greases made with nonsoap thickeners, such as carbon black, were less affected. Micrographs of a carbon black grease showed the same structure (agglomeration of carbon particles dispersed throughout the oil phase) both before and after irradiation.

The stabilization of the thickening structure under irradiation solves the problem of softening or bleeding of the base oil but will not solve the eventual solidification of the grease. This is a function of the base oil, and the solutions discussed under lubrication oils (use of antirad additives or synthetic organics as base fluids) are valid.

The mechanism of change for three greases is shown in Fig. 11-21. In one case the grease had an unstable thickener and progressively softened to

Table 11-8   Radiation-Resistant Fluids, and Potential Applications

Application	Fluid Type	Initial Viscosity, cSt at 100°F	Flash Point, COC°F	Boiling Range, °F at 760 mm	% Viscosity Increase (100°F, cSt) at 10⁹ Rads	
					In Air	In Inert Atmosphere
Hydraulic fluid/spindle lubricant	Synthetic aromatic A	13.6	305	490/745	45	25
	Synthetic aromatic B	34.5	345	645/745	70	45
Bearing lubricant						
moderate stability	Oil/synthetic aromatic	25	345	. . .	152	80 approx
high stability	Synthetic aromatic B	34.5	345	645/745	70	45
	Synthetic aromatic/polymer	74.5	345	645/800	20 approx	58
Gear lubricant						
moderate stability	Oil/synthetic aromatic/disulfide	24	345	. . .	130 approx	70 approx
high stability	Aromatic/disulfide	31	345	645/745	54	29
	Aromatic/polymer/disulfide	72	345	645/800	15 approx	40 approx
Heat transfer						
low temperature	Kerosine/aromatics	2.9	115	320/745	100	60 approx
intermediate temperature	Synthetic aromatic A	13.6	305	490/745	45	25
high temperature	Synthetic aromatic C	38.3	340	630/745	65	40 approx

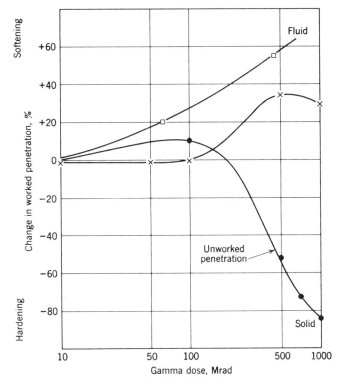

×	Approved UKAEA schedule 1 grease
□	Grease with unstable thickener
●	Grease with unstable fluid

Fig. 11-21   Effect of Radiation on Greases

fluidity. Although such a grease might protect the bearing, the problem of leakage would be great, and incompatibility with reactor components would be an added problem. The second grease gradually decreased in penetration (solidified) after an initial increase or softening. Such a grease would cause failure in the lubricated mechanism. The third grease showed good stability with a slight softening up to 10^9 rad.

Radiation Stability of Thickeners   The selection of the thickener or solid phase of a grease designed for nuclear applications requires consideration of compatibility as well as resistance to radiation, high temperature, mechanical shear, and atmosphere.

Certain elements are unsuitable, for their presence within or close to the reactor core would seriously affect the neutron economy, or react with the fuel element cladding to cause destruction of the canning and release of fission products. Accordingly the UKAEA has restricted lubricant composition (Table 11-9).

Effect of atmosphere can be illustrated by air, which has a serious oxidizing effect, especially when coupled with radiation. Conventional antioxi-

Table 11-9 Elements on Which Restrictions Are Placed For Radiation-Resistant Lubricants Used in United Kingdom-Type Reactors Employing Magnox Fuel Cans

None allowed	Mercury.
0.1% allowed	Barium, bismuth, cadmium, gallium, indium, lead, lithium, sodium, thallium, tin, zinc.
1% allowed	Aluminum, antimony, calcium, cerium, copper, nickel, silver, strontium, praseodymium.

Note: In certain instances the above limits can be exceeded, where it can be shown that the metals are present in a stable compounded form and that practical compatibility tests are satisfied.

dants are destroyed as noted above. Some of the organic modified thickeners have an antioxidant effect and perform a dual function. Hot pressurized carbon dioxide can cause rapid degeneration of conventional soap thickened greases, presumably by carbonate formation.

In a selection of the thickener, the compatibility of thickener and base fluid is of paramount importance, for even an exceptionally radiation resistant thickener, when in combination with certain base fluids, may at best yield weak gels that soften easily; for example, a satisfactory grease structure is extremely difficult to obtain by using an Indanthrene pigment with a paraffinic bright stock.

Various nonsoap thickeners that form good grease structure with both mineral oil and synthetic fluid bases are available. These thickeners may be grouped as follows:

1. *Modified clays and silicas.* Typical of the modified clays are Bentone 34 and Baragel, which are formed by a cation exchange reaction between a montmorillonite clay and a quatenary ammonium salt. This reaction produces a hydrocarbon layer on the surface of the clay which makes it oleophilic. Finely divided silicas may be treated with silicone to render them hydrophobic, or, as with Estersil, the silica may be esterified with n-butyl alcohol.
2. *Dye pigments.* Organic toners or dye pigments are utilized as grease thickeners (e.g., Idanthrene).
3. *Organic thickeners.* Typical of this type are the substituted aryl ureas characterized by the diamide-carbonyl linkage which may be formed in situ by the reaction of diisocyanate with an aryl amine.

The behavior of these thickeners, when used in conjunction with a synthetic fluid, is shown in Fig. 11-22.

As with fluid lubricants, antirads may be added to the grease to increase its radiation stability. The effect of free radical scavengers and aromatic energy absorbers on a modified clay, synthetic fluid grease is shown in Fig. 11-23.

The comparison of radiation resistant greases with conventional soap type, mineral oil greases is shown in Figs. 11-24 and 11-25. The radiation stability of the fluids extracted from these greases is shown in Fig. 11-26.

LUBRICATION RECOMMENDATIONS

The advent of nuclear energy has added a new dimension to the requirements of lubricants and other petroleum products in industrial applications. Equipment in the nuclear industry — research and power reactors; fuel processing machinery; conveyors, manipulators, and cranes in irradiation facilities; viewing windows and shield doors in hot cells; and heat exchange units — all require oils, greases, and organic fluids to perform conventional and special functions in radiation atmospheres.

Nowhere are the operating conditions of radiation, temperature, and atmosphere more demanding than in the power reactor field. At the outset, equipment was designed to operate without conventional petroleum lubricants because little was known of the behavior of petroleum products under irradiation and the exact severity of the application was overestimated. This placed a design and economic burden on nuclear power generation. As experience was gained in the operation of these plants, the original position was reconsidered. First, specific operating parameters of radiation flux, temperature and so on, were obtained, which realistically established the requirements; second, research in the radiation resistance of petroleum

Fig. 11-22 Effect of Thickener on Grease Polymerization

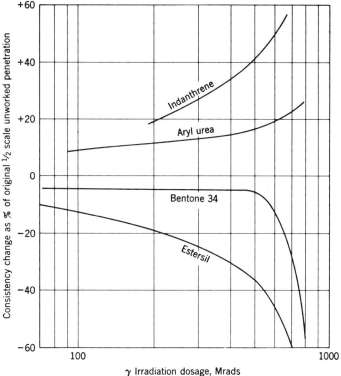

γ Irradiation dosage, Mrads

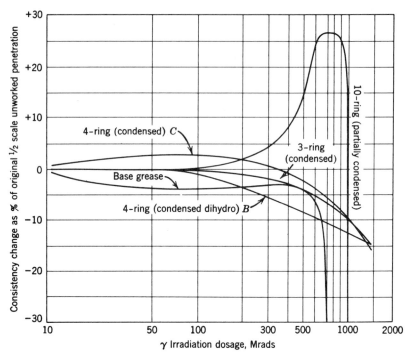

Fig. 11-23   Effect of Antirads on Grease Stability

Fig. 11-24   Effect of Radiation on Conventional Greases

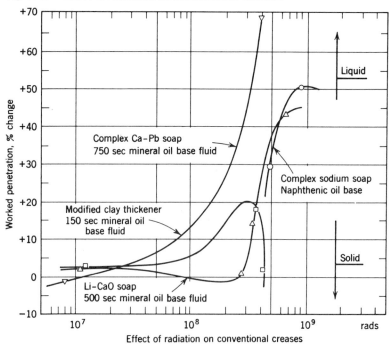

Effect of radiation on conventional creases

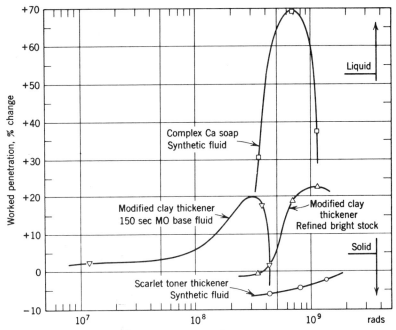

Fig. 11-25   Stability of Radiation Resistant Greases

Fig. 11-26   Radiation Stability of Extracted Base Fluids

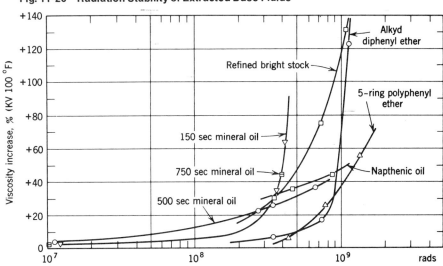

materials showed that conventional lubricants could withstand doses up to 10^7 to 10^8 rad and that new petroleum lubricants could be developed to withstand more than 10^9 rad. As a consequence of this increase in knowledge and experience, nuclear power station owners and operators requested that whenever possible, proved engineering designs employing conventional petroleum greases and lubricating fluids should be adopted. This has lead to an increase, in new generation plants, of the use of petroleum lubricants.

Because the nuclear power industry is still in a developing stage, in which patterns of design and operating conditions are changing, most equipment is unique, and each plant requires separate consideration before proper lubricants and lubrication schedules can be established. Little repetitiveness exists in equipment or operating conditions, and, therefore, the best that can be accomplished in this chapter on lubrication recommendations is to furnish the background experience and establish the guide lines so that lubricating engineers can, after a survey of specific conditions, recommend the best lubricants for a particular application.

General Requirements

All lubricating engineers are familiar with the effect of the environmental factors of speed, load, temperature, and time on the life of gears and bearings, and on the physical and chemical properties of oils and greases exposed to these environmental conditions.

In conventional applications, the effects of speed, load, and temperature are evaluated in making recommendations. Selecting the correct lubricant and service interval is determined by evaluating the lubricant's anticipated performance under the most critical of these conditions; for example, speed may be the determining factor with antifriction bearings and the proper grease must resist excessive softening for the service period. In high loading, extreme pressure properties may be the determining consideration. In most cases, however, temperature is the most critical factor. Operating severity may be measured by the degree of heat, or the temperature, and the extent of exposure, or the time interval. The additional environmental factor of radiation in nuclear applications affects lubricants in much the same manner as heat. Both are modes of energy and, as we have seen in a

Table 11-10 Nuclear Reactor Systems: Estimated Dose to Components — Rads per Year*

Component	Pressurized-Water Reactor	Boiling-Water Reactor	Organic-Moderated Reactor	Liquid Metal-Cooled Reactor	Gas-Cooled Reactor
Control-rod mechanisms	10^5–10^8	10^6 max	10^6–10^8	0–10^5	10^3–10^5
Fuel-handling devices	10^6–10^7	. . .	. . .	0–10^8	10^2–10^9
Primary coolant pumps	10^6–10^7	10^6–10^7	0–10^2	Negligible	10^3 max
Auxiliary pumps	10^4–3×10^6	10^4–3×10^6	0–10^2	0–10^8	10^4 max
Auxiliary motors, etc.	10^5–3×10^7	10^5–10^7	. . .	. . .	10^4 max
Turbogenerator and auxiliaries	Negligible to 10	400–800	Negligible to 25	Negligible to 25	Negligible to 5×10^3

*The rad, a unit of dose, is the absorption of 100 ergs of radiation energy per gram of material.

Table 11-11 Typical Operating Conditions in Components of CO_2 Cooled, Graphite Moderated Reactors

Component	Operating temperature, °C	Radiation dose (rad/yr)	Compatibility Requirements With CO_2	With Reactor Materials	Volatility Requirements	Operating Atmosphere	Pressure of Operating Atmosphere (lb/in²)	Lubricant
CO_2 gas circulators								
bearings	80 max	10^3 max	Yes	Yes	Yes	CO_2 and air	Nil–slight	Oil
seals	30–140 or higher	10^3 max	Yes	Yes	Yes	CO_2	Slight	Oil
CO_2 gas valves								
gears, bearings and hydraulics	70–150 or higher	10^3 max	No	No	No	Air	Nil–slight	Oil or grease
Control-rod mechanisms								
gears and bearings	70–150	10^3–5×10^5	Yes	Yes	Yes	CO_2	150–300	Grease; also oil for sintered bearings
Fueling machines								
bearings and gears	100–400	10^2–3×10^9	Yes	Yes	Yes	CO_2 and air	Partial vacuum up to 150–300	Grease
hydraulic mechanisms	100 max	10^4 max	No	No	No	Air	Nil–slight	Oil
cooling fan bearings	100 max	10^3 max	Yes	Perhaps	Perhaps	CO_2 and air	150–300	Grease or oil
Ancillary machines service machines (when separate from fueling machine), gears, and bearings	70–100	10^5 max under normal conditions	Yes	Yes	Yes	CO_2	Partial vacuum up to 150–300	Grease
Charge chute machines (when separate from fueling machine), gears, and bearings	100–400	10^6–10^8	Yes	Yes	Yes	CO_2	Nil up to 150–300	Grease
Reactor observation equipment (when separate from fueling machine)	100 max	Dependent on use but could be high	Yes	Yes	Yes	CO_2 and air	Nil up to 150–300	Grease
Spent fuel element conveyor systems Chains and bearings	100 max	2×10^8 max under normal conditions	No	Yes	No	Air	Nil	Grease

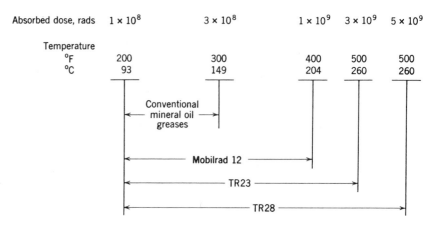

Fig. 11-27 Mobil's Radiation Resistant Greases

previous section, lubricants, like all organic materials, undergo major structural changes when certain thresholds of absorbed energy are reached. We know that petroleum products undergo thermal cracking and polymerization at certain temperatures and, likewise, that cleavage and crosslinking occur at certain radiation dosage thresholds.

In nuclear applications, radiation energy is expressed as the flux or dose rate. The particular types of radiation and the units have already been discussed. It is sufficient to state here that this dose rate applied over a specified interval will yield the absorbed dose for an exposed material. If, for example, a grease can absorb a dose of 5×10^9 rad before suffering physical or chemical changes that will render it useless as a lubricant, this grease can be used for 5000 days if the dose is absorbed at a rate of 1 Mrad a day or for only five days if the dose is 1000 Mrad a day.

Selection of Lubricants

Much has been published and extensive studies have been made that lead to the recommendation of correct lubricants. In this final analysis, however, the selection of the proper lubricant and its application to any particular equipment must be made by a lubrication engineer for each specific instance, based on the operating conditions and the type of bearing, gear, cylinder, etc., requiring lubrication. Nowhere is this more pertinent than in the lubrication of nuclear power plants. Because of the newness of the field, the uniqueness of the designs and the severity of operating conditions of temperature, radiation, and operating atmospheres, each plant is markedly different from every other. Therefore the experience of the lubricating engineer is important, and blanket recommendations serve only as guidelines to these engineers. Mobil engineers for more than 100 years have devoted their efforts to the study of lubrication requirements in thousands of plants all over the world, to the development of lubricants to meet these requirements, and to the perfection of application methods. The accumulated experience with all kinds of machinery has been helpful in numerous plants in the solution of lubrication problems and, in many cases, in the elimination of mechanical problems as well.

In selecting lubricants for a nuclear plant, the engineer should be consulted in the design phase, be cognizant of development work at equipment manufacturers, and participate in practical evaluation of prototype units under the operating conditions. In surveying the plant requirements, particular attention must be paid to the radiation flux profile which has been calculated for the various parts of the plant and compared with actual surveys during operation of similar plants. Extreme conservatism has been the rule in estimations of nuclear plant requirements, often to the detriment of practical solutions.

In estimating the radiation flux to which a lubricant will be exposed, benefit must be taken for the shielding effect of the equipment components which will attenuate the flux estimated for the surroundings. Estimated averages for the general flux levels in various nuclear power plants are shown in Table 11-10; more detailed estimates of a gas cooled, graphite moderated reactor plant of the British type are shown in Table 11-11.

The ability of Mobil lubricants to withstand radiation is shown in the graph plotted (Fig. 11-18) for two turbine oils statically irradiated and in the chart (Fig. 11-27) which shows the ranges for available greases.

12 Automotive Chassis Components

While the engine and power train generally are considered to be part of the automotive chassis, they are presented in detail in other sections of this book. The discussion in this section is concerned only with suspension and steering linkages, steering systems, wheel bearings, and brake systems.

SUSPENSION AND STEERING LINKAGES

A suspension is an arrangement of linkages, resilient members — such as coil or leaf springs or torsion bars — and shock absorbers which attach the wheels to the frame or body of a vehicle. In over-the-road vehicles, the front suspension includes pivots that permit the front wheels to be turned for steering. Most off highway and farm equipment vehicles do not have steerable front wheels. Steering is accomplished either by an articulated joint located between the front and rear sections of the machine, or by disconnecting the power and applying a brake to the wheels on the inside of the turn. Many self-propelled combines and some trucks use steerable rear wheels.

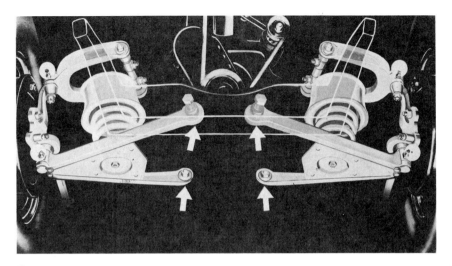

Fig. 12-1 Passenger Car Front Suspension This illustration shows one typical passenger car front suspension. Points that are normally lubricated are indicated by white circles; white arrows indicate points that may be lubricated.

Many suspension designs are used for passenger cars and trucks. Almost universally, passenger cars are equipped with "independent" front suspensions which permit the front wheels to move independently with respect to the frame and body. Although this type of front suspension has been extended to trucks, many medium and heavy trucks still are equipped with rigid front axles. Rigid rear axles are the most common. Independent rear suspensions are used on a number of high performance and luxury cars with rear wheel drive and on all cars with front wheel drive.

In one typical passenger car front suspension (Fig. 12-1), the wheels are pivoted on a pair of ball joints (Fig 12-2) seated in the outer ends of the up-

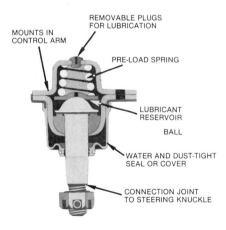

MOUNTS IN CONTROL ARM

REMOVABLE PLUGS FOR LUBRICATION

PRE-LOAD SPRING

LUBRICANT RESERVOIR

BALL

WATER AND DUST-TIGHT SEAL OR COVER

CONNECTION JOINT TO STEERING KNUCKLE

SUSPENSION BALL JOINT

Fig. 12-2 Suspension Ball Joint This type of joint is designed for periodic relubrication, at low pressure, through the removable plug. Some similar joints have a vent in the seal so that they can be lubricated through a fitting.

per and lower control arms. These joints permit the wheels to move up and down with respect to the frame and to be turned for steering. Additional pivot points at the inner ends of the control arms permit them to swing up and down against the restraint of the springs and shock absorbers. Other front suspension arrangements use ball type joints at the pivot points. Where a rigid front axle is used, usually the wheels are mounted on a yoke and kingpin (Fig. 12-3) to permit turning the wheels for steering; see Fig. 12-3.

The usual rear suspension, because there is no requirement for steering, is simpler in design than the front suspension. Either leaf or coil springs may

Fig. 12-3 Rigid Axle Front Wheel Mounting In this design the yoke is on the wheel spindle, steering knuckle assembly. In other designs the yoke is on the axle.

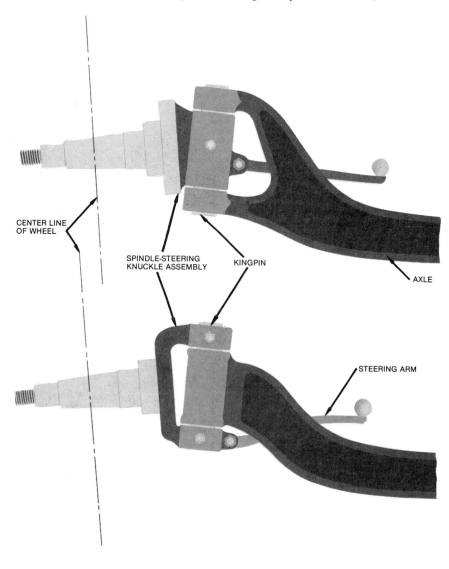

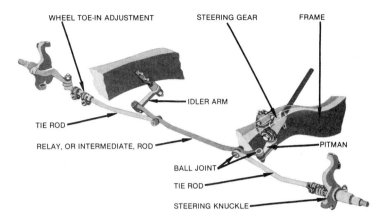

Fig. 12-4 Typical Steering Linkage Some variation of this type of design is used on the majority of passenger cars today. Where a rack and pinion steering gear is used, the rack replaces the intermediate rod and the pitman and idler arm are eliminated.

be used. Some additional complexity may be introduced with independent rear suspension to permit the wheels to move with respect to the frame while still transmitting the driving force to it.

The most common type of steering linkage used on road vehicles is the parallelogram type; see Fig. 12-4. This term is derived from the parallelism maintained between the pitman and idler arm at all linkage positions. The pitman and idler arm are connected to the intermediate rod through pivot bearings, such as those shown at the right in Fig. 12-5, all of the other connections are made through ball joints such as those shown at the left in Fig. 12-5. The pitman is operated by the steering gear which is discussed in the following section.

Fig. 12-5 Typical Steering Linkage Ball Joints The joint on the right is not truly a ball joint and is designed for rotation only in a single plane. It is suitable for connecting the intermediate link to the pitman and idler arm.

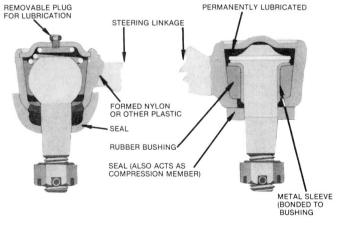

STEERING LINKAGE BALL JOINTS

Rack and pinion steering is now used on many light cars. In this design no pitman or idler arm is used. The intermediate rod is replaced by a rack which meshes with a pinion on the end of the steering shaft. This arrangement results in a simple, direct acting linkage.

Other linkage arrangements may be found on farm and off highway equipment. For example, with the row crop tractors, no linkage is used. The wheels are mounted on a single vertical shaft that is rotated directly by the steering gear. Generally, where steerable front wheels are used, some type of linkage similar in principle to the parallelogram type is used. However, hydrostatic steering without linkages or gears, as such, is also employed.

Factors Affecting Lubrication

A number of factors contribute to the lubrication problems associated with suspension and steering linkage components. Motion of the pivot points is of an oscillating nature through relatively short arcs. This does not permit the formation of fluid lubricating films and also causes wiping off of lubricant films. Since these components are located below the body, they are exposed to water (sometimes containing salt used for melting ice) and dirt. Although effective sealing reduces the possibility of entry of these contaminants, seal leakage or failure is always possible. Loads are usually high and severe shock loads are often present especially in suspension members.

During the last 10 to 15 years, procedures for lubrication of suspension and steering linkage components have changed markedly, particularly for passenger cars. Lubrication intervals have been extended with improved designs, particularly with respect to sealing, and by the use of better quality greases. A number of vehicle manufacturers have eliminated the requirement for periodic lubrication entirely, either by adopting designs that use flexible elastomeric bushings, or by coating the rubbing surfaces with a wear resistant, low friction coating. Nevertheless, most ball joints are equipped with some means of relubrication.

Two types of seals are used on ball joints designed for relubrication: umbrella seals and balloon seals. Umbrella seals permit excess lubricant to escape fairly readily. Thus, joints with umbrella seals can be equipped with fittings and lubricated with the usual high pressure grease dispensing equipment. Balloon seals without a pressure release feature can be damaged if grease is pumped into the joint under high pressure. Thus, many joints are equipped with plugs. To relubricate the joint, the plugs are removed and a special fitting is installed. Grease is then applied with a low pressure adapter. This type of seal is now being superseded by balloon seals with a pressure relief feature. With this design, lubrication fittings usually are factory installed.

Lubricant Characteristics

Most suspension and steering linkage pivot points designed for lubrication are lubricated with grease. Fluid lubricants are used in some types of off highway equipment. These applications may involve the use of central lubrication systems to replenish the lubricant automatically. Lubricants used include automotive gear lubricants of API GL-4 or GL-5 quality and semifluid greases. Rheopectic greases have also been used. These greases, which are semifluid as manufactured and dispensed, stiffen when subjected to mechanical shearing in bearings.

The grease used in suspension and steering linkage components, particularly where extended lubrication intervals are recommended, should provide:

1. Good oxidation and mechanical stability
2. Protection against corrosion by both fresh and salt water
3. Resistance to water washout
4. Good wear protection under the conditions of loading and motion involved

In addition, this grease should resist pounding and squeezing out. The ability to reduce friction decreases steering effort and provides smoother riding characteristics.

Several grease formulations are used and provide acceptable performance in these applications. Most of these formulations are similar to NLGI No. 2 in consistency and are made with an oil component approximating SAE 40 viscosity (ISO viscosity grade 150). They are intended for multipurpose use. Usually, this oil viscosity is considered optimum for wheel bearing greases. In addition, special applications, such as wheels equipped with disc brakes, require greases that are resistant to high temperatures. Lithium soap greases, which are used to a considerable extent, are modified by the addition of extreme pressure and antiwear agents to provide better load carrying ability and lower wear rates. Fine particle size solids such as molybdenum disulfide or polyethylene are often added, particularly when the grease is intended for use in passenger car suspensions. These materials generally reduce both friction and wear. Calcium complex greases, which have inherently good extreme pressure properties, are commonly used as are calcium complex greases containing molybdenum disulfide. Barium soap greases containing molybdenum disulfide are used primarily for initial lubrication of ball joints during assembly since they are difficult to dispense and may stiffen at low temperatures.

STEERING GEAR

The steering gear, which transmits motion from the steering wheel and steering shaft to the pitman of the steering linkage, generally employs some type of worm acting against short levers or a gear segment connected to the pitman shaft. In cam and lever type steering gears, the worm is referred to as the cam. It meshes with pins on a lever attached to the pitman shaft. The pins may be either solid or mounted on roller bearings to reduce friction. In other variations, a segment of, or a complete worm wheel meshing with the worm, or a throated worm meshing with a roller in the shape of a short worm may be used. Probably the most common steering gear design used in passenger cars is the recirculating ball type in which the worm drives a ball nut through a series of steel balls; see Fig. 12-6. The path of the balls includes an external guide tube that permits them to recirculate freely in either direction. This arrangement provides low friction between the steering shaft and ball nut, helping to keep steering effort low.

Many cars today are equipped with rack and pinion steering; see Fig. 12-7. In this design, a small pinion on the end of the steering shaft meshes

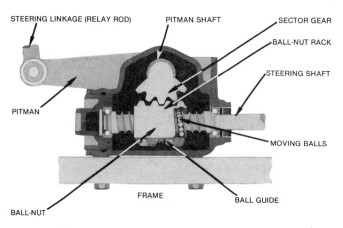

STEERING LINKAGE (RELAY ROD)   PITMAN SHAFT   SECTOR GEAR

BALL-NUT RACK

STEERING SHAFT

PITMAN

MOVING BALLS

FRAME   BALL GUIDE

BALL-NUT

Fig. 12-6   Recirculating Ball Type Steering Gear   As indicated, the recirculating balls act as a low friction method of transferring the motion of the steering shaft to the ball nut.

with a rack mounted in a guide tube. The tie rods are connected directly to the rack through a mounting bracket. This simple linkage produces better "road feel" for the driver and may reduce steering effort.

Power steering, or power assisted steering, is now widely applied to all types of automotive equipment. Generally, two types are used: the linkage type (Fig. 12-8) and in the integral type (Fig. 12-9). However, in some cases, rack and pinion power steering may also be used. Although these illustrations show passenger car applications, the principles are applicable to systems used on other equipment.

Fig. 12-7   Rack and Pinion Steering Gear

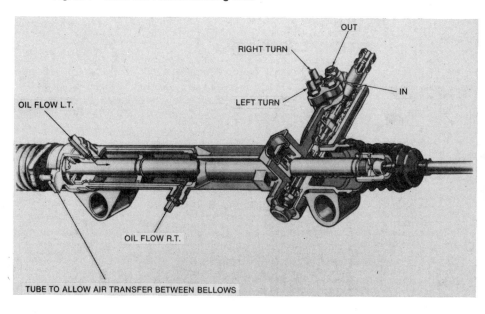

OUT

RIGHT TURN

IN

OIL FLOW L.T.

LEFT TURN

OIL FLOW R.T.

TUBE TO ALLOW AIR TRANSFER BETWEEN BELLOWS

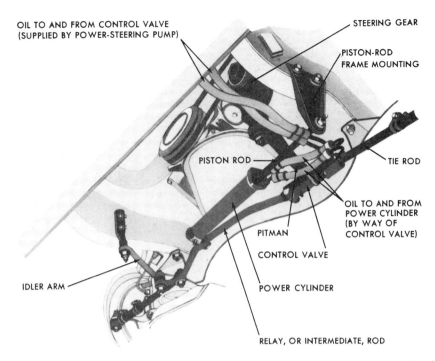

OIL TO AND FROM CONTROL VALVE
(SUPPLIED BY POWER-STEERING PUMP)

STEERING GEAR

PISTON-ROD
FRAME MOUNTING

PISTON ROD

TIE ROD

OIL TO AND FROM
POWER CYLINDER
(BY WAY OF
CONTROL VALVE)

PITMAN

CONTROL VALVE

IDLER ARM

POWER CYLINDER

RELAY, OR INTERMEDIATE, ROD

Fig. 12-8 Linkage Type Power Steering The control valve may be incorporated in the cylinder, rather than being a separate unit as shown here. The valve opening is proportional to the force applied by the pitman, so the amount of hydraulic assist provided is proportional to the turning force applied to the steering wheel.

In linkage type power steering, a double acting hydraulic cylinder is attached between a point on the frame and the intermediate rod. A control valve is mounted so that it is operated by the pitman arm. The control valve is connected to the pitman arm through a ball joint. Steering load deflects the spool valve allowing power steering fluid to flow into the appropriate reaction chamber in the power cylinder. The amount of power assistance (pressure) is proportionate to the spool valve deflection which increases with the steering effort. High pressure hydraulic fluid is supplied from a small pump mounted on and driven by the engine. When the pitman moves in one direction, as a result of turning the steering wheel, the control valve allows fluid under pressure to flow into the correct end of the cylinder. This provides the hydraulic assist which forces the intermediate rod in that direction. Turning in the other direction applies assist in that direction. When the steering wheel is centered, the valve is also centered and fluid can flow freely in and out of either end of the cylinder. With this arrangement the steering gear itself is a conventional type, although the ratio may be somewhat lower than when the hydraulic assist is not installed. The steering gear is lubricated separately with a special lubricant. The power steering fluid serves only as the hydraulic medium.

Integral type power steering systems are the most common type in use today. In these systems, when developed from a recirculating ball type

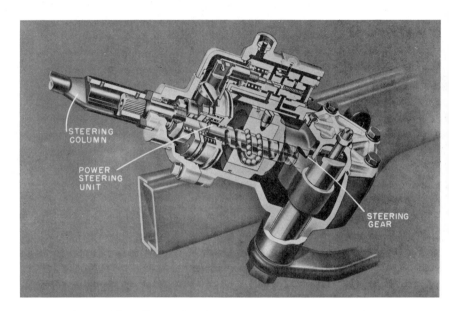

STEERING
COLUMN

POWER
STEERING
UNIT

STEERING
GEAR

Fig. 12-9 Integral Type Power Steering In this design, longitudinal reaction forces from the worm act on reaction members to control the opening of the control valve and the amount of assist provided. In the other common design, the control valve is actuated by a torsion bar in which the amount of twist is proportional to the turning force applied to the steering wheel. With both designs, the vehicle can be steered if no power is available since there is a mechanical connection between the steering shaft and the ball nut.

steering gear, the ball nut is also a double acting piston. Hydraulic fluid under pressure is admitted to one end, or the other, by means of a control valve. The control valve can be actuated by reaction forces along the steering shaft, or by torsional forces acting on a torsion bar which rotates a spool valve. In either case, the amount the valve opens, and the power assist provided, is designed to be proportional to the amount of force applied to the steering wheel. In this way, the driver's 'feel' of the road is retained.

With both types of power steering, the steering shaft is connected mechanically through the steering gear. Thus, if fluid pressure is not available, either because the engine is not running or a system failure has occurred, the vehicle can still be steered.

In the usual automotive practice, hydraulic fluid is supplied from an integrated pump and reservoir mounted on the engine. Some use has been made of the central hydraulic system to supply a number of mechanisms such as the power steering, brakes, and windshield wipers. A number of projections suggest, that this approach may be more widely used in the future. Power steering systems on farm and construction equipment may be supplied from a separate system, but generally they are supplied from the main hydraulic system of the machine.

Factors Affecting Lubrication The lubricant in the steering gear is required to lubricate the gears and bearings. It must withstand shock loads which are transmitted to the steering gear from the wheels hitting bumps and obstructions, and it must resist the wiping action of the gear teeth. At the lowest expected operating temperatures, it must not exhibit excessive resistance to motion, yet it must have enough viscosity to lubricate properly at the highest operating temperatures reached.

The fluid in integral power steering systems serves both as the hydraulic fluid and as the lubricant for the gears. Thus, it must perform all of the functions of the lubricant in conventional steering gears with the additional requirement of being a highly stable, shear resistant hydraulic fluid.

The conditions under which the power steering pump operates make the hydraulic fluid service relatively severe, particularly in passenger cars. The operating speed of the pump at maximum engine speed may be 6 to 8 times what it is at idle speed, yet at idle speed the output must be sufficient to provide whatever power assist is required. This means that at high speeds the pump output far exceeds the requirements of the system and the excess fluid must be recirculated internally through a flow control valve and the reservoir. Since the quantity of fluid must be kept low to keep the system compact, in operation the fluid temperature may be quite high and severe shearing of the fluid may occur. Under these conditions, wear of the pump vanes or rollers may be a problem.

Lubricant Characteristics

Most U.S. steering gear manufacturers fill their units with a semifluid grease to minimize leakage during shipment and installation. Some automotive units, however, are filled with a multipurpose gear lubricant of API GL-4 or GL-5 quality. For tractors and similar equipment, the lubricant may be a multipurpose fluid designed for use in tractor hydraulic systems, transmissions, and final drives. Except in the latter case, the lubricant used for makeup in the field is usually a multipurpose gear lubricant of about SAE 80W or 90 (ISO viscosity grades 68 or 150).

Automatic transmission fluids were originally used in passenger car power steering systems. However, in some cases these fluids permitted excessive pump wear and in other cases seal compatibility problems were encountered. Special fluids were developed to overcome these problems, and the majority of units are now operated on one of these fluids. Generally, a small amount of automatic transmission fluid may be used as field makeup, but if a complete refill is required, the special power steering fluid should be used. For makeup and refill, car manufacturers' recommendations should be followed.

WHEEL BEARINGS

Wheel bearings for automotive equipment are of the rolling element type. Since the bearings of steerable wheels must carry considerable thrust loads in addition to the radial loads, they are of angular contact ball, taper, or spherical roller type. Usually, the bearings are installed in pairs so that each bearing is subjected to thrust loads in only one direction. Bearings for non-steerable wheels are not subjected to as high thrust loads, thus deep groove ball bearings or cylindrical roller bearings may be used and, in many, only a single bearing is required.

Most wheel bearings are designed for grease lubrication which is done by periodic repacking, as in the case of many passenger car rear wheel bearings, by being "packed for life" on assembly. Some wheel bearings on auxiliary motive equipment have fittings for relubrication. The bearings of driving wheels are sometimes lubricated by gear lubricant from the final drive housing, and no periodic relubrication from an external source is required. Oil lubrication of nondriving wheel bearings is also used to some extent on trucks, trailers, and off highway equipment. Lubrication with oil requires careful attention to sealing to prevent leakage which might find its way into the brakes and cause brake failure. Use of a grease reduces the leakage tendencies and simplifies the sealing problem.

Front wheel bearings of rear wheel drive cars, trucks, and buses, as well as the rear wheel bearings of front wheel drive cars normally are designed to be removed periodically for cleaning and relubrication. Repacking can be done by hand or with a bearing packer. Generally, the latter is preferred since it is quicker, somewhat less wasteful of grease, and less skill is required to accomplish a satisfactory packing job. On assembly, the shaft and housing should be coated lightly with grease as a corrosion prevention measure; however, the housing should never be filled with grease as this may cause overheating due to churning.

However, only specially trained personnel should perform the entire wheel bearing repacking. The final adjustment of the wheel bearing, running clearance, or preload, is critical and specifications vary from one vehicle to another. Further, removal of the disc brake caliper when necessary requires mechanical expertise.

Lubricant Characteristics

Greases for wheel bearings are expected to provide acceptable performance over long service intervals. They must resist softening and leakage, and hardening which could cause increased rolling resistance and heating. They should also provide good rust and corrosion protection as well as protection against friction and wear.

In stop-and-go city driving, the use of disc brakes may cause higher operating temperatures for the wheel bearings than with drum brakes. As a result, greases with higher dropping points and better high temperature stability may be required for disc brake equipped vehicles.

Traditionally, wheel bearing greases were specialty short fiber, sodium soap products. Today, many of the multipurpose automotive greases have proven extremely satisfactory for wheel bearing lubrication. Generally, single purpose wheel bearing greases are now disappearing from use. All of the types of multipurpose greases mentioned as being suitable for suspension and steering linkage components are being used for wheel bearing lubrication, usually in the NLGI No. 2 or No. 3 consistency. Where higher temperature capability is required for disc brake equipped vehicles, both lithium complex and calcium complex greases are used.

Oil lubricated wheel bearings usually are lubricated with a multipurpose gear lubricant of SAE 90 or 140 (ISO viscosity grades 150 or 460).

Two excellent publications on wheel bearing lubrication, "Recommended Practices for Lubricating Automotive Front Wheel Bearings" and "Recommended Practices for Lubricating Truck-Wheel Bearings," are available from the National Lubricating Grease Institute, 4635 Wyandotte Street, Kansas City, MO 64112.

BRAKE SYSTEMS

Most service brake systems are hydraulically operated; see Fig. 12-10. U.S. National Highway Traffic Safety Administration (NHTSA) regulations require a dual or two-sided system so that some braking capability is retained if a failure occurs on either side of the system. Common practice is to operate the front wheel brakes from one side of the system and the rear wheel brakes from the other side. Also used are "dual diagonal" systems in which a front wheel and the diagonally opposite rear wheel are paired.

Operation of the brake pedal forces fluid from the master cylinder under considerable pressure to the wheel cylinders. In turn, the wheel cylinders act through connecting links to force the brake shoes out against the surfaces of the brake drums, or to force brake pads against the sides of a disc. A mechanical linkage to the rear wheel brakes usually is provided to serve as a parking brake.

Disc brakes are widely used on the front wheels of passenger cars and are being installed on all four wheels on an increasing number of models. Compared to drum brakes, disc brakes are most resistant to brake "fade" or loss of effectiveness due to heat buildup with repeated applications and, usually, provide straighter stopping. Since higher application pressure is required, disc brakes are nearly always equipped with a power assist system to reduce the pedal pressure required.

Generally, power assist systems involve a double acting piston, or diaphragm, coupled to the master cylinder piston. Normally, both sides of the piston are under vacuum supplied by engine manifold vacuum through a reservoir. When the brake pedal is depressed, a valve is opened to allow

Fig. 12-10   Hydraulic Brake System

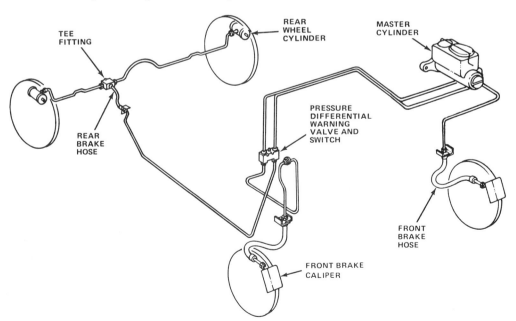

atmospheric pressure to act on one side of the piston assisting the pedal action in moving the master cylinder piston.

Hydraulic retarders for dynamic braking are used on some heavy vehicles equipped with automatic transmissions. A torque converter does not transmit power well in the reverse direction, so with an automatic transmission, the engine cannot be used as effectively for dynamic braking. With a hydraulic retarder, which operates as a torque converter coupled in the opposite direction, energy from the wheel is converted into heat in the fluid in the retarder, and then dissipated to the atmosphere.

Another type of assistance called power boost or "hydroboost" is a hydraulically operated power brake. The hydraulic booster consists of an open center spool valve and a hydraulic cylinder combined in a single housing. A dual master brake cylinder bolted to the booster is actuated by a push rod projecting from the booster cylinder. The power steering pump provides the hydraulic fluid under pressure to the booster cylinder. The master brake cylinder and the braking system use conventional brake fluid.

Oil immersed or "wet" brakes are used on some crawler tractors as well as on many other types of tractors and heavy equipment. These brakes operate submerged in the lubricant for the final drive. Actuation is hydraulic, using fluid from the main hydraulic system on the machine.

Fluid Characteristics   Primarily, the fluid in brake systems is a hydraulic fluid but it must also lubricate the elastomer seals in the wheel cylinders, and protect system metal parts against rust and corrosion. In practice, it has been found to be nearly impossible to exclude moisture completely from brake systems. To prevent this moisture from collecting and freezing in cold weather or causing rust and corrosion, which can cause brake failure, the usual approach is to use brake fluids that are miscible with water. Petroleum base fluids do not meet this requirement, so most brake fluids are based on glycols. Although some vehicle manufacturers have designed brake systems to operate on mineral oil or silicone fluids, the practice is not widespread.

In operation, considerable heat is transmitted from the friction surfaces of the brakes to the fluid. If this raises the temperature of the fluid sufficiently to cause vaporization of some of the fluid, braking effectiveness will be reduced because of the compressibility of the vapor. The problem may be more severe with disc brakes because the smaller area and higher pressures of the friction pads may result in higher operating temperatures. With higher brake pad temperatures, more heat is transmitted through the wheel cylinders to the brake fluid.

The absorption of moisture by a brake fluid lowers the boiling point of the fluid, since water boils at a lower temperature than glycols used as a base for brake fluids. For this reason, brake fluid specifications usually include both an equilibrium reflux boiling point, which is the boiling point under reflux conditions of the pure fluid, and a wet equilibrium reflux boiling point, which is the boiling point of the fluid when it is contaminated with a specified amount of water. To some extent, the amount that these boiling points can be raised is restricted because higher boiling glycols also have higher viscosities at low temperatures and may not provide satisfactory brake performance in cold weather.

The most widely accepted brake fluid specification is the U.S. Federal Motor Vehicle Safety Standard (FMVSS) No. 116 for Grade DOT 3 fluids. Fluids meeting this standard generally are suitable for all normal brake systems designed for nonpetroleum fluids. Some manufacturers specify a higher boiling fluid for vehicles with disc brakes; and higher boiling fluids may also be required in certain types of severe service, such as mountain operations, particularly in hot climates, and road racing.

The SAE Standard J1703e "Motor Vehicle Brake Fluid" is generally similar to the DOT 3 Standard, but is somewhat less restrictive in its requirements.

Where mineral oil or silicone base fluids are used, usually the fluids are developed specifically to meet requirements set out by the brake system manufacturer.

In service, it is important that every precaution be taken to keep brake fluids as clean and free of moisture as possible. Containers should be kept sealed when not in use and should never be reused. Master cylinder reservoirs should be kept filled to the proper level since this minimizes the amount of breathing that can occur. Reservoir caps should be replaced properly after the fluid level is checked or fluid is added. A number of manufacturers now use translucent plastic reservoirs so the fluid level may be observed without removing the cap. This method reduces the possibility of contaminants being introduced into the system.

MISCELLANEOUS COMPONENTS

The "fifth wheel" of tractor-trailer combinations requires effective lubrication to assure proper vehicle control. The grease used must have good adhesion and water resistance to resist wiping and removal by exposure to the elements. A variety of greases are used; however, the best results are obtained with greases having antiwear and EP characteristics. The addition of molybdenum disulfide to the grease is generally beneficial.

Refrigerator trucks and trailers have small refrigeration units, usually driven by a small gasoline engine. The refrigeration compressor is of the sealed type and does not require periodic oil additions. The gasoline engine is usually lubricated with the same oil as used in the vehicle engine.

13 Automotive Power Transmissions

In automotive equipment, the power developed by the engine must be transmitted to the drive wheels to propel the vehicle. This is accomplished by the power train, the elements of which vary from application to application. In probably its simplest form, a bicycle equipped with an auxiliary engine, the power train may consist only of a belt drive with an idler pulley that can be actuated to engage or disengage the drive. At the other extreme, the power train may consist of a clutch or coupling, a transmission, transfer case, inter-axle differential, propeller shafts, front and rear differentials — possibly in tandem at the rear, driving axles — and planetary reducers at the wheel ends of the axles. Almost any arrangement between these two extremes will be encountered in some type of automotive equipment.

In addition to transmitting the power from the engine to the drive wheels, the power train performs several other functions. For example, the power train provides:

1. Mechanism for engaging and disengaging the drive so the vehicle can be started and stopped with the engine running

2. Torque multiplication (speed reduction) so that sufficient torque is available to start from rest, accelerate, climb hills, and pull through soft ground
3. Mechanism for reversal of direction
4. Mechanism that allows one wheel to be driven at a higher or lower speed than the other when the vehicle is negotiating curves and turns
5. Except in vehicles with transversely mounted engines, the change in direction of torque and power flow to couple a longitudinally mounted engine to the transverse drive axle

The components of power trains can be considered conveniently starting from the engine and progressing toward the drive axle.

CLUTCHES

In order to engage and disengage the various gear ratios in a mechanical transmission, provision must be made to disconnect the engine from the power train. This is accomplished with a clutch.

Most road vehicles with mechanical transmissions are now equipped with a single plate, dry disc clutch. A machined surface on the flywheel serves as the driving member. The driven member is usually a disc with a splined hub that is free to slide lengthwise along the splines of the transmission input shaft (clutch shaft), but which drives the input shaft through these same splines. About 65 percent of the clutch plate area is faced on both sides with friction material. A typical passenger car clutch plate, for example, with a 15 cm (10 in.) diameter would have a band of friction material about 6 cm (4 in.) wide along its outer portion. When the clutch is engaged, the pressure plate (driven members) is clamped against the driving member by a diaphragm type spring or an arrangement of coil springs; see Fig. 13-1. The entire mechanism usually is called the pressure plate and cover assembly. To disengage the clutch, pressure is applied through mechanical linkage or a hydraulic system on the end of the throwout fork that pivots on a support in the clutch housing. When release pressure is applied, the fork transmits the force to the release levers in the cover assembly which compresses the springs and retracts the pressure plate from the driven member.

On models with hydraulic clutch activator, the system reservoir should be kept filled to the correct level. Usually, conventional brake fluid is specified. Always follow manufacturer's recommendations. The only lubricated part of this type of clutch is the throwout bearing. In most cases, these bearings are "packed for life" on assembly and do not require periodic relubrication. In a few instances, these bearings are equipped with a fitting and require periodic relubrication, usually with a multipurpose automotive grease. Also, various pivot points in the actuating linkage may require periodic lubrication, with either a small amount of engine oil or a multipurpose automotive grease.

Multiple plate clutches, usually of oil immersed or "wet" type, are used in some tractors and offhighway machines. With these clutches, the frictional properties of the fluid surrounding the clutch are extremely important if the clutch is to engage smoothly, resist slipping, and provide extended service life. These fluids are discussed in this chapter.

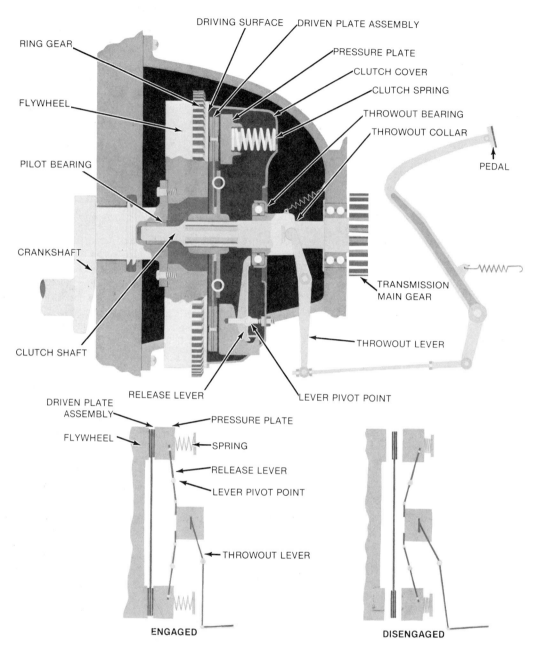

Fig. 13-1 Single Plate, Dry Disc Clutch

Single plate, double acting, dry disc clutches are used with some torque converter transmissions for buses and similar applications. Two driven members are used, one on each side of the driving member. When engaged in one direction, the clutch connects the drive to the torque converter which in turn drives the input shaft of the transmission. When engaged in the other direction, it connects the drive to a through shaft which provides a mechanical drive to the transmission, either to the input or output shaft depending on the arrangement. Again, the only lubrication required is for the throwout bearing, which is located so that it is lubricated by the torque converter fluid.

TRANSMISSIONS

The primary functions of the transmission are to:

1. Provide a method of disconnecting the power train from the engine so the vehicle can be started and stopped with the engine running.
2. Provide torque multiplication for conditions where greater driving torque is required at the wheels than is available from the engine.
3. Provide a method of reversing the drive.

Since torque multiplication is accompanied by speed reduction at the output end, the transmission also permits the operator to select different travel speeds for any given engine speed.

There is probably more variation in transmission design and application than in any other automotive component. For ease of discussion, transmissions can be considered as mechanical, automatic, semiautomatic, or hydrostatic.

Mechanical Transmissions

A mechanical transmission is an arrangement of gears, shafts, and bearings in a closed housing such that the operator can select and engage sets of gears that give different speed ratios between the input and output shafts. In most cases, a set of gears that can be engaged to drive the output shaft in the opposite direction is also included. For a constant power, torque increases as speed is decreased; the transmission provides a series of steps of torque multiplication.

In an elementary sliding element transmission (Fig. 13-2) one of each pair of gears is splined onto its shaft in such a way that it can be moved along it by a shift fork into and out of mesh with its mating gear. Drive is from the input shaft (also called the clutch shaft) through the main gear to the countershaft. The output shaft ends in a pilot bearing in the main gear, which is free to revolve at a different speed or in a different direction than the input shaft. Thus, the output shaft is driven from the countershaft by whichever pair of gears is engaged, or directly from the main gear if the direct drive gear is engaged with the internal gear in the main gear.

Sliding element transmissions are used now only in low speed applications, such as tractors. In this application, the clutch must be disengaged and the vehicle at a complete stop before the gears can be engaged or the gear ratio changed. For other applications, synchromesh or synchronized transmissions are used. In this type of transmission, all gears are always in mesh, except the reverse gear. One of each pair of gears is free to revolve on its shaft unless locked to it by a clutching mechanism called a synchronizer; see Fig. 13-3. The synchronizer, which is keyed or splined to the shaft, consists of a friction clutch and a dog clutch. As the shift fork moves the synchronizer toward the gear, the friction cones make contact first to bring the shaft to the same rotational speed as the gear. The outer rim of the clutch

Fig. 13-3 Elementary Synchronizer After the friction cones make contact to bring the gear that is to be engaged up to synchronous speed, the clutch sleeve slides past the ball lock to engage with the dogs on the gear. Once engaged, the ball lock holds the clutch sleeve in position until it is moved on disengagement by pressure from the shift fork.

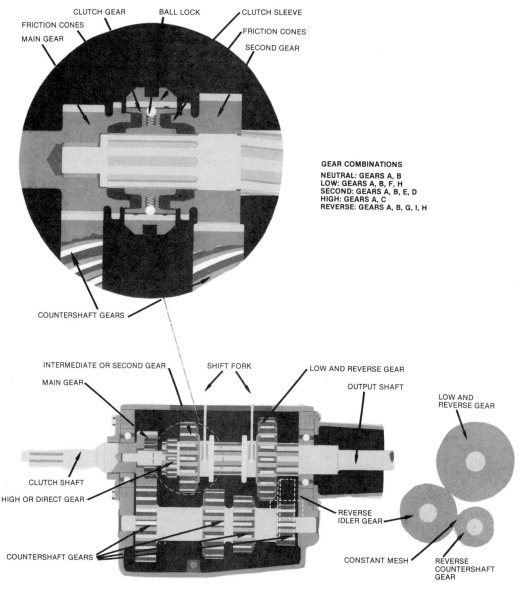

GEAR COMBINATIONS
NEUTRAL: GEARS A, B
LOW: GEARS A, B, F, H
SECOND: GEARS A, B, E, D
HIGH: GEARS A, C
REVERSE: GEARS A, B, G, I, H

Fig. 13-2 Elementary Sliding Element Mechanical Transmission The shift forks move the gears into and out of mesh along the splined main shaft. When the C-D combination is moved to the left, gear C meshes with the internal gear in the main gear A so that the output shaft is driven directly by the input shaft. If this gear combination is moved to the right so that gear D meshes with gear E, drive is from the main gear A to countershaft gear B, and from countershaft gear E to gear D. The main shaft will then be driven at a lower speed than the input shaft speed. Similarly, gear H can be moved to the left to engage with gear F for a greater speed reduction, or to the right to mesh with the reverse drive gears.

gear then slides over its hub causing a set of internal teeth to engage with a set of teeth (dogs) on the side of the gear. This then provides a positive mechanical connection between the gear and shaft.

Usually, synchronizers are equipped with a blocking (also called baulking) system to prevent engagement of the dog clutches until the gear and shaft speeds are fully synchronized. Generally, this is a spring loaded mechanism which keeps the teeth on the synchronizer from lining up with the teeth on the gear as long as there is any slip in the friction clutch.

Mechanical transmissions are built with up to about six gear ratios. Where more ratios are required, as in the case of heavy trucks equipped with diesel engines, they are usually obtained by means of a two or three speed, auxiliary transmission mounted behind the main transmission. With this arrangement, for each ratio in the main transmission there are two or three ratios in the auxiliary; so, for example, a four speed main transmission with three speed auxiliary becomes a 12 speed transmission. Heavy duty transmissions are often built with twin countershafts to decrease gear tooth loading.

Sliding element transmissions are built with straight spur gears. Synchromesh transmissions for over-the-road vehicles are usually built with helical gears, both because they provide greater load carrying capacity and because they operate more quietly. Transmission for off highway equipment may be built with either type of gearing.

Automatic Transmissions   Early passenger car automatic transmissions were built with a fluid coupling and a hydraulically operated power shift gear box. The fluid coupling permitted enough slip with the engine idling so the vehicle could be stopped with the gears engaged. Power transfer efficiency was also good. However, since a fluid coupling does not multiply torque, the gear box required four forward speeds to provide a smooth progression of gear ratios and this resulted in considerable complexity. As efficient hydraulic torque converters were developed they replaced the fluid couplings. Gradually, the gear boxes have been standardized on an arrangement with three forward speeds and a reverse speed. Truck automatic transmissions may have more forward speeds, and may also have an arrangement to lock out or bypass the torque converter when the transmission is in any gear except first or reverse. Transit coach transmissions may have only a drive through the torque converter or direct drive. All of these transmissions are sometimes referred to as "hydrokinetic" transmissions since engine power is transmitted by the kinetic energy of the fluid flowing in the torque converter.

Torque Converters   The simplest single stage, torque converter consists of three elements: a centrifugal pump, a set of reaction blades called a stator, and a hydraulic turbine; see Fig. 13-4. These three elements are installed inside a case filled with a hydraulic fluid. The pump is driven by the engine, and the turbine drives the input shaft of the gear box. The pump blades are shaped so that they discharge the fluid at high speed and in the correct direction to drive the turbine. As the fluid flows out of the turbine, it strikes the fixed stator blades and is redirected into the inlet side of the pump, where any velocity it still retains is added to the velocity imported to the fluid by the pump. With this arrangement, most of the power delivered to the

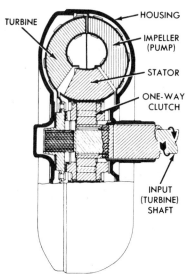

TURBINE

HOUSING

IMPELLER (PUMP)

STATOR

ONE-WAY CLUTCH

INPUT (TURBINE) SHAFT

Fig. 13-4 Three Element Torque Converter

pump is available to drive the turbine (some power is lost due to fluid friction), and as long as the turbine is running at a lower speed than the pump, torque multiplication will occur. Most single stage, torque converters are designed for maximum torque multiplication ratio of slightly more than 2:1 which, due to maximum load conditions, occurs under "stall" conditions when the turbine is stationary.

Torque converters can be built with more than one stage, that is, additional pumps and stators in pairs to give greater torque multiplication. However, this is done less frequently now, and the majority of units being built is single stage.

A torque converter does not transmit power very efficiently when the speed of the turbine reaches approximately the speed of the pump. To improve this power transfer efficiency, the stator is mounted on a one-way or overrunning clutch; see Fig. 13-5. With this addition, when the "coupling stage" is reached, the stator revolves ("free wheels") with the turbine and the whole assembly performs as fluid coupling.

The efficiency of power transfer through the converter when it is operating normally in the coupling phase is fairly satisfactory. However, to gain some percentage points in fuel economy, a number of car manufacturers have introduced torque converter "lockup" devices that eliminate all slippage in the coupling phase. The lockup mechanisms are designed to be effective when the transmissions are in direct drive and converter torque multiplication is not required.

At low engine speeds, the torque transmitted by a torque converter is low enough that it is insufficient to move the vehicle or will cause only a small amount of creep. This feature permits the vehicle to be stopped without disconnecting the engine from the power train.

Planetary Gears Passenger car automatic transmissions are built with planetary and neutral gear sets to provide the additional torque multiplication, reversal of direction, and neutral. This type of gearing is also used in some truck and heavy equipment automatic transmissions. Planetary gearsets have several advantages for these applications:

1. Ratio changes and reversal of direction can be accomplished through these constant mesh gears by locking or unlocking various elements of the gearset.
2. The gears are coaxial; thus they provide a compact arrangement.
3. In addition, good load carrying ability can be obtained from a relatively small gearset. The coaxial construction of planetary gears contains most of the operating loads. This allows the use of thin, light-weight aluminum die cast housings because they do not have to withstand extreme mechanism loads. A simple planetary gearset is shown in Fig. 13-6, and some of the various possible arrangements are shown in Fig. 13-7.

At A in Fig. 13-7, the gearset is shown with the planet carrier locked to the sun gear. With this arrangement, the sun gear, acting as the input, will drive the gear carrier and in turn the planet gears will drive the annular gear (shown as the output) all at the same speed so the unit operates as a solid coupling. The rule is that if any two members of a planeting gearset are coupled, the gearset will rotate as a unit at a 1:1 ratio. At B in Fig 13-7, the annular gear is locked to the case and the sun gear is driving (input). The carrier will be driven (output) in the same direction as the sun gear but at a lower speed. The unit then operates as a speed reducer or torque multiplier. The same type of operation can be obtained by locking the sun gear while using the annular gear as input. Now the planet carrier will also rotate in the same direction at a lower speed, but at a different ratio. At C with the planet carrier locked to the case and the drive through the annular gear, the planet pinions, which are free to revolve on their shafts, will drive the sun gear in

Fig. 13-5 Overrunning Clutch When a clockwise force is applied to the movable member, the rollers wedge on the ramps to prevent rotation. When the force is released, the rollers move back down the ramps permitting the movable member to rotate freely in the clockwise direction.

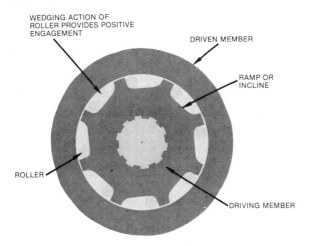

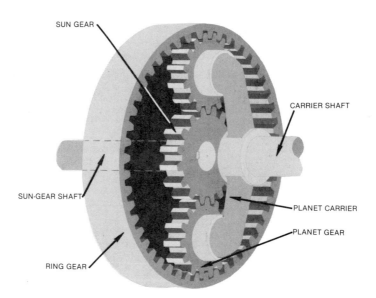

SUN GEAR

CARRIER SHAFT

SUN-GEAR SHAFT

PLANET CARRIER

PLANET GEAR

RING GEAR

Fig. 13-6 Planetary Gear Set In this arrangement only two planet gears are used. Automatic transmission gear sets commonly have either three or four planet gears.

the opposite direction to the annular gear and at a lower speed. The unit then operates as a reverse speed reducer. In automatic transmission applications, various arrangements of oil-immersed plate type or one way clutches and brake bands are used to engage the drive to the various elements, or lock elements to the case.

A single planetary gearset can provide direct drive, two stages of forward speed reduction, and reverse. Though not shown in the illustration, a planetary gearset can also be operated in the overdrive phase. By locking the sun gear and using the planet carrier as input, the annular gear output rotation speed is increased. On earlier model cars, this overdrive effect for increasing fuel economy has been very successful. This overdrive effect has not been successful on automatic transmissions. With the mounting pressures of today's energy shortage, at least one U.S. car company is now providing an automatic transmission with the overdrive feature. More are likely to follow this trend in the near future. To provide the three forward speeds (including direct drive) now used on most passenger car automatic transmissions, either two planetary gearsets or a compound planetary gearset with two sets of planet pinions and carriers are used. A typical, current, three speed transmission is shown in Fig. 13-8.

Transmission The gear box of an automatic transmission is a "power shift" gear box. That is, it can be engaged, or the gear ratio changed, while engine power transmitted by the torque converter is being applied continuously to the input shaft. As indicated in the preceding discussion of planetary gears, these operations are performed by engaging and disengaging clutches in the drive lines to various planetary elements and by applying and releasing brakes to lock or unlock elements. The clutches and brakes are operated by hydraulic servo mechanisms, which are controlled by a complex valve ar-

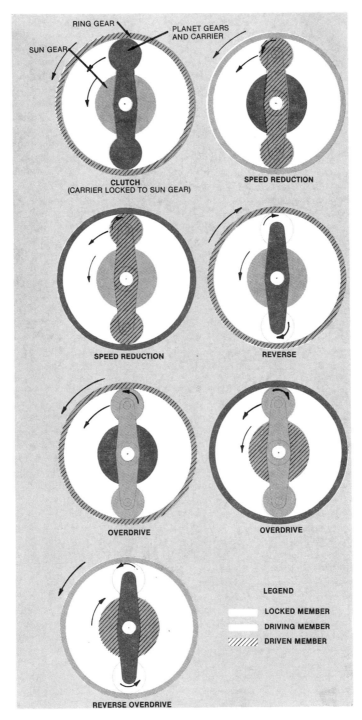

Fig. 13-7 Planetary Gear Operating Modes When any two members are locked together, the unit acts as a solid coupling. Other arrangements involve locking (grounding) one of the elements to the case to prevent it from rotating.

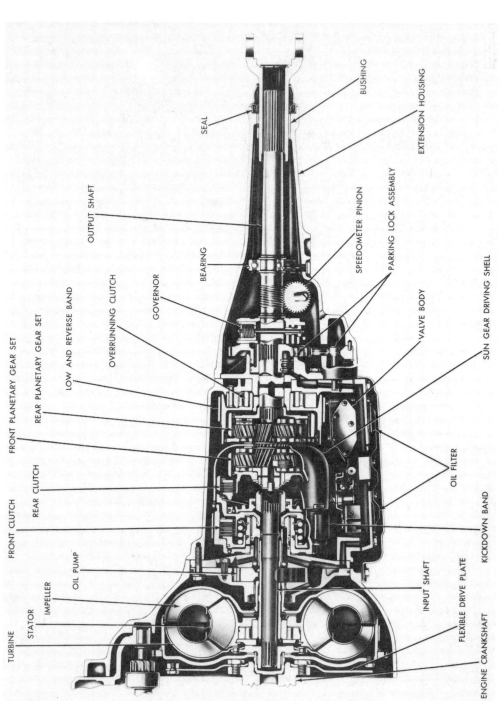

SEAL

BUSHING

EXTENSION HOUSING

OUTPUT SHAFT

BEARING

SPEEDOMETER PINION

PARKING LOCK ASSEMBLY

GOVERNOR

OVERRUNNING CLUTCH

LOW AND REVERSE BAND

REAR PLANETARY GEAR SET

FRONT PLANETARY GEAR SET

FRONT CLUTCH

REAR CLUTCH

VALVE BODY

SUN GEAR DRIVING SHELL

OIL FILTER

KICKDOWN BAND

INPUT SHAFT

OIL PUMP

IMPELLER

STATOR

TURBINE

FLEXIBLE DRIVE PLATE

ENGINE CRANKSHAFT

Fig. 13-8 Cutaway of Typical Automatic Transmission

rangement. Hydraulic pressure to operate the servos is provided by an auxiliary pump, usually of the internal-external gear type, mounted in the front end of the gear box.

In operation, the operator selects a driving range and the gear ratio changes are made automatically within that range. Basically, these shifts are determined by a speed sensing device on the output shaft of the gear box. Throttle position and, in some models, engine vacuum are used to modulate the shift speeds. The farther the throttle is depressed, the higher the speeds the shifts occur at. Normally, a forced downshift is provided so the operator can downshift the transmission for additional acceleration by "flooring" the accelerator pedal.

Semiautomatic Transmissions

A number of arrangements are used to reduce the operator effort required to select and engage gear ratios. Clutch operation may be made automatic, or the gear box arranged so that a clutch is not needed. Gear ratio selection and engagement are still performed by the operator, thus these arrangements are considered semiautomatic.

In one type of semiautomatic passenger car transmission, Fig. 13-9, a torque converter is installed ahead of an electro-pneumatic clutch and three speed, manually shifted gear box. Electrical contacts in the shift lever knob are closed when slight downward pressure is applied to the knob. This completes the circuit to the vacuum valve, allowing vacuum from a vacuum storage tank to act on the servo and disengage the clutch. Gear shifts can then be made in the normal manner and when the knob is released, the clutch engages again. The torque converter provides enough multiplication so that

Fig. 13-9 Semiautomatic Passenger Car Transmission The clutch used in this arrangement is a conventional single plate, dry disc clutch. The arrangement of transmission and final drive in a single housing is often called a transaxle.

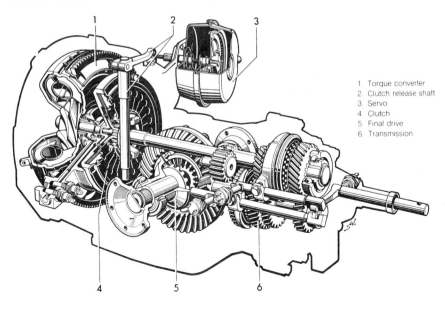

1. Torque converter
2. Clutch release shaft
3. Servo
4. Clutch
5. Final drive
6. Transmission

a three speed gear box is adequate with a small engine. It also eliminates the need for some shifting and helps to cushion shocks that may occur when the clutch engages at the end of a shift.

Increasing numbers of farm and construction machines are equipped with "power shift" transmissions. The gear boxes of these transmissions are somewhat similar in principle to synchromesh transmissions, except that the synchronizers are replaced by hydraulically operated, oil immersed clutches. The hydraulic circuit is in turn controlled by a shift lever. In some cases, two levers are used, one for gear ratio selection and one for direct shifting from forward to reverse. No clutch is required since the gears can be engaged and disengaged under power.

Another type of arrangement has a power shift gear box in series with a conventional clutch and a conventional gear box. The clutch and conventional gear box are used to select a range. Shifts within that range may then be made with the power shift gear box without using the clutch.

Hydrostatic Transmissions

Hydrostatic transmissions are now used on many selfpropelled harvesting machines and garden tractors, as well as significant numbers of large tractors and construction machines. Applications in trucks for highway operation are also being developed. In the sense that no clutch is used and no gear shifting is involved, this type of transmission could be called an "automatic," but in all other respects the hydrostatic transmission has no similarity to the hydrokinetic automatic transmission.

The hydrokinetic transmission transfers power from the engine to the gear box by first converting it into kinetic energy of a fluid in the pump. The kinetic energy in the fluid is then converted back to mechanical energy in the turbine. In the hydrostatic system, engine power is converted into static pressure of a fluid in the pump. This static pressure then acts on a hydraulic motor to produce the output. While the fluid actually moves through the closed circuit between the pump and motor, energy is transferred primarily by the static pressure rather than the kinetic energy of the moving fluid. The relatively incompressible fluid acts much like a solid link between the pump and motor.

The pump in a hydrostatic system is of the positive displacement type. It may be either fixed or variable displacement, but for mobile equipment applications, it usually is a variable displacement type. Axial piston pumps are the most common, although some radial piston pumps are used for small transmissions. In the variable displacement, axial piston pump, Fig. 13-10, the cylinder block and pistons are driven from the input shaft. Piston stroke, pump displacement, and direction of fluid flow are controlled by the reversible swash plate, which in this case is moved by a pair of balanced, opposed servo pistons. The servo pistons are, in turn, controlled by a speed control lever. On smaller units, where the forces acting on the swash plate are not as great, its position is controlled directly by the speed control lever. With radial piston pumps, a moveable guide ring is used to control piston stroke instead of a swash plate.

With the pump in Fig. 13-10, when the speed control lever is in neutral, the swash plate is perpendicular to the pistons and no pumping occurs. As the speed control lever is moved in one direction, the swash plate is tilted, piston stroke is gradually increased, and fluid is pumped from one of the out-

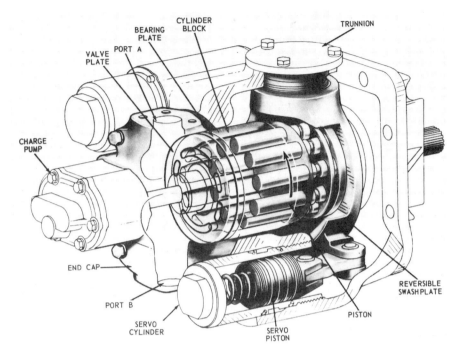

Fig. 13-10 Variable Displacement Pump The servo pistons tilt the reversible swash plate on the trunnion to vary the displacement from maximum in one direction through zero to maximum in the other direction. In many cases motion in the reverse direction is limited so that the maximum reverse speed is one-half or less of the maximum forward speed.

let ports. If the speed control lever is moved in the opposite direction, the swash plate is tilted in the opposite direction, and the piston stroke is moved 180 degrees around the case. Fluid is then pumped from the other outlet port. In combination with a reversing motor, this permits a continuously variable range of speeds from full forward to full reverse with only the single lever for control.

The motor in a hydrostatic system can be any type of positive displacement hydraulic motor. Axial piston motors usually are used for larger drives, and are also for some smaller drives. Both gear motors and radial piston motors are used for low power drives. The motor is usually of the fixed displacement type, Fig. 13-11, but may be variable displacement. As noted, the motor is reversible with the direction of rotation dependent on the direction of flow in the closed loop circuit to the pump.

In addition to the pump and motor, connecting lines, relief valves, and a charge pump are required. The connecting lines may be passages where the pump and motor are in the same housing, or may be hoses where the motor is mounted away from the pump. The charge pump provides initial pressurization of the motor and replaces any fluid lost due to internal leakage. On small tractors it may also be used to supply fluid for remote hydraulic cylinders. A typical small tractor schematic diagram is shown in Fig. 13-12.

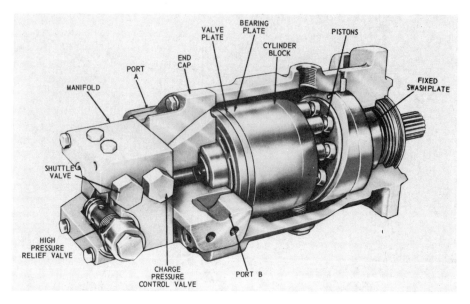

Fig. 13-11 Fixed Displacement Motor This axial piston motor has a fixed swash plate. Similar design motors are available with a movable swash plate.

Various arrangements of the pump and motors are used. A variable displacement pump with a variable displacement motor may be used. Thus, the swash plates must be linked and synchronized so that the motor displacement decreases as the pump displacement is increased. Because motor displacement is maximum when pump displacement is low, motor speed will be low and torque will be high. Conversely, motor displacement will be minimum when pump displacement is maximum so the maximum speed will be high. The arrangement gives high starting torque and the widest range of speeds for any given size of pump and motor.

Another variation has a variable displacement pump and a two piston swash plate or guide ring on the motor. The latter is controlled by a range lever. In the low range, motor displacement is greater, thus starting torque is higher and maximum speed lower.

Most drives in mobile type equipment have a variable displacement pump in combination with a fixed displacement motor. This type of circuit gives a constant torque output, with the power output increasing as the pump displacement is increased.

The fact that the pump and motor do not need to be connected directly together permits considerable flexibility in arrangement. The pump may be connected directly to the engine output shaft and motors located at the driving wheels. In another arrangement, two pumps and two motors may be used with each pump and motor driving one wheel. One wheel can then be driven forward with the other in neutral or reverse for spin turns.

One of the main disadvantages of hydrostatic drives is that they permit the operator to select any travel speed up to the maximum without varying the engine speed. The engine can be operated at governed speed to provide

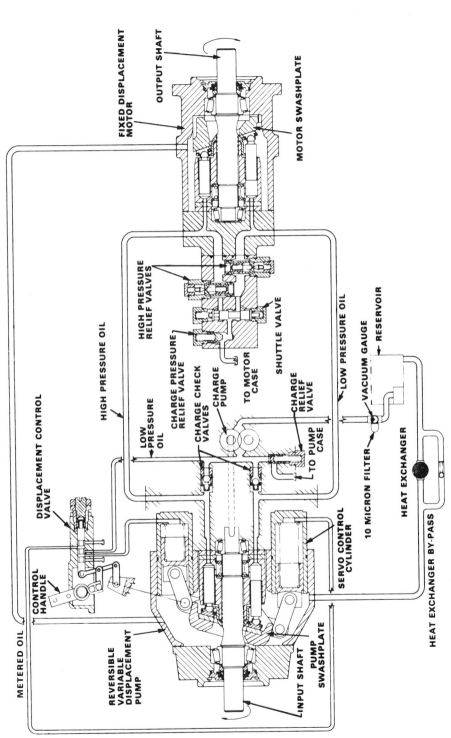

Fig. 13-12 Hydrostatic Drive Schematic Diagram In this arrangement, as in many similar small tractor systems, the low pressure makeup pump is also used to supply auxiliary hydraulic units. The main drive system consists of a variable volume pump with a fixed displacement motor. Fluid is drawn through a strainer from the reservoir and excess fluid not required to charge the main pump flows through the filter back to the reservoir.

METERED OIL

CONTROL HANDLE

DISPLACEMENT CONTROL VALVE

REVERSIBLE VARIABLE DISPLACEMENT PUMP

INPUT SHAFT

PUMP SWASHPLATE

SERVO CONTROL CYLINDER

HEAT EXCHANGER BY-PASS

HEAT EXCHANGER

10 MICRON FILTER

VACUUM GAUGE

RESERVOIR

LOW PRESSURE OIL

CHARGE RELIEF VALVE

TO PUMP CASE

CHARGE PUMP

TO MOTOR CASE

SHUTTLE VALVE

CHARGE CHECK VALVES

CHARGE PRESSURE RELIEF VALVE

LOW PRESSURE OIL

HIGH PRESSURE RELIEF VALVES

HIGH PRESSURE OIL

FIXED DISPLACEMENT MOTOR

OUTPUT SHAFT

MOTOR SWASHPLATE

proper operating speed for elements such as the threshing section of a combine, but a full range of travel speeds is available to adjust to terrain or crop conditions. Operation is also greatly simplified.

Factors Affecting Lubrication

The differences in the lubrication requirements of the various types of transmissions require separate consideration of the factors affecting lubrication.

Mechanical Transmissions The elements in mechanical transmissions requiring lubrication are the bearings, gears, and sliding elements in the synchronizers. Bearings may be either plain or rolling element. As noted, gears are usually either straight spur or helical, gear loads are moderate to heavy. Normally, lubrication is by bath and splash, but some large transmissions may have integral pumps to circulate the lubricant.

Most mechanical transmissions are designed to be lubricated by fluid products. Soft or semifluid greases may be used in small units, such as the transmissions of some motorcycles and scooters.

Generally, the lubricant in a mechanical transmission is expected to remain in service for an extended period of time; normally, many passenger car manufacturers do not recommend periodic draining and refilling. Thus, the lubricant must have the chemical stability to resist oxidation and thickening under conditions of agitation and mixing with air. Operating temperatures may also be quite high. Plain bearings and synchronizer components are often of bronze or other copper alloys. Thermal degradation of the lubricant can result in formation of materials that are corrosive to these components. Severe agitation also occurs, therefore, the lubricant must have good resistance to foaming.

A lubricant selected for mechanical transmissions must have adequate fluidity to permit immediate circulation and easy shifting when a vehicle is started in cold weather. At the same time, the lubricant must have high enough viscosity at operating temperature to maintain lubricating films and to cushion the gears so that acceptably quiet operation results.

A variety of lubricants is recommended by mechanical transmission manufacturers. Straight mineral gear lubricants suitable for API Service GL-1 (see discussion of automotive gear lubricants in this chapter) are recommended by a number of manufacturers. Some manufacturers will accept multipurpose gear lubricants, but only of API Service GL-4 quality, while others will accept either GL-4 or GL-5 quality lubricants. At least one manufacturer recommends DEXRON®* Automatic Transmission Fluid, but permits the use of SAE 90 or SAE 140 gear lubricants if there is objectionable noise when operating on the DEXRON® fluid. Manufacturers of farm and construction machines frequently install the transmission in a common sump with the final drive; the sump may also serve as the reservoir for the central hydraulic system on the machine. Special fluids designed for service as combination heavy duty gear lubricants and hydraulic fluids are usually required for these applications.

Automatic Transmissions In some installations the torque converter is located in a separate housing with its own supply of hydraulic fluid. However,

---

*®General Motors Company registered trademark.

most of the passenger car automatic transmissions, the torque converter, and gear box operate from a common fluid reservoir.

In a torque converter, the fluid serves mainly as a power transfer fluid. It also lubricates the bearings and transfers heat resulting from fluid friction and power losses to a cooler or to the case for dissipation in the atmosphere. Power transfer efficiency increases with decreasing viscosity. Heat transfer efficiency also generally increases with decreasing viscosity. These factors dictate that a torque converter fluid has as low a viscosity as is practical. On the other hand, high operating temperatures and the need for long service life of the fluid require that it can resist oxidation well. Since extremely low viscosity mineral oil base fluids generally have inferior oxidation stability, some compromise on viscosity is usually necessary.

Where the torque converter operates from the same fluid supply as the gear box, the physical characteristics of the fluid also must be compromised in order to meet the lubrication requirements of the gear box.

The fluid in the gear box portion of an automatic transmission performs several functions.

1. It lubricates the gears and bearings of the planetary gear sets.
2. It serves as a hydraulic fluid in the control systems.
3. It controls the frictional characteristics of the oil immersed clutches and brakes.

These functions must be performed under a variety of operating conditions that tend to make the service severe.

Automatic transmissions are expected to engage and shift properly at low temperatures when a vehicle is started in cold weather. In operation, temperatures in the order of 250 to 300°F (121–149°C) may be reached. Gear loads are relatively heavy, and the fluid is exposed to severe mechanical shearing both in the gears and in the hydraulic circuit. Some breathing of air occurs due to changes in temperature inside the unit and this tends to promote oxidation of the fluid, particularly when operating temperatures are high. Where the gear box operates on the same fluid as the torque converter, the severe churning in the torque converter tends to cause foaming. Seal compatibility of the fluid is also an important consideration.

In order to satisfy these requirements for automatic transmission fluids, various highly specialized products have been developed. They are discussed in detail in this chapter.

Semiautomatic Transmissions Since the passenger car type of semiautomatic transmissions discussed is a combination of torque converter and a mechanical transmission, the earlier discussions of the lubrication requirements of these units apply to it.

Power shift transmissions used in heavy equipment have lubrication requirements not unlike the gear box section of automatic transmissions. Because of the higher torques transmitted, somewhat higher pressure may be required in the hydraulic system in order to obtain proper engagement of the clutches. This in turn may apply somewhat higher mechanical shear stresses to the fluid. Again, frictional characteristics of the fluid are critical if the clutches are to engage smoothly and firmly.

Hydrostatic Transmissions A hydrostatic drive is a high pressure hydraulic system so the basic fluid requirements correspond closely with those of industrial hydraulic systems. Good oxidation stability is required, as well as good resistance to foaming and good entrained air release. Antiwear properties are also required since operating pressures are usually in excess of 2500 psi (17.2 MPa).

In addition, the fact that hydrostatic drives must operate over a wide range of temperatures generally dictates that very high viscosity index (VI) fluids with good low temperature fluidity must be used. Since the fluid is a major factor in proper sealing of the pump pistons, the high temperature viscosity of the fluid is important, and excellent shear stability is required to maintain this viscosity in spite of the severe shearing that occurs in the pump and motor.

Many hydrostatic drives are operated from a common reservoir with the differential or final drive. In these cases, the fluid used must also provide satisfactory lubrication of the gears and bearings.

The hydrostatic drives used on garden tractors usually are designed to operate on automatic transmission fluids. These fluids are readily available and generally provide the combination of performance characteristics required. They may also be recommended for hydrostatic drives on larger machines. Engine oils, often in SAE 10W-30 viscosity may also be recommended. Where a hydrostatic drive on a larger machine is operated from a common reservoir with other drive elements, the fluid recommended is usually one of the special fluids discussed later under "Multipurpose Tractor Fluids."

PROPELLER SHAFTS AND UNIVERSAL JOINTS

Road vehicles have the wheels connected to the body and chassis through springs, but the engine and transmission are mounted directly on the chassis. Where a rigid (live) axle is used and the springs flex, the position of the axle with respect to the engine and transmission changes. Thus, there must be provision in the power connection between the transmission and drive axle to accommodate these changes. This is accomplished in the propeller shaft and universal joints.

Typically, a propeller shaft consists of a tubular shaft with a universal joint at each end. The universal joints allow for angular changes and a slip joint at one end allows for changes in length. Some long drive shafts are made in two parts with a center support bearing to minimize whip and vibration. Three universal joints are then used.

Universal joints are usually of the cross or cardan type; see Fig. 13-13. Ball and trunion type universal joints were used to some extent in the past; see Fig. 13-14. If there is any angular misalignment between the driving shaft and driven shaft, both of these types of joints will transmit rotation with fluctuating angular velocity. The amount of fluctuation increases with increasing misalignment, rising from about 7 percent at 15 degrees misalignment to over 50 percent at 40 degrees. Since this fluctuation in velocity may be accompanied by vibration, the propeller shafts, in which these joints are

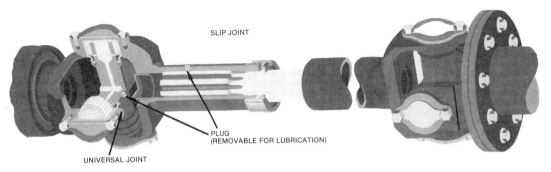

SLIP JOINT

PLUG
(REMOVABLE FOR LUBRICATION)

UNIVERSAL JOINT

Fig. 13-13 Cross Type Universal Joint This type of joint, which is often called a cardan joint, may have needle roller bearings or plain bearings at the ends of the cross. Fittings may be used for relubrication, but the weight of the fitting can cause unbalance so a more common arrangement is to use special flush fittings, or plugs that must be removed and replaced with fittings during lubrication.

used, are designed for minimum misalignment. Another approach is to use what are called "constant velocity" joints.

One type of constant velocity joint consists of two cross type joints in a tandem assembly. Several other designs are available. Constant velocity joints are now being used to some extent in propeller shafts and are used in the drive axles of front engine, front wheel drive or rear engine, rear wheel drive vehicles, and conventional arrangement vehicles with independent rear suspension. In some cases, rather than constant velocity joints, cross type joints may be used at each end of the axle shaft, positioned so that the changes in angular velocity cancel out.

Lubrication In some designs, the universal joint and slip joint at the transmission end of the propeller shaft are lubricated by the transmission lubricant. In other

Fig. 13-14 Propeller Shaft with Ball and Trunnion Type Universal Joint Misalignment is accommodated by the spherical blocks on the arms of the trunnion moving longitudinally in the grooves cut in the body. The spherical blocks are mounted on needle bearings to reduce friction. This type of joint is always enclosed and must be disassembled for repacking.

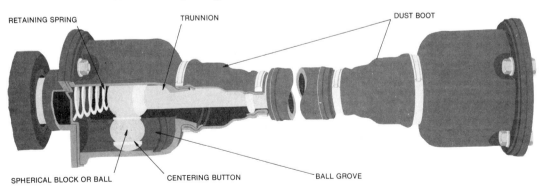

RETAINING SPRING

TRUNNION

DUST BOOT

SPHERICAL BLOCK OR BALL

CENTERING BUTTON

BALL GROVE

designs, the joints are lubricated with grease. Many joints are now packed for life on assembly and require servicing only if other repairs are required. Some joints require periodic disassembly and repacking, while some are equipped with a fitting or plug for periodic relubrication. The plug can be replaced with a special fitting while relubrication is being performed. In most cases, the grease used is a multipurpose automotive grease, sometimes with the addition of molybdenum disulfide.

Drive Axles

The drive axle usually contains one or more stages of gear reduction such as the gears in the universal which enable the wheels to be driven at different speeds. Also, in vehicles with a longitudinally mounted engine, the drive axle provides the gears with the capability to produce a 90 degree change in direction of power flow — to couple the transverse axle shafts to the longitudinal transmission output shaft. In passenger cars and most trucks, the gear reduction in the drive axle is the final stage of gear reduction in the power train. In heavy trucks, and farm and construction equipment, additional stages of speed reduction, usually called final drives, may be used at the wheel ends of the drive axle.

In the most common passenger car and light truck arrangement (Fig. 13-15), the propeller shaft couples through a universal joint to the front end of a pinion shaft. The pinion gear at the rear of this shaft meshes with the ring gear which is bolted or riveted solidly to the differential case. The differential, in turn, drives the half axle shafts.

In most drive axles of this type, hypoid gears are used for the reduction stage. This type of gear design has high load carrying capacity in proportion to the size of the gears and operates quietly. In addition, the offset position of the centerline of the pinion, with respect to the centerline of the gear, permits the propeller shaft to be located lower. This helps to lower the center of gravity of the vehicle and reduces the tunnel through the floor of the passenger compartment that covers the propeller shaft.

With front engine, front wheel drive or rear engine, rear wheel drive cars, spiral bevel gears are usually used for this reduction stage. If the engine is mounted transversely with either of these arrangements, the 90 degree change in direction of power flow is not required and straight spur or helical gears are used. In many heavy trucks, two stages of reduction are used in the drive axle. The first stage of reduction is usually a set of spiral bevel gears and the second stage either straight spur or helical gears. Some trucks are equipped with worm gears where a large total reduction is required.

Differential Action

As a vehicle turns, the wheels on the outside of the turn follow a longer path than those on the inside of the turn. In order to provide for this and other differences in rolling distances between the driving wheels, a differential is used.

The principle of operation of a differential is shown in Fig. 13-16. In this illustration, the arm represents the differential case which is bolted to the ring gear. The differential pinion is free to rotate on its shaft, and meshes with the side gears which drive the axle shafts. When driving resistance is equal at both wheels, the pinion acts as a simple lever to drive the axle shafts at the same speed as the gear ring. If greater rolling resistance is en-

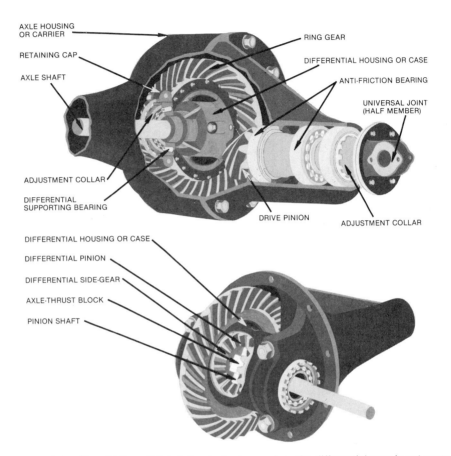

AXLE HOUSING OR CARRIER
RETAINING CAP
AXLE SHAFT
ADJUSTMENT COLLAR
DIFFERENTIAL SUPPORTING BEARING
RING GEAR
DIFFERENTIAL HOUSING OR CASE
ANTI-FRICTION BEARING
UNIVERSAL JOINT (HALF MEMBER)
DRIVE PINION
ADJUSTMENT COLLAR

DIFFERENTIAL HOUSING OR CASE
DIFFERENTIAL PINION
DIFFERENTIAL SIDE-GEAR
AXLE-THRUST BLOCK
PINION SHAFT

Fig. 13-15 Hypoid Type Drive Axle In the lower view, the differential case is cut away to show one of the side gears.

Fig. 13-16 Elementary Differential In an actual differential at least two pinions are used and the arm is replaced by a case which more or less completely encloses the pinions and side gears.

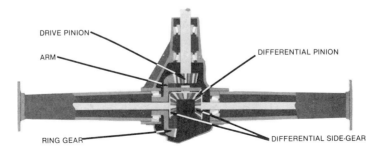

DRIVE PINION
ARM
DIFFERENTIAL PINION
RING GEAR
DIFFERENTIAL SIDE-GEAR

countered at one wheel than at the other, the unbalanced reaction forces acting on the pinion will cause it to rotate on its shaft. The wheel encountering the least resistance will then be driven faster and the wheel encountering the most resistance will be driven slower. The increase in speed of one wheel will be exactly equal to the decrease in speed of the other.

Differentials are built with either two or four pinions. Two pinions are mounted on a shaft which runs across the case and is mounted in a bearing at each end. Four pinions are mounted on a cross shaped member called a spider, and the case is split to permit assembly. Normally, straight bevel gears are used for the pinions and side gears.

Limited Slip Differentials

Conventional differentials have a major drawback in that exactly the same torque is delivered to both wheels regardless of traction conditions. Thus, if one wheel is on a surface with low enough traction for the applied torque to exceed the traction, that wheel will break loose and increase in speed until it is revolving at twice the speed of the ring gear and the other wheel would stop revolving. All the power will then be delivered to the spinning wheel and no power will be delivered to the wheel with traction. Limited slip, or torque biasing, and locking type differentials have been developed to overcome this problem.

The limited slip differentials used in passenger cars are all similar in principle. Clutches are inserted between the side gears and the case. When these clutches are engaged they lock the side gears to the case and prevent differential action. Either plate or cone type clutches may be used. A typical unit using cone type clutches is shown in Fig. 13-17. Initial engagement pressure for the clutches is provided by the springs. As torque is applied to the unit, normal gear reaction forces tend to separate the side gears which apply more pressure to the clutches. The more torque is applied, the more closely the unit approaches a solid axle. When differential action is required the changes in torque reaction at the wheels tend to reduce the pressure on the clutches, permitting them to slip. Coil springs, dished springs, and Belleville springs are all used to provide the initial engagement pressure.

A variation of the unit shown in Fig. 13-17 has the cones reversed so that increasing torque input reduces the engagement pressure on the clutches. This is referred to as an "unloading cone," spin resistant, differential. It has been found useful for the interaxle differential of four wheel drive vehicles and some high performance cars. Both torque biasing and locking differentials are used for trucks and offhighway equipment. Some locking differentials lock and unlock automatically, while others are arranged so the operator can lock them when full traction at both driving wheels is needed. Because of the higher torque inputs involved with these machines, more positive locking arrangements than the clutches used in passenger cars are required for the torque biasing differentials. One type uses cam rings and a set of blunt nosed wedges that operate much in the manner of an overrunning clutch. Other types use special tooth profiles on the pinions such that a torque bias in favor of the wheel with the best traction is always provided.

Factors Affecting Lubrication

The hypoid gears used in drive axles are among the most difficult lubricant applications in automotive equipment. The high rate of side sliding between the gear teeth tends to wipe lubricant films from the tooth surfaces, and the

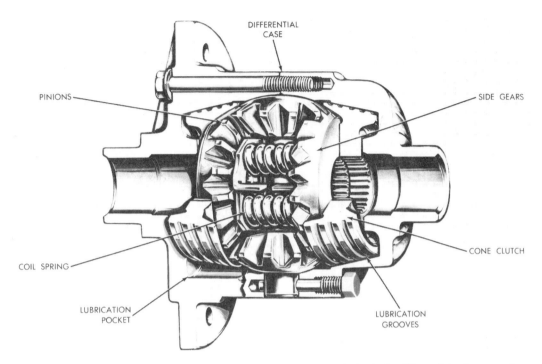

Fig. 13-17 **Limited Slip Differential** A typical limited slip differential using cone type clutches.

amount of sliding in proportion to rolling increases as the offset between the shaft center lines is increased. The gears must transmit high torques, and shock loads are often present. The drive axle is normally not sprung; thus, to keep the unsprung weight low it is desirable to make the gears as compact and lightweight as possible. This has resulted in the use of high strength, hardened steels for these gears, and tooth loading is usually high.

Where spiral bevel gears are used in the drive axle, conditions are more favorable to the formation of lubricant films and tooth loading is generally lower. Worm gear axles present special problems because of the high rate of sliding between the teeth and the metallurgy that must be employed. Where spur gears are used, lubrication conditions are generally not severe.

Many manufacturers no longer recommend periodic lubricant changes for passenger car drive axles. The lubricant for these axles must be suitable for extended service, often at high temperatures.

Breathing of moisture into drive axles frequently occurs. Combined with the severe churning action of the gears, this can contribute to foaming of the gear lubricant and can also promote rust and corrosion.

An important consideration for the drive axles of all vehicles operated in cold weather is the ability of the lubricant to flow to the teeth to provide lubrication of the gears, and to be carried up by the gears in sufficient quantities to lubricate the pinion shaft bearings.

In farm and construction equipment, the drive axle and differential are often installed in a common sump with the transmission which may also serve as the reservoir for the hydraulic system. Oil immersed clutches and brakes may also be involved. In these cases, the lubricant requirements of these other elements must be considered in the selection of the lubricant for the drive axle.

Limited slip axles present special lubrication problems because of the clutches. The clutches must engage firmly so that proper torque biasing is obtained, but the clutches must slip smoothly when differential action is required. If the clutches do not release properly, or stick and slip, then chatter will result and in extreme cases one wheel may be forced to break traction in high speed turns. This can be an unsafe condition. In order to minimize these problems, lubricants for limited slip axles must have special frictional properties, which may conflict with their ability to protect the gears against wear scuffing and scoring. In most cases a small amount of chatter or noise while turning corners is considered normal for these units.

TRANSAXLES

Where a transmission and drive axle are combined in a single housing the unit may be referred to as a transaxle; see Fig. 13-18. The arrangement is common for front engine, front wheel drive or rear engine, rear wheel drive cars.

Factors Affecting Lubrication The drive axle reduction gears used in a transaxle are either spiral bevel or, with a transverse engine, spur or helical gears. As a result, the lubrication requirements are not as severe as with hypoid gears. At the same time, the

Fig. 13-18 Typical Transaxle Cutaway

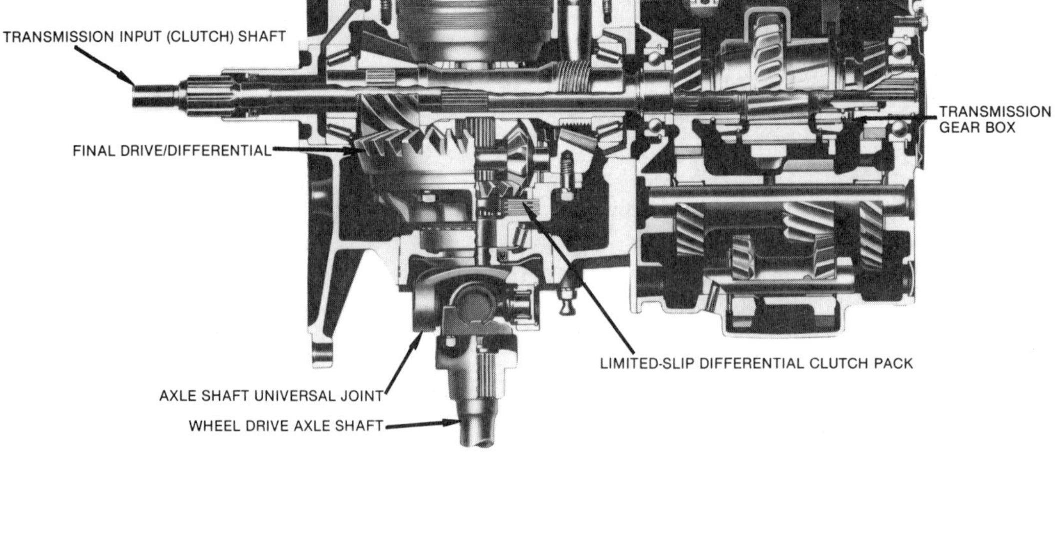

lubricant must meet the requirements of the transmission portion of the transaxle, which often has synchronizer elements that are sensitive to active extreme pressure agents designed for use in hypoid axle gears. Most builders of transaxles now recommend either specially formulated lubricants, or lubricants for API Service GL-4.

Engine/transaxle combinations with automatic transmission, whether at the front or rear of the vehicles, are usually equipped with separate compartments for the transmission and final drive. The transmission compartment contains automatic transmission fluid; the final drive compartment contains suitable gear lubricant for the spiral bevel gears. The modern trend is toward the use of front wheel drive transaxles with transverse engines. This design is usually provided with a common compartment that contains automatic transmission fluid for both the transmission and final drive.

OTHER GEAR CASES

A number of other gear cases may be used in various types of automotive equipment. Some of the more important are discussed in the following sections.

Auxiliary Transmissions

Auxiliary transmissions are used with mechanical transmissions to provide a larger choice of reduction ratios, see "Factors Affecting Lubrication" in this chapter. These are usually two or three speed gear boxes. Shifting is by means of a range control lever which, on trucks, usually controls a hydraulic circuit to perform the actual shifting. On low speed equipment, shifting may be by means of a mechanical linkage. In some cases, the auxiliary may be built into the same housing as the main transmission, in other cases it may be a separate unit. Gears in auxiliary transmissions may be either spur or helical type. Bearings may be either plain or rolling element.

Transfer Cases

A transfer case is required with most four wheel drive vehicles in order to provide a second output shaft to drive the second axle. A transfer case may also provide a power take-off to drive accessory equipment such as a hoist. In some heavy equipment, a transfer case is not required because the main transmission is provided with both front and rear outputs.

With the conventional four wheel drive arrangement, operation at highway speeds can cause stresses to build up in the drive line due to the different distances travelled by the front and rear wheels. If these stresses become excessive, they may cause one wheel to break traction and "hop." To prevent this, the drive to the front wheels normally is disconnected for highway travel. A recent approach is to use a third differential in place of the transfer case. Differentiation in this unit prevents the build up of stresses so the front wheel drive can be left engaged all the time.

Overdrives

An overdrive is an arrangement that drives the transmission output shaft at a higher speed than the input shaft. At cruising speeds this reduces engine rpm and improves fuel economy. Two general approaches are used.

Many of the four speed mechanical transmissions used in small cars have an overdrive fourth gear. Third gear is made with a ratio only slightly greater than 1:1 and fourth gear has a ratio slightly less. The normal progression of ratios in the gear box is maintained, but in fourth gear the engine rpm will be somewhat below the rpm of the output shaft of the transmission.

With three speed transmissions an auxiliary unit is added to the rear of the main transmission. This unit has an arrangement of planetary gears that provides a step-up ratio. The unit may be controlled electrically or hydraulically. When not engaged, it acts as a solid coupling. When the operator moves the control lever to the engaged position, the overdrive is activated but does not engage until a predetermined cut-in speed is reached and the operator momentarily releases the accelerator pedal. The overdrive will then remain engaged until car speed drops below the cut-in speed, or until the operator forces disengagement by fully depressing the accelerator pedal. Separate overdrive units are usually bolted to the rear of the main transmission and provision is made for lubricant to flow from one case to the other. Separate drain plugs are usually provided, and special care in filling may be necessary to ensure that both units are properly filled.

Final Drives

Several types of drive units are used at the wheel ends of the drive axles to obtain additional reduction or rotate the wheels. Planetary reducers are used on many large, offhighway trucks. They are also used on many tractors, and some self-propelled harvesting machines. Planetary speed increases are used at the front wheels of conventional tractors equipped with power front wheel drive to match the travel speed of the smaller front wheels to that of the larger rear wheels. Various types of chain drives are used on self-propelled harvesting machines. Chains or gears are also used to couple a single drive axle to tandem driving wheels on certain types of heavy construction machines. Drop housings, which may or may not involve speed changes, are used on farm tractors to increase the clearance under the tractor for row crop work.

In many cases, these final drives have a separate lubricant supply. In other cases, they are lubricated from the drive axle or a common sump that supplies other units.

Factors Affecting Lubrication

Auxiliary transmissions, transfer cases, and overdrives are generally similar to mechanical transmissions in their lubricant requirements. As noted, auxiliary transmissions and overdrives frequently are coupled to the main transmission so that the same lubricant supply serves both units. Transfer cases are usually independent, but do not present any special lubrication problems. If an interaxle differential of the limited-slip type is used instead of a transfer case, the frictional characteristics of the lubricant are extremely important in the proper operation of the clutches.

Final drives present a range of lubrication problems because of the diversity in design of these units. Some are lubricated from the drive axle, thus, they are designed to operate on the type of lubricant that is suitable for the axle. To simplify lubrication, many final drives with independent lubricant reservoirs are also designed to operate on one of the lubricants required for other parts of the machine. Chain drives may present special problems. Some are fully enclosed and run in a bath of oil, but others are enclosed in relatively loose fitting dust shields. In the latter case, hand oiling or a drop feed oiler may be used. In extremely dusty conditions; it may be necessary to let the chains run dry. Also, it may be desirable to remove the chains periodically and soak them in a bath of oil so that some lubrication will be present inside the rollers on the pins.

AUTOMOTIVE GEAR LUBRICANTS

In automotive gear units, gears and bearings of different designs and materials are employed under a variety of service conditions. The selection of the lubricant involves careful consideration of these factors and their relationship to performance characteristics of the lubricant. Some of the more important performance characteristics of automotive gear lubricants are discussed in the following sections. The reader is also referred to the discussion of Physical and Chemical Characteristics, Additives, and Performance and Evaluation Tests in Chap. 2.

Load Carrying Capacity One of the most important performance characteristics of a gear lubricant is its load carrying capacity, that is, its ability to prevent or minimize wear, scuffing, or scoring of gear tooth surfaces. Some gears are operated under conditions such that the load carrying capacity of straight mineral oils is adequate, however, most gears require lubricants with higher load carrying capacity. This higher capacity is provided through the use of additives. Lubricants of this type are generally referred to as extreme pressure (EP) lubricants.

In order to provide differentiation between automotive gear lubricants with different levels of EP properties, the American Petroleum Institute (API) prepared a series of five lubricant service designations for automotive manual transmissions and axles. These service designations describe the service in which various types of lubricants are expected to perform satisfactorily. In addition, for the two service designations intended for use in hypoid axles, antiscore protection must be equal to or better than that of certain reference gear oils, and the lubricants must have been subjected to the test procedures and provide the performance levels described in ASTM STP-512, "Laboratory Performance Tests for Automotive Gear Lubricants Intended for API GL-4 and GL-5." This latter publication describes tests for other performance characteristics in addition to those for load carrying capacity.

API Lubricant Service Designations The gear lubricant service designations, effective February 1976, are as follows:

API-GL-1 Designates the type of service characteristics of automotive spiral bevel and worm gear axles and some manually operated transmissions* operating under such mild conditions of low unit pressures and sliding velocities, that straight mineral oil can be used satisfactorily. Oxidation and rust inhibitors, defoamers, and pour depressants may be utilized to improve the characteristics of lubricants for this service. Frictional modifiers and extreme pressure agents shall not be utilized.

---

*Automatic or semiautomatic transmissions, fluid couplings, torque converters, and tractor hydraulic systems usually require special lubricants. For the proper lubricant to be used, consult the manufacturer or lubricant supplier.

API-GL-2 Designates the type of service characteristic of automotive type worm gear axles operating under such conditions of load, temperature, and sliding velocities, that lubricants satisfactory for API-GL-1 service will not suffice.

API-GL-3 Designates the type of service characteristic of manual transmissions and spiral bevel axles operating under moderately severe conditions of speed and load. These service conditions require a lubricant having load carrying capacities greater than those which will satisfy API-GL-1 service, but below the requirements of lubricants satisfying API-GL-4 service.

API-GL-4 This classification is still used commercially to describe lubricants, but the equipment required for the antiscoring test procedures to verify lubricant performance is no longer available.

Designates the type of service characteristic of gears, particularly hypoid* in passenger cars and other automotive type equipment operated under high speed, low torque, and low speed, high torque conditions.

API-GL-5 Designates the type of service characteristic of gears, particularly hypoid in passenger cars and other automotive equipment operated under high speed, shock load; high speed, low torque; and low speed, high torque conditions.

Lubricants suitable for this service are those which provide antiscore protection equal to or better than that defined by CRC Reference Gear Oil RGO-105 and have been subjected to the test procedures and provide the performance levels described in ASTM STP-512 dated April 1972. Lubricants that are suitable for more than one service classification may be so designated.

Reference Gear Oils RGO-105 and RGO-110 are members of a series available from Coordinating Research Council, Inc. (CRC). The series ranges from RGO-100, a solvent refined straight mineral oil, to RGO-115, a very high load carrying capacity lubricant made from the straight mineral oil and 15 percent of a particular EP additive. Intermediate levels are made with lesser amounts of this additive, as indicated by the last two digits of the number, RGO-105 containing 5 percent and RGO-110 containing 10 percent. Reference Gear Oil L-1000, a product meeting Ford Motor Company specification ESE-M2C105A, is available from Southwest Research Institute.†

The performance tests detailed in ASTM STP-512 for API-GL-4 and GL-5 services are those required for qualification against the former U.S. Military Specifications MIL-L-2105 and MIL-L-2105B (the latter are the same as those presently required for MIL-L-2105C). Gears for the tests required for GL-4 are no longer available, therefore, it is now usual to estimate this performance level on the basis of more or less equivalent tests or from the known performance of a particular additive system. The gear tests used for

---

*Limited-slip differentials generally have special lubrication requirements. The lubricant supplier should be consulted regarding the suitability of his lubricant for such differentials. Information helpful in evaluating lubricants for this type of service may be found in ASTM STP-512 dated April 1972.
†P.O. Drawer 28510, San Antonio, Texas 78284.

GL-5 are the CRC L-37 and CRC L-42. The CRC L-42 test is primarily a scoring test and it is in this test that a satisfactory lubricant must provide antiscore protection equal to or better than RGO-110. The CRC L-37 test has broader criteria for acceptability in that tooth distress may be evident as rippling, ridging, pitting, scratching, wear, or combinations of these. In general, an acceptable lubricant should control wear so that the original tool marks are visible over most of the tooth surface.

These lubricant service designations have generally provided satisfactory levels of performance for current automotive gear units. Some difficulties have been experienced in low temperature service as a result of the fact that operating temperatures may not be high enough to fully activate the various EP agents. This is usually overcome by using a higher dosage level of additive in lubricants intended specifically for arctic type service.

Viscosity The viscosity of a gear lubricant has some effect on load carrying capacity, leakage, and gear noise. At low temperatures, it also determines ease of gear shifting and has considerable influence on flow to gear tooth surfaces and bearings.

The viscosities of automotive gear lubricants are usually reported according to the SAE Axle and Manual Transmission Lubricant Viscosity Classification (Chapter 2). This classification was last revised in 1974,* primarily to provide for measurement of low temperature viscosities according to ASTM D 2983, "Standard Method of Test for Apparent Viscosity of Gear Oils at Low Temperatures Using the Brookfield Viscometer." The 150,000 cP (150 Pa · s) value selected for the definition of low temperature viscosity in this revision was based on a series of tests in a specific axle design which showed that pinion bearing failures could occur if the lubricant viscosity exceeded this value. However, other axle designs may operate safely at higher viscosities or fail at lower viscosities, so it is the responsibility of the axle manufacturer to determine the viscosity required by any particular axle design under low temperature conditions.

Other gear applications may have different limiting viscosities. For example, for satisfactory ease of shifting, many manual transmissions require a lubricant viscosity not exceeding 20,000 cP (20 Pa · s) at the shifting temperature. However, this does not necessarily mean that a lubricant with a viscosity in excess of 20,000 cP at the lowest expected starting temperature cannot be used satisfactorily. At low temperatures, it is usually necessary to idle the engine for a short period before driving away. During this period, the main gear and countershaft in the transmission will be revolving if the clutch is engaged. As long as the lubricant does not channel, some of it will be picked up and circulated by the gears. The resulting fluid friction may warm the lubricant sufficiently for satisfactory shifting. Gear or bearing failures usually do not occur during this period since the loads are relatively low, being only those resulting from fluid friction in the lubricant. In a drive axle, on the other hand, as soon as the drive is engaged, loads are relatively high and the need for adequate flow of lubricant is critical.

*A further revision, converting the temperatures for viscosity determinations from degrees Fahrenheit to degrees Celsius, with the high temperature viscosities determined at 100°C rather than 210°F, was published during 1977.

Multigrade lubricants such as 80W-90 and 85W-140 can be formulated fairly readily under this system using conventional gear lubricant base stocks. Lubricants in these viscosity grades are generally suitable for a wide range of operating temperatures in automotive gears. Multigrade lubricants covering an even wider range, such as 75W-90 or 80W-140 can also be formulated, but these usually require either special high VI base oils such as some of the synthetics, or the use of VI improvers. VI improvers must be selected with extreme care since, under the severe mechanical shearing that exists in gears, the contribution of the VI improver to high temperature viscosity can be lost rapidly if an unsuitable material is used.

Channeling Characteristics

Under low temperature conditions, a lubricant may "channel," that is, it may become solid enough that when gear teeth cut a channel through it, the lubricant does not flow back rapidly enough to provide fresh material for the gear teeth to pick up.

The channeling temperature of a lubricant is somewhat related to its pour point, and also to its low temperature viscosity. However, the pressure head tending to cause flow into a channel cut in a lubricant may be greater than in the pour point test, and the conditions in the Brookfield viscometer are sufficiently different that it can provide a viscosity measurement for a lubricant that might not flow under other conditions. In effect, the Brookfield viscometer will predict if a lubricant can be carried up by the gears and distributed, but it does not necessarily predict if the lubricant will flow to the gear teeth to be picked up.

Presently, no test is considered to give a reliable prediction of the channeling characteristics of a gear lubricant. The Channel Point test, CRC L-15 or FTM-3456.1, has been found to give little or no correlation with field performance. Work is underway on the subject, but it may be some time before a suitable procedure is developed. In the meantime, tests such as Pour Point, Channel Point, and Brookfield viscosity can be used to provide some rough guidance as to the lowest temperature at which a lubricant can be used safely.

Storage Stability

In extended storage, particularly if the storage temperature is excessively high or low, some of the additive materials may separate from gear lubricants, or reactions may occur which can change the properties of these materials. With some of the older additive systems, this was occasionally a severe problem, but the newer additive systems generally show improved oil solubility and a reduced tendency for the components to react with each other. Proper storage and control of inventory to prevent excessively long storage are, however, still important; see Chapter 15.

Oxidation Resistance

Operating temperatures of drive axles and manual transmissions in normal passenger car service are usually moderate so that the oxidation resistance of the lubricant may not be a major consideration. In heavy duty service, such as trailer towing and in many commercial vehicles, high operating temperatures may be encountered. In these applications, the lubricant must have adequate oxidation resistance to prevent excessive thickening or the development of sludges that could restrict oil flow. Oxidation may also result in the development of materials that are corrosive to some of the metals used.

The oxidation resistance of automotive gear lubricants is usually evaluated in the thermal oxidation stability test (TOST), FTM 2504, and is used in the MIL-L-2105B and MIL-L-2105C qualification programs. The test uses a small gear set which is operated at a high temperature (325°F, 163°C). A copper catalyst coupon is immersed in the oil, and air is bubbled through it. The test is either run for a specified number of hours or until the viscosity has increased by a specified percentage. After the test, the gears and case are examined for deposits and other abnormalities, and the lubricant is examined for insoluble materials. Lubricants that are rated as satisfactory in the TOST have generally provided satisfactory oxidation stability in service, although there are indications that improved oxidation stability may be desirable for heavy duty type operations.

Foaming The churning in gear sets, combined with contaminants such as moisture that enter the case through breathers, tend to promote foaming of the gear lubricant. Foaming may cause overflow with loss of lubricant, and may interfere with lubricant circulation and the load carrying ability of lubricant films.

Defoamers are used in gear lubricants to reduce the foaming tendency. One of the requirements for a defoamer to function effectively is that it must be insoluble in oil. As a result, defoamers are carried in suspension. If the density of the defoamer is significantly greater than that of the oil, the defoamer can settle out during extended storage. As a check on this, gear lubricants submitted for U.S. military approval must pass a foam test both when freshly prepared and after six months storage. In order to meet this requirement, organic defoamants rather than silicone defoamants are now generally used in gear lubricants.

There is considerable evidence that gear lubricants that pass the laboratory foam test may not always be satisfactory in service. Some form of full-scale testing in automotive gear sets is usually used to check on this aspect of the performance of new gear lubricant additive systems and gear lubricant formulations.

Chemical Activity or Corrosion Extreme pressure agents function by reacting chemically with metal surfaces. This reaction normally is initiated when local overheating occurs due to rubbing of surface asperities through the oil film. If the reactivity of the extreme pressure agents is too high, some chemical reaction may occur at normal temperatures, resulting in corrosion and metal loss. Copper and its alloys are particularly susceptible to this type of corrosion and, although copper alloys are not normally used in drive axles, they are frequently used in manual transmission components.

Rust Protection Some moisture enters gear sets through normal breathing, and heavy water contamination may occur in certain types of offhighway equipment. The gear lubricant must provide adequate rust protection to prevent rusting that might result in gear or bearing damage, particularly during shutdown periods.

Seal Compatibility

In order to maintain control of leakage, gear lubricants must be formulated to be compatible with the elastomeric materials used as seals. This means that the lubricant must not cause excessive swelling, or shrinkage and hardening. A slight amount of swelling is usually considered desirable in that it helps to keep the seals tight.

Frictional Properties

As pointed out earlier, the frictional properties of the lubricant are critical in limited slip drive axles, particularly the types that use clutches to accomplish lockup. It is generally considered that for slip in the clutches to be initiated smoothly, the lubricant must have a lower coefficient of friction at low sliding speeds than it has at high sliding speeds. This requires the addition of special friction modifiers to the lubricant, or formulation of additive system with one or more components that have these special frictional properties. However, due to the strong polar attraction of the friction modifiers to metal surfaces, they may reduce access to these surfaces by the extreme pressure agents. This may reduce the load carrying ability of the lubricant. Also, the frictional properties of the friction modifiers may be lost fairly rapidly in service.

In order to minimize these difficulties with limited slip axles, some manufacturers apply special coatings to their axle gears to reduce the severity of the service on the gear lubricant. Regular lubricant changes are also recommended by some manufacturers.

Friction modification of the lubricant may also offer benefits in conventional axles, particularly during break in or during severe service. The reduction in sliding friction between the gear tooth surfaces reduces the amount of heat generated at these surfaces. Since less heat is generated, less heat will be rejected to the lubricant and the operating temperature of the lubricant will be lower. With lower operating temperatures, oxidation and thermal degradation of the lubricant may not be as serious. During break in, surface temperatures high enough to affect the metallurgy of the gear tooth surfaces have been experienced in some axle designs. The reduction in friction from the use of friction modifiers can provide some relief from this problem too.

Identification

In addition to the API lubricant service designations and the SAE viscosity classification, automotive gear lubricants are frequently identified by U.S. Military specifications. Some manufacturer specifications also exist, but products meeting them are usually used only for factory fill of new machines, or for sale through the service parts organization of the sponsoring company.

U.S. Military Specifications Only one gear lubricant specification is now current. Three others may be encountered as performance references:

U.S. Military Specification MIL-L-2105C This current specification requires performance essentially equivalent to API-GL-5. It covers SAE 75W, SAE 80W-90, and SAE 85W-140 viscosity grades. All grades must pass the gear performance tests. The SAE 75W grade replaces the former MIL-L-10324A Sub-Zero Specification, and the 80W-90 and 85W-140 grades replace the 80, 90, and 140 grades of MIL-L-2105B.

U.S. Military Specification MIL-L-2105B This obsolete specification differed from MIL-L-2105C in that it covered viscosity grades corresponding to an earlier SAE viscosity classification. Generally, only the SAE 90 grade was fully tested in the qualification program.

U.S. Military Specification MIL-L02105 This obsolete specification described lubricants essentially equivalent to API-GL-4.

U.S. Military Specification MIL-L-10324A This obsolete specification described one low viscosity grade intended for use in arctic regions. These requirements are now supplied with SAE 75W products meeting MIL-L-2015C.

TORQUE CONVERTER AND AUTOMATIC TRANSMISSION FLUIDS

The specialized fluids used in passenger car, truck, bus, and offhighway automatic and semiautomatic transmissions are all manufactured to meet specifications developed by equipment manufacturers, mainly General Motors Corporation and Ford Motor Company. At the present time, there are no industry-wide specifications for any of these fluids.

Torque Converter Fluids Separately housed torque converters are used with some passenger car automatic transmission arrangements, but the main application is in transit coach drives and some construction equipment drives. Separately housed torque converters for passenger cars are operated on automatic transmission fluids. Some early torque converters for vehicles with diesel engines were designed to operate on diesel fuel. The fuel was recirculated from the vehicle tank to the torque converter, so that the bulk fuel in the tank served as a heat sink to absorb heat resulting from power losses in the torque converter. This arrangement was not too successful because thermal degradation of the fuel could result in sludge and deposits in the torque converter.

To provide better oxidation stability, low viscosity mineral oils were adopted. These oils were about 60 SUS (10.5 cSt) at 100°F (38°C). Again, some problems with thermal degradation were encountered, particularly in extended service and in newer, larger vehicles where operating temperatures were higher. Some quantities of these fluids are still used in older equipment, mainly in countries outside the United States, but the most commonly used products now are oxidation inhibited fluids of about 100 SUS (20.5 cSt) at 100°F (38°C).

Two manufacturer specifications covering these 100 SUS fluids are generally recognized: the General Motors Corporation Truck & Coach Division Specification No. C-67-I-23 for Type 2 V-Drive Fluids, and the Dana M-2003 specification for torque converter fluid. These specifications have the same physical requirements, and the performance requirements are similar. Viscosity and pour point requirements are shown in Table 13-1.

Automatic Transmission Fluids Commercially available passenger car automatic transmission fluids are all developed to meet either the General Motors Corporation or Ford Motor Company specifications.

Table 13-1   Torque Converter Fluids[a]

Characteristic	Limits
Viscosity	
cSt @ 38°C	19.35–24.0
cSt @ 99°C	2.92–4.80
SUS @ 100°F approx.	95–115
SUS @ 210°F approx.	36–42
Pour Point, °F (°C)	−10 (−23) max

[a]In addition to the torque converter applications, these fluids are also recommended for some hydraulic retarders.

Most other manufacturers accept one or the other of these types of fluids, although some may require additional testing to ensure compatibility with the requirements of their transmissions. For heavy truck and off highway automatic and semiautomatic transmissions, one fluid type is standardized.

Both the General Motors and Ford fluids have evolved over the years as car weights have increased and decreased and operating conditions have made the service on the fluids more severe. Other factors such as extension or elimination of drains and designs in which only the fluid in the gear box is drained when the fluid is changed have also contributed to the need for products that are more resistant to deterioration in service.

The various versions of these specifications are outlined briefly in the following paragraphs.

General Motors Type A Fluid   This specification and qualification procedure as introduced in 1949. Products qualified under the program were assigned AQ-ATF-XXX numbers where the X's represent the digits in a three digit number. These products came to be known as "Type A" fluids. They provided satisfactory performance in the transmissions of that time, and were recommended by a number of manufacturers in addition to General Motors.

General Motors Type A, Suffix A Fluid   In 1956, the Type A specification was revised to require improved quality. Fluids qualified under this program were assigned AQ-ATF-XXXX qualification numbers. They were suitable for use in transmissions where Type A fluids had been originally recommended and, at least for the first few years, were acceptable to most other automatic transmission manufacturers. While this type of fluid has disappeared from the U.S. market, some demand for it may still exist among certain European manufacturers.

General Motors DEXRON® Fluid*   In 1967, General Motors issued the DEXRON specification which represented a further improvement in quality, mainly in the areas of high temperature stability, low temperature fluidity, and retention of frictional properties in service. Products qualified against

*®General Motors Registered Trademark.

this specification were assigned B-XXXXX qualification numbers. DEXRON fluids were intended to be suitable where Type A or Type A, Suffix A fluids had been recommended originally, but, as noted, some European manufacturers continued to show preference for the older Type A, Suffix A fluids. In addition to the General Motors applications, DEXRON fluids were acceptable to Chrysler, American Motors, and some Borg-Warner affiliates.

General Motors DEXRON II Fluid   The DEXRON II specification was intended originally as a multipurpose fluid suitable for passenger car automatic transmissions, the lubrication of rotary engines, and as a replacement for the Type C-2 fluids recommended by the Detroit Diesel Allison Division of General Motors for automatic and semiautomatic transmissions used in trucks and offhighway equipment. Some early fluids qualified on this basis were assigned C-XXXXX qualification numbers. Unfortunately, these fluids proved to be deficient in corrosion protection for the braze alloys used in certain transmission fluid coolers. The specification was revised to include a braze alloy corrosion test, and the rotary engine lubrication requirements were dropped. Fluids qualified under this version of the specification are assigned D-XXXXX qualification numbers and are generally suitable for applications where any of the earlier General Motors approved fluids were recommended originally.

Compared to the DEXRON fluids, the DEXRON II fluids have somewhat lower low temperature viscosity, and the oxidation stability is improved. Oxidation stability for the DEXRON II fluids is tested in a modern, three speed, automatic transmission which is probably more severe than the older two speed transmission used in the DEXRON test. Additionally, the rate of air flow into the transmission during the test is twice as great in the DEXRON II test. A humidity cabinet rust test is also required in the DEXRON II specification.

Ford Fluids   In the past, specifications for the Ford automatic transmission fluids were issued under the designation ESW-M2C33 with a suffix letter indicating different revisions. They are usually referred to as "Type F" fluids. M2C33-D was issued in 1959 and was suitable for use in all Ford automatic transmissions manufactured prior to 1967. Products qualified against it were assigned 1P-XXXXX qualification numbers. It was superceded by M2C33-E/F ("E" covered the natural colored product, while "F" required a red dye) which was suitable for use in all Ford automatic transmissions through 1976, as well as all 1977 transmissions except the C-6. Products qualified against M2C33-F were assigned 2P-XXXXX qualification numbers. A number of other manufacturers, including some of the Borg-Warner affiliates, also recommend this type of fluid.

More recently, M2C33-G was circulated for information, but has not been adopted in the United States, and probably won't be. It has, however, been adopted by the European Ford affiliates for both factory fill and service refill. Compared to M2C33-F, the transmission oxidation test is more severe as a result of the air flow through the transmission during the test being doubled, the allowable low temperature viscosity is decreased, and the anti-wear characteristics are improved.

Early in 1977, a new Ford specification, M2C138-CJ was issued and only this type of fluid is recommended for the C-6 and "JATCO" automatic transmissions, the former is the largest Ford passenger car automatic transmissions manufactured. These fluids are sometimes referred to as "CJ" fluids. They differ from the earlier fluids in that they contain friction modifiers. It is expected that these fluids will be extended to all Ford automatic transmissions at some time in the future.

Fluid Comparison   The major difference between the General Motors fluids and the M2C33 Ford fluids is in the frictional properties. The Ford fluids have a higher coefficent of friction at low sliding speeds than they do at high sliding speeds, while the General Motors fluids are the opposite; see Fig. 13-19. This results in faster and more positive lockup of the clutches and bands at low sliding speeds with the Ford fluids.

The Ford fluids are physically compatible with the General Motors fluids, but should not be mixed. Any mixing changes the frictional properties which may result in poor transmission shifting or, in extreme cases, failure of the transmission to engage.

The viscosities and low temperature fluidity requirements of all the fluids are similar. As shown in Table 13-2, there has been a gradual improvement in low temperature fluidity and the high temperature viscosity after use is becoming the more important criterion.

Detroit Diesel Allison Division Type C Fluids   This division of General Motors offers both fully automatic and semiautomatic transmissions for trucks and off highway equipment. Early designs were intended to operate on SAE 10W heavy duty engine oils, but variations in the performance of some of these products led to the issuance of the Type C-1 specification. Type C-1 were SAE 10W heavy duty engine oils that had performed satisfactorily in Allison

Fig. 13-19 Automatic Transmission Fluid Frictional Characteristics   As shown, the friction coefficient of the Ford fluids decreases with increasing sliding speeds, while the DEXRON fluids increase.

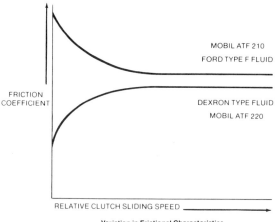

Variation in Frictional Characteristics

Table 13-2 Automatic Transmission Fluid Fluidity Requirements

Requirement	General Motors Fluids			Ford Fluids		
	Type A, Suffix A	DEXRON*	DEXRON* II	M2C33-F	M2C33-G	M2C138-CJ
New Fluid						
Viscosity,						
cSt @ 210°F (99°C), min	7.0	7.0	—	7.0	6.35	7.0
cP @ 0°F (−18°C), max	—	—	—	1,400	1,400	1,400
cP @ −10°F (−23°C), max	4,500	4,000	4,000	—	—	—
cP @ −40°F (−40°C), max	64,000	55,000	50,000	55,000	40,000	50,000
Pour Point, °F (°C) max	—	—	—	−40 (−40)	—	—
Used Fluid						
Viscosity,						
cSt @ 210°F (99°C), min	—	5.5	5.5	6.2	6.35	6.2
cP 0°F (−18°C), max	—	—	—	—	1,400	1,400

transmissions. In 1967 the Type C-2 specification was issued to replace the Type C-1. Type C-2 was intended only as a transmission fluid and included tests such as a transmission oxidation test (operated at a somewhat lower temperature than the test used in the DEXRON II program) and a power steering pump wear test. Some engine oils met this requirement, as did some DEXRON fluids, and a number of special fluids were developed specifically to meet it.

Recent model Allison automatic transmissions for highway vehicles are designed to operate on either DEXRON or DEXRON II fluids. When the DEXRON II specification was developed, there was some thought that it would be applicable for all Allison transmissions, permitting the eventual elimination of the Type C-2 fluids. However, the fact that DEXRON II fluids are somewhat more costly than Type C-2 fluids has caused objections from fleet operators with large numbers of Allison transmissions, and the DEXRON II fluids have also been unsatisfactory from the standpoint of retention of frictional properties under the severe operating conditions encountered in offhighway equipment. Since some Type C-2 fluids may also be unsatisfactory in this latter respect, a new specification, Type C-3, which contains a test for retention of frictional properties has been introduced to replace Type-C2. In addition, the Type C-3 specification covers both SAE 10W and SAE 30 viscosity grades.

MULTIPURPOSE TRACTOR FLUIDS

Probably the area of lubrication where there has been the greatest proliferation of proprietary manufacturer specifications is in fluids for the drive and hydraulic systems of farm and industrial tractors and self-propelled farm equipment. Currently, at least 20 to 25 different fluids are used by the equipment manufacturers for initial fill or are marketed by them through their parts departments. In many cases these fluids must be used during the warranty period if the warranty is to be maintained. This can cause considerable complexity in stock and application for the operator with several different makes of equipment. The risk of misapplication may also be high under these conditions.

Many reasons can be given for the large number of these fluids and their variations in characteristics. Some of the differences among the products result from differences in design philosophies and construction material among the manufacturers. The number and types of machine elements served by the fluid may be extensive, and only minor differences in materials of construction, applied loads or operating characteristics can have a marked effect on the lubricant properties required for satisfactory performance. When these systems were first being developed, there were no commercially available fluids that provided the type of performance desired by the equipment manufacturers. Thus, these manufacturers tried to develop fluids that were specific to their needs.

There are continuing attempts to classify these fluids so that some simplification can be achieved, but progress has been slow. In addition to the differences in product characteristics, it may be that commercial considerations are also a factor in this (sale of the fluids may be a significant source of profit for the parts departments of the equipment manufacturers). However, a factor that is increasingly focusing attention on the need for standardization is the intermingling of fluids that occurs when a tractor is attached to a towed implement. Within a short period of time, the fluid from the tractor and the fluid in the implement are completely intermingled. Under these conditions, it is impossible to maintain the integrity of a special fluid for the tractor systems.

Fluid Characteristics Designations and application of the more widely used multipurpose tractor fluids are shown in Table 13-3. In addition to the usual physical and chemical requirements, most of these fluids have requirements for seal compatibility. The tests for this characteristic vary widely both in the types of seal elastomers used for testing and in the test conditions. Various other bench type tests for such properties as wear sensitivity or tolerance, filterability, compatibility with other similar fluids, and cold oil flowability are also used in some of the specifications. Many of the specifications also include performance tests, including such tests as automatic transmission oxidation tests, transmission durability tests, cold engagement tests, various pump tests, wet brake, clutch chatter and durability tests, friction tests, frictional property retention tests, and dynamometer or full scale field tests.

In addition to these products, for oil cooled friction compartments, Caterpillar Tractor Company recommends products that are referred to as CD/TO-2 oils. While related to the other tractor fluids in application, the CD/TO-2 oils are API Service CD (MIL-L-2104C) engine oils that have passed the Caterpillar TO-2 test. This test is designed to evaluate the retention of frictional properties of the oil after repeated engagement of oil immersed clutches. Many API Service CD or CD, SE oils are now being formulated to meet this requirement.

Some of the more important properties of the multipurpose tractor fluids are discussed briefly in the following paragraphs.

Table 13-3 Multipurpose Tractor Fluids

Specification	Title	Application
Ford ESN-M2C41-A	Hydraulic and Automatic Transmission Oil — Tractor Quality	Combination power-transmission medium, hydraulic control fluid, heat transfer medium, bearing surface lubricant, and gear lubricant for automatic transmissions where specified. Other tractor applications are: power steering units, hydraulically operated loaders and backhoes.
Ford ESEN-M2C86-A	Oil — Mild EP — Wet Brake Lubricant	Tractor rear axles equipped with wet brakes, and tractor hydraulic systems.
Ford ESN-M2C134-A	Lubricant — Gear — Wet Brake and Hydraulic	Tractor transmissions, rear axles, hydraulic systems, and tractors equipped with wet brakes.
Massey-Ferguson M-1127	Transmission-Hydraulic Oil	Designed to serve the multiple function of a friction controlled PTO clutch and wet brake fluid, pressure hydraulics fluid and gear lubricant under condition of high and low temperature, moderately high gear tooth contact pressure and/or high shear where power train gears, clutches, wet brakes, pressure hydraulics, power steering components and other devices are contained within a single housing or served by a common reservoir.
	M-1127A — High Temperature Grade	The high temperature grade provides superior break-in gear lubrication through increased viscosity and is intended for initial factory fill of mechanical drive train gears and service of all vehicles in mainland USA.
	M-1127B — Low Temperature Grade	The viscosity of the second grade is designed to provide optimum low temperature fluidity and is intended for initial factory fill of hydrostatic transmissions and service of all vehicles in Canada.
Massey-Ferguson M-1129-A	Extreme Pressure Transmission Fluid	Intended to serve the complex function of gear lubricant and hydraulic fluid under high gear tooth sliding contact pressures, low −20°C (−4°F) to moderately high 85°C (185°F) reservoir temperatures and moderately high fluid shear conditions resulting from flow through a piston pump, valves and other hydraulic lift systems components at 17,000 kilopascals (2500 psi) fluid pressure and provide high fluidity at low temperature necessary for machine startup under severe cold conditions.
John Deere JDM J14	Combination Hydraulic and Transmission Oils	J14B is used in systems having a common oil for hydraulic systems, wet clutch, transmissions, and/or wet brakes where squawk or chatter is a problem. J14C is used in systems having a common oil for hydraulic systems, wet clutch, transmissions, and/or wet brakes where squawk or chatter is not a problem.
John Deere JDM J20A	Transmission and Hydraulic Oil; Anti-Chatter	Intended primarily for use in the transmission and/or hydraulic system of John Deere vehicles.
John Deere JDM J21A	All-Weather Hydrostatic Fluid	To provide satisfactory durability, life, cold weather performance, and compatibility with water contamination.
International Harvester B-6	Combination Hydraulic and Transmission Fluid	Suitable for use in certain hydraulic, transmission, power steering, wet brake and other systems.
White Q-1722	Hytr-Trans Fluid	—
Allis-Chalmers Part No. 257541	Oil — Multi-Purpose	Transmissions, Final drives, oil lubricated brakes and clutches, and hydraulic systems.

Viscosity and Viscosity Index The type of equipment in which these fluids are used is often operated on a year round basis. This means that the fluids must have adequate low temperature fluidity to flow to parts requiring lubrication, and to the inlet of hydraulic pumps at normal winter starting temperatures. All of the specifications prescribe a maximum allowable viscosity at 0°F (−18°C), and some also limit the viscosity at even lower temperatures. Since these limits are determined by the equipment manufacturers on the basis of tests on their own equipment, it is probable that the fluids will give satisfactory low temperature performance in the equipment for which they are recommended. However, there may be some compromises in the maximum allowable low temperature viscosity in order to meet a desired viscosity at high temperatures.

In high temperature operation, the viscosity of the fluid should be sufficient to permit adequate maintenance of hydraulic pressure and to protect gears and bearings from excessive wear. While many of the fluids are similar in viscosity at 210°F (99°C) or 100°C (212°F), there are some divergences that may result from different attitudes toward VI improvers. The degradation products of some VI improvers are reported to cause glazing of certain types of facing materials used on oil immersed (so-called "wet") clutches and brakes. This condition can cause slippage and other problems. Therefore, some equipment manufacturers require that VI improvers not be used in their fluids, so the viscosity characteristics of those fluids are limited to what can be obtained from conventional high VI mineral oil bases.

In general, VI improved fluids are desirable to maintain adequate high temperature viscosity while maintaining the low temperature viscosity in a range that will permit satisfactory operation at the lowest ambient temperatures encountered in service. Where used, careful selection of the VI improvers is necessary to obtain materials that will resist the mechanical shearing in hydraulic pumps and gears.

Foam and Air Entrainment Control Churning and pumping through the system can promote foaming, which can seriously degrade the performance of hydraulic components, and may interfere with the formation of proper lubricating films. Since the rest time of the fluid in the reservoir is low, the fluid must not only resist foaming but also permit rapid collapse of any foam that does form. Excessive entrainment of air in the fluid can also cause difficulties in hydraulic components, so the fluid must release entrained air as rapidly as possible. Careful selection of the components of the fluid and use of an effective defoamer are necessary to meet these requirements.

Some breathing of moisture into these systems always occurs. Since moisture can promote foaming, some of the manufacturers require that their fluids have good foam resistance when they are contaminated with a small amount of water.

Rust and Corrosion Protection Much construction and farm equipment operates at least part of the time under wet or humid conditions. Breathing of this moisture into the lubrication systems can cause rust and corrosion of ferrous parts, particularly above the oil level during shutdown periods. Nearly all of the specifications have some requirement for protection against rust and corrosion, with some type of humidity cabinet test the most common. Quite effective rust inhibitors are necessary to pass this type of requirement.

As with gear lubricants, the extreme pressure agents used in multipurpose tractors fluids function by reacting chemically with the metal surfaces. If the reactivity is too high, then corrosion may occur. Nearly all of the specifications contain a copper strip corrosion test, although the time and temperature vary considerably. Since oxidation of a fluid in service can increase its corrosivity, therefore, in at least one case, a combination oxidation/corrosion requirement must be met.

Oxidation and Thermal Stability   Operating temperatures often are high in these systems as a result of heavy loads and continuous operation. Thermal and oxidative degradation of the fluid can result in thickening, which can reduce hydraulic pump capacity and the formation of varnish and sludge. These latter materials can plug filters and leave deposits on clutch facings where they can cause slippage, or in hydraulic control valves where they can cause erratic control.

Some type of beaker oxidation test is used in nearly all of the specifications. These tests can be interpreted in various ways. Excessive evaporation loss usually results from the use of volatile, low viscosity components in the base oil blend. The amount of viscosity increase is somewhat indicative of the amount of oxidation of the fluid. Sludge or sediment may indicate thermal decomposition of additive components or some form of chemical reaction between additive components at high temperatures.

Various performance tests are also used in many of the specifications to evaluate the stability of fluids in long term service or under simulated service conditions. These tests may be field or dynamometer tests or specialized tests such as transmission oxidation tests.

Frictional Characteristics   The growing use of oil immersed clutches and brakes has led to increased emphasis on the frictional characteristics of the fluid and the retention of these characteristics in service. Generally, friction modifiers must be included in the fluid if the clutches and brakes are to engage smoothly and resist chatter and squawk. Detergent-dispersants may be desirable to help keep the facing surfaces clean. Good oxidation stability is important to reduce the amount of varnish type materials that might deposit on the friction surfaces. The additive materials must also be carefully selected since some are reported to cause rapid deterioration of some types of friction materials. The overall problem is complicated by the varying levels of friction performance required in different systems, and the many types of friction materials involved.

Extreme Pressure and Antiwear Properties   Multipurpose tractor fluids must provide adequate protection against wear and gear tooth surface distress of a variety of gearing that operates under conditions of loading, varying from mild to severe. Shock loads are often present. In hydraulic systems, protection against wear of pumps operating at pressures that may be in the order of 2700 psi (19 MPa) or higher must be provided. The situation is further complicated by the fact that some of the specifications permit, or require, zinc dithiophosphates, which give good antiwear performance in hydraulic pumps and certain types of gear applications, while other specifications prohibit the use of these materials.

In addition to a considerable number of bench tests for EP properties, there are several specialized gear tests used by the equipment manufacturers. Differences in gear types and operating conditions make it difficult to compare the gear load carrying capacity of the various products.

Proprietary Fluids

In spite of the differences in the specification requirements and applications, a number of oil companies market multipurpose fluids that they recommend for all or most of the applications. These products are not usually recommended for use during the warranty period. Obviously these products can't meet all of the specifications, so are the product of a series of compromises. However, overall performance of these products appears to be satisfactory and, in some cases, they may actually provide better performance than the proprietary specification products. Possibly of the greatest importance, when applied intelligently, these products simplify stocking and reduce the risk of misapplication in fleets with equipment of more than one manufacturer.

14
Compressors

Compressors are manufactured in several types and for a variety of purposes. Lubrication requirements vary considerably, depending not only on the type of compressor but also on the gas being compressed. In general, air and gas compressors are mechanically similar so that the main difference is the effect of the gas on the lubricant. Refrigeration and air conditioning compressors require special consideration because of the recirculation of the refrigerant and mixing of the lubricant with it.

Compressors are classified as either positive displacement or dynamic. The positive displacement class includes reciprocating (piston) types and several rotary types. Dynamic compressors are usually of either the centrifugal or axial flow type, although some mixed flow machines that combine some elements of both of these types are used.

Excluding the refrigeration and air conditioning applications, in terms of number of machines, more compressors are used to compress air for utility use than for any other purpose. These include both the portable compressors used on construction projects, in mining and other outdoor applications, and the stationary compressors used to provide plant air in applications ranging from service stations to industrial plants of all types. In the past, these applications were largely the province of reciprocating compressors, but due to various factors, large numbers of other types of compressors are now being used.

In plant air application, the vibration associated with medium and larger size reciprocating compressors generally necessitates either heavy, vibration absorbing mountings, or special isolating mountings. The pulsation in the air delivery may require pulsation dampers. Rotary and centrifugal compressors are now being supplied as packaged units which can be installed with minimal mounting requirements from the vibration absorption point of view. Compared to reciprocating units, these packaged rotary units are more compact, offer nonpulsating flow, have fewer wearing parts, and require less maintenance. In the case of the centrifugal units, they also deliver oil free air.

In the construction field, the development of helical lobe (screw) compressors to efficiencies approaching those of reciprocating compressors, combined with the lower noise levels associated with the rotary machines, has increased the use of rotaries.

When considering compressor lubrication, it is necessary to recognize that compressing a gas causes its temperature to rise. The more the gas is compressed, the higher will be its final temperature. When high discharge pressures are required, compression is carried out in two or more stages with the gas being cooled between stages so as to limit temperatures to reasonable values. This also improves compressor efficiency and reduces power consumption. For the range of temperatures that can be reached, see Table 14-1. This table is based on adiabatic* compression of air with an intake pressure of 14.7 psi (101.3 kPa) and temperature of 60°F (15.6°C) with intercooling between stages to the same temperature. These temperatures, although higher than those reached in actual practice since some of the heat is removed by cooling the cylinder walls, are indicative of the temperatures that must be considered when selecting compressor lubricants.

RECIPROCATING AIR AND GAS COMPRESSORS

Reciprocating compressors are used for many different purposes involving extremes of pressure and volume requirements. As a result, a great variety of designs is commercially available. Most reciprocating compressors are of the single or two stage type, with smaller numbers of multistage machines — three or more stages. From the lubrication point of view, single and two stage machines generally are similar, while multistage have somewhat different requirements.

The principal parts common to all reciprocating compressors are pistons, piston rings, cylinders, valves, crankshafts, connecting rods, main and crankpin bearings, and suitable frames. In double acting machines, (which compress on both faces of the pistons) piston rods, packing glands, crossheads and crosshead guides are required; the connecting rods are connected to the crossheads by crosshead pins. Crossheads and associated parts are

---

*Adiabatic compression assumes that all of the mechanical work done during compression is converted to heat in the gas, i.e., the cylinder is perfectly insulated so that no heat is lost from it.

Table 14-1 Effect of Staging on Discharge Temperatures

Discharge Pressure Gage		Discharge Temperature					
		1 Stage		2 Stages		3 Stages	
psi	kPa	°F	°C	°F	°C	°F	°C
70	483	398	203	209	98		
80	552	426	219	219	104		
90	621	452	233	226	109		
100	689	476	247	238	114		
110	758	499	256	246	119		
120	827	519	271	254	122	182	83
250	1724			326	163	225	108
500	3447			404	207	269	132

also used in some multistage single acting compressors, but the majority of single acting compressors are of the trunk piston type where the connecting rods are connected directly to the piston by piston pins (wristpins). For lubrication purposes, all parts associated with the cylinders, including pistons, rings, valves, and rod packing, are considered as cylinder parts, and all parts associated with the driving end, including main, crankpin, crosshead pin or wristpin bearings, and crossheads and guides, are considered running parts, or running gear.

Every reciprocating compressor is provided with cooling facilities in order to limit the final discharge temperature to a reasonable value and to minimize power requirements. The cylinder walls and head are cooled and in the case of two stage and multistage machines, the gas being compressed is cooled between stages in intercoolers. Cooling can be by air or water, but in the larger machines water cooling is usually required. Frequently the gas being discharged is cooled in aftercoolers. In the case of air, this removes water and, thus, prevents or minimizes condensation of moisture in the air distribution system. Aftercoolers also act as separators to assist in removing oil that may be carried over from the cylinders.

A considerable amount of moisture can be condensed in intercoolers. For example, in a two stage compressor taking in air at atmospheric pressure, 70°F (21°C), and 75 percent relative humidity and discharging at 120 psi (758 kPa), about 3 ¾ gal (14L) of water per hour will be condensed in the intercooler for each 1000 cubic feet per minute (1700 m³/h) of free air compressed. This fact has an influence on the lubrication of subsequent stages of the compressor.

Methods of Lubricant Application

In reciprocating compressors, the cylinders and running gear may be lubricated from the same oil supply, or the cylinders may be lubricated separately.

Cylinder Lubrication Except where cylinders are open to the crankcase, oil generally is fed directly to the cylinder walls at one or more points by means of a mechanical force feed lubricator. In a few cases, the main oil feed is supplemented by an additional feed to the suction valve chambers. For some small diameter, high pressure cylinders of multistage machines, oil is fed only to the suction valve chambers. Essentially, all the oil fed to cylinders, which are not open to the crankcase, is carried out of the cylinders by the discharging gas and collects in the discharge passages and piping.

Cylinders, which are open to the crankcase, are lubricated by means of oil thrown from a reservoir by means of scoops or other projections on the connecting rods or cranks. Where this splash lubrication method is used, the pistons are provided with oil control rings, which are designed to prevent excessive oil feeds to the cylinders.

Compressor valves require very little lubrication. Usually the small feed of oil required spreads to the valves from the adjacent cylinder walls or is brought to them in atomized form by the air or gas stream. However, when air compressors operate under extremely moist suction conditions, it is sometimes necessary to provide supplementary lubrication by means of force feed lubricator connections to the suction valve chambers.

Valve operators may hold valves open or closed for certain types of pressure regulation systems that require very little lubrication. Usually the small feed of oil required spreads from the adjacent cylinder walls or is brought to the valve operators in atomized form by the gas stream.

When metallic piston rod packing is used on double acting machines, stuffing boxes are lubricated by means of force feed lubricators. Usually when nonmetallic packing is used, stuffing boxes are adequately lubricated by oil from the compressor cylinders. In some cases, however, mechanical force feed lubricators or drop feed cups may be used.

Bearing (Running Gear) Lubrication In practically all reciprocating compressors, the oil for the lubrication of the running gear is contained in a reservoir in the base of the compressor. Oil from the bearings, crosshead, or any cylinders open to the crankcase drains back to the reservoir by gravity. However, a variety of methods and combinations of methods are used to deliver oil from the reservoir to the lubricated parts.

Oil may be delivered to lubricated parts entirely by splash. Where this is done, a portion of, or projection from, one or more cranks or connecting rods dips into the oil and throws it up in a spray or mist that reaches all internal parts.

Many horizontal compressors have a flood system for bearing and crosshead lubrication; see Fig. 6-17. Oil is lifted from the reservoir by discs on the crankshaft and is removed from the discs by scrapers. The oil is then led to the bearings by passages, or allowed to flow down over the crosshead bearing surfaces.

A full pressure circulation system is often used for running gear lubrication. A positive displacement pump draws oil from the reservoir and delivers it under pressure to the main bearings (if plain) and crankpin bearings, then through drilled passages to the wristpin bearings and crosshead (if used). Where rolling element main bearings are used, the small controlled feed of oil required for them is commonly supplied by a drip or spray from the cylinder walls or rotating parts. Sometimes a jet of oil is directed toward these bearings. In some compressors, oil is supplied under pressure to the crankpin bearings from which it is thrown by centrifugal force to the cylinder walls. The wristpins are then lubricated by oil scraped from the cylinder walls and led to the pin bearing surfaces by drilled passages in the pistons.

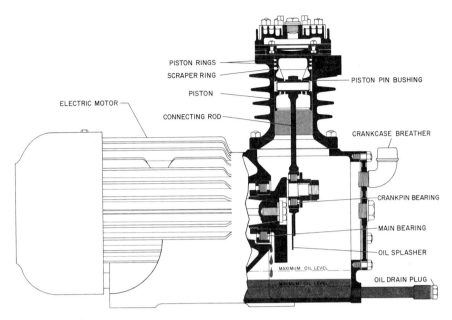

Fig. 14-1 Vertical Single Acting Air Compressor In this 3 hp machine designed for a discharge pressure of 100 psi (690 kPa) the motor and compressor are combined in a single compact unit. Lubrication is by a simple splash system.

Single and Two Stage Compressors Industrial single and two stage compressors range in size from fractional horsepower units to large machines of 20,000 hp (14,900 kW) or more. The smallest compressors are of the vertical single acting type, usually air cooled; see Fig. 14-1. Larger compressors are built in a variety of arrangements, including vertical single acting (air or water cooled), vertical double acting water cooled (Fig. 14-2), and horizontal balanced opposed compressor (Fig. 14-3). The largest machines are of the latter type. Machines may be single cylinder or multicylinder with cylinders in tandem, opposed, or in a "V" type (Fig. 14-4), or a "W" type (Fig. 14-5). Single stage machines are available for pressures up to about 150 psi (1030 kPa), while two stage machines are available for pressures up to about 1000 psi (6.9 MPa), although pressures this high are comparatively rare. Most two stage compressors are designed for pressures in the range of 80 to 125 psi (550 to 860 kPa).

Factors Affecting Cylinder Lubrication The operating temperature in compressor cylinders is an important factor, both because of its effect on oil viscosity and on oil oxidation, and the formation of deposits. At high temperatures, oil viscosity is reduced, so that when the operating temperature is high, higher viscosity oils are required in order to provide adequate lubricating films.

The thin films of oil on discharge valves, chambers, and piping are heated by contact with hot metal surfaces and are continually swept by the heated gas as it leaves the cylinders after compression. This is a severe oxidizing condition, and all compressor oils oxidize to an extent that depends on their

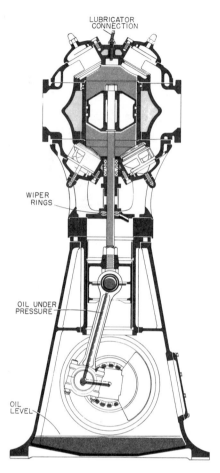

LUBRICATOR
CONNECTION

WIPER
RINGS

OIL UNDER
PRESSURE

OIL
LEVEL

Fig. 14-2 Vertical Double Acting Compressor In this single stage machine, rolling element main bearings are used. A pressure circulation system supplies oil to the crankpin bearings, and through drilled holes in the connecting rods to the crosshead pin bearings and crosshead guides. Oil draining back from the crosshead bearing is broken into a mist which lubricates the main bearings. A mechanical force feed lubricator supplies oil to the cylinders and wiper rings.

ability to resist this chemical change. Oil oxidation is progressive. At first the oxidation products formed are soluble in the oil, but as oxidation progresses these materials become insoluble and are deposited, mainly on the discharge valves and in the discharge piping, which are the hottest parts. After further baking, these deposits are converted to materials that are high in carbon content. These materials, along with contaminants that adhere and bond with them, are commonly called carbon.

Deposits on discharge valves may interfere with valve seating and permit leakage of hot, high pressure gas back into the cylinders. This high temperature gas heats the gas taken in on the suction stroke so that the temperature at the start of compression is raised and the final discharge temperature is also raised. This is called recompression. Since the action is repeated on each stroke, the effect is cumulative, although for a constant rate of leakage there is a tendency for the temperature to level off at some higher value. If leakage increases, there will be a further increase in temperature.

Deposits also restrict the discharge passages and cause cylinder discharge pressures to increase and overcome the pressure drop caused by the restriction. Again, the increased discharge pressure is accompanied by an increased discharge temperature.

Abnormally high discharge temperatures resulting from these effects result in more rapid oil oxidation and, therefore, contribute to further accumulation of deposits and temperature rise. This cycle may eventually result in a fire or explosion.

A variety of contaminants is often present in the gas being compressed. Hard particles abrade cylinder surfaces and may interfere with ring and valve seating. Some contaminants promote oil oxidation by catalytic action, and some entrained chemicals may react directly with oil to form deposits. Solid particles adhere to oil wetted surfaces and contribute to deposit build-up on discharge valves and in discharge passages. In many cases, the deposits found in compressors are largely composed of contaminants with a relatively small proportion of carbonaceous materials from oil oxidation.

All the oil fed to compressor cylinders is subjected to oxidizing conditions. Since most of the oil fed to compressor cylinders eventually leaves through the discharge valves where the temperatures are highest, keeping the amount of oil fed to a minimum helps to minimize the formation of deposits on these areas. Recommended rates of oil feed are shown in Table 14-2.

Moisture is a factor in single and two stage air compressors principally because of condensation that occurs in the cylinders during idle periods when the cylinders cool below the dew point of the air remaining in them. The water formed tends to displace the oil films, exposing the metal surfaces to rusting. The amount of rust formed during any single idle period may be

Fig. 14-3 Horizontal Balanced Opposed Compressor This construction is used for very large capacities. There may be from two to eight cylinders, each pair being arranged as shown. Oil is supplied to the cylinders and packings by a mechanical force feed lubricator (not shown), and to the running parts by a pressure circulation system. A gear pump, driven from the crankshaft or a separate motor, delivers oil to a main distribution line and from there to the precision insert main and crankpin bearings, the bronze crosshead pin bushings, and the crosshead slippers.

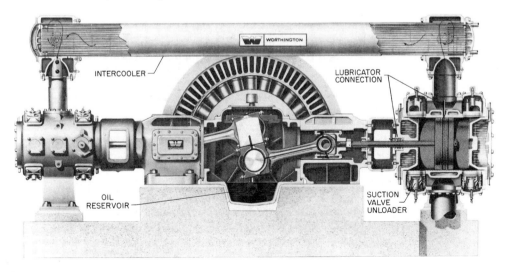

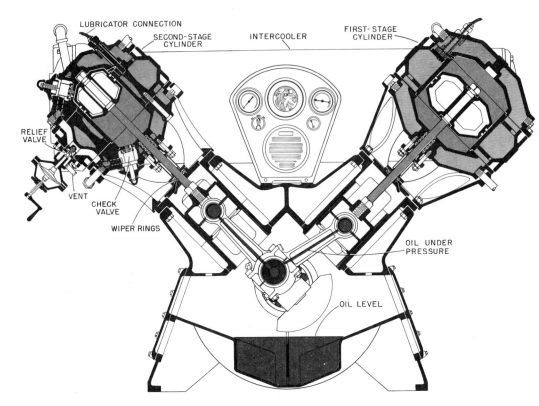

Fig. 14-4   V-Type Two Stage Double Acting Compressor   The cylinders are lubricated by means of a mechanical force feed lubricator that is geared to the crankshaft. Oil is delivered to the bearings through drilled passages by a pressure circulation system.

minor, and this may be scuffed off as soon as the compressor is started again, but in time the process will result in excessive wear. In addition, rust tends to promote oil oxidation, and the rust particles contribute to the formation of deposits.

Factors Affecting Running Gear Lubrication   In general, the factors that affect compressor bearing lubrication — loads, speeds, temperatures, and the presence of water and other contaminants — are moderate, so that the main requirement is that the oil be of suitable viscosity at operating temperature.

Much of the oil in circulation in compressor crankcases is broken up into a fine spray or mist by splash or oil thrown from rotating parts. Thus, a large surface of oil is exposed to the oxidizing influence of warm air, and oxidation will occur at a rate and to an extent that depends on the operating temperature and the ability of the oil to resist this chemical change. However, conditions that promote oil oxidation in crankcases are mild in comparison to the oxidizing conditions in compressor cylinders.

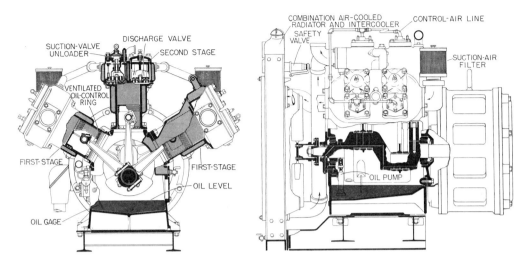

Fig. 14-5 W-Type Stage Single Acting Compressor The first stage has two cylinders in each bank, while the second stage has only one. Two banks with six cylinders form the complete compressor. The radiator contains two sections: an air cooled intercooler and a cooler for the cylinder jacket water. The crankpin bearings are lubricated by pressure from a plunger type pump. The tapered roller main bearings and cylinder walls are lubricated by oil thrown from the crankpin bearings.

Table 14-2   Rates of Oil Feed

Cylinder Diameter, in. (cm)	Drops Oil Per Minute (total for cylinder and packing)* Discharge Pressure of Cylinder, psi (kPa) gage					
	25–75 (172–517)	75–150 (517–1034)	150–250 (1034–1724)	300–600 (2068–4137)	600–1500 (4137–10.3 MPa)	1500–3000 (10.3–20. MPa)
6–8 (15–20)	3–6	5–8	6–9	7–12	9–16	12–18
8–10 (20–25)	4–7	6–9	8–11	10–15	12–20	
10–12 (25–30)	5–9	7–11	9–13	12–18		
12–14 (30–36)	6–10	9–12	11–16	14–21		
14–16 (36–41)	7–12	10–15	13–18			
16–18 (41–46)	8–13	12–17	15–20			
18–20 (46–51)	9–15	13–19	17–22			
20–22 (51–56)	10–17	15–21	19–24			
22–24 (56–61)	11–18	17–23	21–26			
24–26 (61–66)	12–20	18–25	23–28			
26–28 (66–71)	13–21	20–26	25–30			
28–30 (71–76)	14–23					
30–32 (76–81)	16–24					
32–34 (81–86)	17–26					
34–36 (86–91)	18–27					
36–38 (91–97)	19–28					
38–40 (97–102)	20–30					

*The figures given are for gravity and vacuum type sight-feed lubricators. For glycerine sight-feed lubricators, the figures given should be divided by three. Feeds to cylinder bores should never be less than one drop per outlet per minute under any conditions. The quantities given are based on an average of 8,000 drops per pint (16,900 drops per liter) of oil at 75°F (24°C).

Source: Reprinted from the "COMPRESSED AIR AND GAS HANDBOOK," Third Edition, Copyright 1966.

Multistage Compressors

Three stage compressors for continuous duty are available up to about 2500 psi (17 MPa). Four, five, six, and seven stage machines are available up to 60,000 psi (414 MPa) or higher. High discharge temperatures may or may not accompany these high discharge pressures. When possible, the machines are designed for compression ratios of 2.5 to 3.5 per stage, and with 60°F (15.6°C) suction temperature and adequate cooling. This practice limits discharge temperatures from each stage to less than 300°F (149°C). However, compression ratios as high as six are often employed, and with 60°F (15.6°C) suction temperature the calculated adiabatic discharge temperature is then approximately 400°F (204°C). Further, with a suction temperature of 110°F (43°C), which is not uncommon for the later stages, the discharge temperature may approach 500°F (260°C). Even higher compression ratios, and discharge temperatures, may be encountered in compressors designed for intermittent duty.

Factors Affecting Lubrication The factors of operating temperature, contaminants, and oil feeds discussed in connection with single and two stage compressors also affect the lubrication of cylinders in multistage compressors. In addition, the lubrication of multistage compressors is affected by the fact that water and oil are often entrained in the suction gas to the higher pressure stages, and by the high cylinder pressures.

Water in the form of droplets is often entrained in the air leaving intercoolers and carried into the higher pressure stages of multistage air compressors. This water, moving at high velocity into relatively small cylinders, tends to wash the lubricant film from the cylinder walls. With an unsuitable oil in service, this results in inadequate lubrication and excessive wear, incomplete sealing against leakage, and exposure of metal surfaces to rusting.

Some of the oil carried into the intercoolers usually is entrained with the water and carried into the higher pressure cylinders. While this oil contributes to lubrication of the succeeding stages, the results are not necessarily beneficial. This oil has already been exposed to oxidizing conditions in the lower pressure stages. Thus its total exposure is increased two or more times. As a result, rapid buildup of deposits may occur even though high temperatures are not necessarily involved.

High cylinder pressures acting behind the piston rings increase the rubbing pressure between the rings and cylinder walls. In addition, in trunk piston compressors, the connecting rods force against the cylinder wall with considerable pressure which increases with increasing compression pressure. The rubbing surfaces involved are parallel; movement between them does not tend to form thick oil films; and as the pressure increases, there is a greater tendency to wipe away the thin films that are formed.

The lubrication of the running gear of multistage compressors presents essentially the same problems as single and two stage machines.

Lubricating Oil Recommendation

As mentioned at the outset, the lubrication of compressors is not only a function of the type of compressor but also of the gas being compressed. In general, gases can be considered to be of four types: air, inert gases, hydrocarbon gases, and chemically active gases.

There are marked differences between compressor cylinder and bearing lubrication, but, in many cases, it is possible to use a single oil for the lubrication of both. In some cases, the oil required to meet the cylinder lubrication requirements may not be suitable for bearing lubrication. For example, it may be too high in viscosity, may contain special compounding that is not compatible with materials used in some crankcases, or may be of a special type that is not suitable for the extended service expected of bearing lubricating oils. Except in the case of air compressors, the following discussion pertains to the characteristics of oils for cylinder lubrication. The oils used for bearing lubrication of air compressors are usually suitable for the lubrication of bearings of compressors handling other gases. Where a problem might exist, the crankcase system is usually adequately isolated from contamination by the gas or the cylinder oil.

Air Compressors   The oils recommended for, and used in, air compressors vary considerably. Such factors as discharge pressure and temperature, ambient temperature, whether the air is moist or dry, and design characteristics of the machine must all be considered in the selection.

Single and two stage trunk piston type stationary compressors operating at moderate pressures and temperatures with dry air are generally lubricated with premium, rust and oxidation inhibited oils. In portable service, these compressors are now frequently lubricated with detergent-dispersant engine oils, typically oils for API Service SC, SD, SE, CA, CB or CC. The engine oils are also being used in many stationary compressors where the air is moist, or deposit or wear troubles have been experienced with circulation oils. The viscosity grade used is frequently ISO Viscosity Grade 68 or SAE 20, but both lower and higher viscosity oils are used depending on ambient temperature and machine requirements.

Under mild conditions, rust and oxidation inhibited turbine oils are also used as cylinder lubricants for crosshead type compressors. Under wet conditions, compounded oils are used. The compounding may be either a fatty oil or synthetic materials. Depending on the pressure and cylinder size, oil viscosities may range from ISO Viscosity Grade 68 to as high as ISO Viscosity Grade 460. During the breaking-in period particularly, higher viscosity oils are used and oil feed rates are increased.

In many cases where high discharge temperatures have resulted in rapid buildup of valve and receiver deposits with conventional lubricating oils, blends of mineral oils with synthetic materials or straight synthetic oils (usually diester based) are used. In the case of the blends, the synthetic material helps to dissolve and suspend potential deposit forming materials so that the buildup of deposits is slowed. The straight synthetics offer both the ability to dissolve and to suspend potential deposit forming materials, and offer better resistance to thermal degradation. These lubricants are usually of ISO Viscosity grade 100, although higher viscosity blends are used for extremely high pressure cylinders, or where severe moisture conditions are encountered.

In a number of cases where receiver fires have been experienced, fire resistant compressor oils are used. These oils are phosphate ester based. For larger machines and higher pressures, or with moist air, the viscosity is usually ISO Viscosity Grade 100. Lower viscosity grades may be used under moderate conditions with dry air.

Where there is a need for oil free air, reciprocating compressors, which operate without cylinder lubrication, are now being offered. These compressors have polytetrafluoroethylene (PTFE), carbon, or filled composition rings. In some newer designs, the pistons may also be filled plastic composition.

The running gear of crosshead compressors is lubricated normally with premium, rust and oxidation inhibited, circulation oils. In some cases where a synthetic oil is being used as the cylinder lubricant, the same oil is recommended for the running gear with considerably extended oil drain intervals.

Inert Gas Compressors Inert gases are those gases that do not react with lubricating oils, and that do not condense on the cylinder walls at the highest pressures reached during compression. Examples are nitrogen, carbon dioxide, carbon monoxide, helium, hydrogen, neon, and blast furnace gas (largely nitrogen, carbon monoxide, and carbon dioxide). Ammonia is relatively inert but some special considerations apply to it.

These gases generally introduce no special problems and the lubricants used for air compressors are suitable for compressors handling them. However, since carbon dioxide is slightly soluble in oil, it tends to reduce the viscosity of the oil. If moisture is present, carbonic acid, which is slightly corrosive, will form. To minimize the formation of carbonic acid, the system should be kept as dry as possible. To counteract the dilution effect, higher viscosity oils than are used in air compressors are desirable.

Ammonia is usually compressed in dynamic compressors, but occasionally it may be compressed in reciprocating machines. In the presence of moisture, it can react with some oil additives and oxidation products to form soaps. It may also dissolve to some extent, diluting the oil. Highly refined straight mineral oils are usually used. Synthesized hydrocarbon base lubricating oils are also being used because of their low solubility for ammonia.

When compressing gases for human consumption, such as carbon dioxide for use in carbonated beverages, carryover of conventional lubricating oils is undesirable. Generally, medicinal white oils are required for cylinder lubrication under these circumstances. This does not apply when air is being compressed for the manufacture of oxygen for human consumption. The subsequent liquification of the air and distillation to separate the oxygen will leave the oxygen free of lubricating oil carryover. However, care in selection of the lubricant, controlling the rate of oil feed, and keeping the system clean is required to minimize "burned oil" odor that can be carried over.

Petroleum hydrocarbons in the lungs can cause suffocation and possible pulmonary disease. Therefore, air compressors for scuba diving equipment should use teflon rings which require no lubrication.

Under some circumstances, conventional petroleum oils cannot be used in inert gas compressors. This is the case in some process work where traces of hydrocarbons cannot be tolerated in the process gas, or some constituents of lubricating oils might poison catalysts used in later processes. Compressors similar to those used to produce oil-free air are used where hydrocarbon carryover cannot be tolerated. Both special low sulfur, straight mineral oils and polybutenes are used where carryover might poison catalysts.

In some cases, a compressor may be used alternately to compress an inert gas and a chemically active gas; for example, hydrogen and oxygen. Petroleum products form an explosive combination with oxygen. Therefore, they must not be used where oxygen is being compressed. The lubricant or lubrication system used with the oxygen must be used with the inert gas.

Hydrocarbon Gas Compressors   More horsepower is consumed compressing natural gas than any other gas except air. When the volumes of other hydrocarbons that are compressed for the chemical and process industries are considered, the total horsepower consumed in compressing hydrocarbons is extremely large. Where the hydrocarbons must be kept free of lubricating oil contamination, dynamic compressors usually are used, but where high pressures are required, reciprocating compressors must be used.

While natural gas is mainly methane, other gases such as ethane, carbon dioxide, nitrogen, and heavier hydrocarbon gases usually are present in some proportion. These heavier hydrocarbons are similar in many respects to the hydrocarbons that are compressed for process purposes.

The temperature at which a material will condense from the gaseous state to the liquid state (also the temperature at which it will pass from its liquid state to the gaseous state, i.e., its boiling point) increases with increasing pressure. With the higher boiling, heavier hydrocarbon, the condensation temperature may be above the cylinder wall temperature at the pressure in the cylinders. The condensate formed:

1. Tends to wash the lubricant from the cylinder walls
2. Dissolves in the lubricating oil reducing its viscosity

Viscosity reduction may also result when hydrocarbon gases go directly into solution in the lubricating oil.

The dilution effect can generally be compensated for by using an oil that is somewhat higher in viscosity than would be used for air under the same operating conditions. Generally, compounded oils help to resist washing where condensed liquids are present in the cylinders. It is usually advisable also to operate with somewhat higher than normal cooling jacket temperatures in order to minimize condensation. This also requires the use of higher viscosity oils.

As it comes from the well, natural gas contains sulfur compounds and is referred to as "sour." Compressors handling sour gas are usually lubricated with detergent-dispersant engine oils. These oils provide better protection against the corrosive effects of the sulfur. The viscosity grade used most frequently is SAE 30, but if the gas is wet, i.e., carrying entrained liquids, oils of SAE 50 or heavier are used. The compressor of integral engine-compressor units is usually lubricated with the same oil used in the engine.

Chemically Active Gas Compressors   Among the chemically active gases that must be considered most frequently are oxygen, chlorine, hydrogen chloride, sulfer dioxide, and hydrogen sulfide.

Petroleum oils should not be used with oxygen since they form explosive combinations. Oxygen compressors with metallic rings have been lubricated with soap solutions. Compressors with some types of composition rings have been lubricated with water. Compressors designed to run without lubrication are also being used. Some of the inert synthetic lubricants, such as the chlorofluorocarbons, can be used safely and provide good lubrication.

Petroleum oils should not be used for the lubrication of chlorine and hydrogen chloride compressors. These gases react with oil to form gummy sludges and deposits. If the cylinders are opened to remove these deposits, rapid corrosion takes place. Concentrated sulfuric acid is sometimes used to lubricate chlorine compressors. Compressors designed to run without lubrication are being used.

Sulfur dioxide dissolves in petroleum oils, reducing the viscosity. It may also form sludges by reacting with additives in the presence of moisture, or by selective solvent action. The system must be kept dry to prevent the formation of acids. Highly refined straight mineral oils or white oils from which the sludge forming materials have been removed, either by acid treating or hydrogen processing, are used. Oil feed rates should be kept to a minimum.

Hydrogen sulfide compressors must be kept as dry as possible since hydrogen sulfide is corrosive in the presence of moisture. Compounded oils are usually used, and rust and oxidation inhibitors are considered desirable.

ROTARY COMPRESSORS

The four main types of rotary positive displacement compressors are: straight lobe, helical lobe, rotary vane, and liquid piston. In addition, variations of the helical lobe type are now offered.

Straight Lobe Compressors Straight lobe compressors are built with identical two or three lobed impellers that rotate in opposite directions inside a closely fitted casing; see Fig. 14-6. The impellers are driven by timing gears outside the case which maintain the relative positions of the impellers. The impellers do not touch each other or the casing, and no internal lubrication is required. No compression occurs in the case since the impellers simply move the gas through. Compression occurs due to back-pressure from the discharge side. Compression ratios are low and for this reason, these machines are often referred to as blowers rather than compressors. Straight lobe compressors are available in capacities up to about 30 000 cfm (51 000 m³/h) and for single stage discharge pressures up to about 25 psi (172 kPa).

Lubricated Parts The principal parts of straight lobe compressors requiring lubrication are the timing gears and the shaft bearings.

The bearings may be either plain or rolling element (usually roller) type. Timing gears are of precision spur or helical type. The bearings and gears are generously proportioned to minimize unit loads and wear since the small clearances between the impellers and casing must be maintained for efficient operation. Most bearings are lubricated by means of integral circulation systems. Some rolling element bearings at the end opposite the drive may be grease lubricated.

Lubricating Oil Recommendations Lubricating oils used in straight lobe compressors are usually turbine quality, rust and oxidation inhibited oils. The use of oils with antiwear properties may be desirable to provide improved gear life. Viscosities usually recommended are ISO Viscosity Grade 150 for normal ambient temperatures, 220 for high ambient temperatures, and 100 or even lower where extreme low temperatures are encountered.

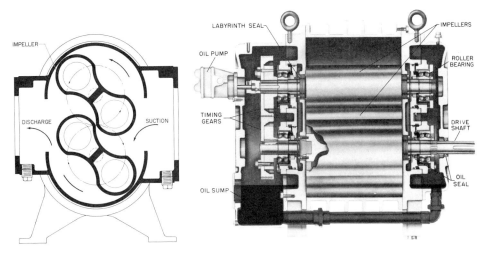

Fig. 14-6 Rotary Straight Lobe Compressor As the impellers rotate, air (or gas) is drawn in at the suction opening, trapped between the impellers and casing, then forced into the discharge area against the pressure that exists there. Lubricating oil is supplied under pressure to the bearings and timing gears by means of an integral circulation system.

Helical Lobe Compressors

These compressors, commonly called screw compressors, are available in two impeller and single impeller designs. The single impeller type was introduced quite recently and is still not in wide use.

In the two impeller type, a four lobed male rotor meshes with a six lobed female rotor (Fig. 14-7). The rotors may be individually driven by timing gears, or the male rotor may drive the female rotor. Gas is compressed by the action of the two meshing rotors as illustrated in Fig. 14-8.

Two variations of two impeller screw compressors are used. In the "flood lubricated" type, oil is injected into the cylinder to absorb heat from the air or gas as it is being compressed, and act as a seal between the rotors. Since oil is available in the cylinder to lubricate the rotors, these machines are now usually built without timing gears. They require an external circulation system to control the temperature of the oil and an oil removal system to remove most of the oil from the discharge air or gas; see Fig. 14-9.

In "dry" screw compressors, no oil is injected. Since the rotors are not lubricated, timing gears are required. These compressors can be used to deliver oil-free air or gas. However, because there is no oil seal between the rotors, operating speed must be relatively high to minimize gas leakage. Water cooling of the cylinder and such features as oil cooled impeller shafts (Fig. 14-7) are usually required.

Two impeller screw compressors are available in capacities up to about 25,000 cfm (42,500 m³/h) and for discharge pressures up to 125 psi (860 kPa) single stage and 250 psi (1720 kPa) two stage.

The single screw compressor (Fig. 14-10) has a driven conical screw which rotates between toothed wheels. The teeth of the wheels sweep the thread cavities of the screw, compressing the air as it is moved up the progressively decreasing volume of the cavity. The wheels are made of two materials:

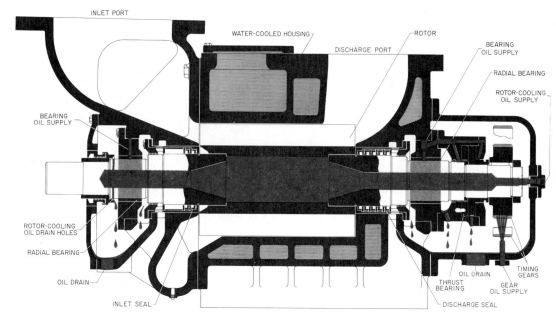

Fig. 14-7 Two Impeller Helical Lobe Compressor This large stationary machine has timing gears and no oil is used inside the casing. The machine is water cooled, with oil cooled impeller shafts. Multiple carbon ring shaft seals are shown. Shaft bearings are plain, and a tilting pad thrust bearing is used.

1. A metal backing which absorbs the stresses imposed on the wheels
2. A wider plastic facing which contacts the sides of the thread cavities

Oil is injected into the case for lubrication, cooling, and sealing. As with flood lubricated, two impeller, screw compressors, oil cooling and oil removal equipment are required as part of the compressor running gear.

Lubricated Parts   The lubricated parts of dry helical lobe compressors are the gears and bearings. Shaft bearings may be either plain or rolling element, and thrust bearings may be either tilting pad or angular contact rolling element. Gears may be either precision cut, spur, or helical type.

In flood lubricated machines, in addition to the gears and bearings, the oil lubricates the contacting surfaces of the rotors. In both types, oil lubricated seals may be used where gas leakage must be kept to a minimum.

Lubricant Recommendations   Flood lubricated screw compressors in portable and plant air applications are lubricated with oils of about SAE 10W or ISO Viscosity Grade 32 (some builders recommend an SAE 20 grade). Heavy duty engine oils suitable for API Service SC, SD, SE, CA, CB, or CC are used widely, as are automatic transmission fluids (both DEXRON®* and Ford type). Premium, rust, and oxidation inhibited oils are used where low ambient temperatures, cyclic operation, or very humid conditions are en-

*Registered trademark.

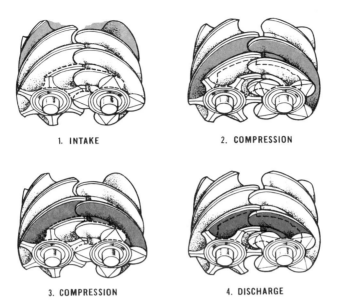

1. INTAKE   2. COMPRESSION

3. COMPRESSION   4. DISCHARGE

Fig. 14-8 Compression in Screw Compressor As the screws unmesh at the rear and below the centerline of the view labeled "intake," the expanding space created draws gas in through the inlet, filling the grooves between the screws. As the screws advance, they seal off this air and carry it forward, compressing it against the discharge end head plate.

Fig. 14-9 Lubrication System for Flood Lubricated Screw Compressor When the oil is cold, the temperature control valve is open, bypassing the oil cooler. As the oil warms up the valve gradually closes, and is completely closed when the oil temperature reaches about 180°F (88°C). All of the oil then passes through the oil cooler. The arrangement provides rapid warmup of the oil to minimize condensation, but limits the maximum oil temperature to provide long oil life.

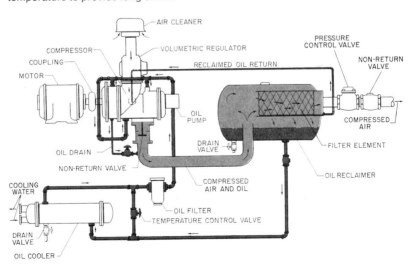

Fig. 14-10 Single Screw Compressor The gear wheels are shown resting in the screw to indicate the manner in which they sweep the thread.

countered. Under any of these conditions, the condensation that occurs in the cylinder may cause troublesome emulsions with the engine oils, so the circulating oils with good water separating ability are preferred.

Even with the engine oils and automatic transmission fluids, problems with varnish and sludge can be encountered in heavy duty operations. Diester based synthetic lubricants are now being used in some of these applications. Synthesized hydrocarbon based lubricants, or blends of synthesized hydrocarbons and synthetic esters are also being developed for use in these applications.

Flood lubricated screw compressors are not being used extensively for compressing gases other than air, although some applications of this type are being developed. Because of the intimate mixing of the oil with the gas, oil selection is even more critical than in reciprocating compressors. In handling hydrocarbon gases, polyglycol based synthetic lubricants are used. The polyglycols have low solubility for hydrocarbons, so dilution of the lubricant is not a problem, and good separation of the oil and gas can be obtained.

Dry screw compressors require only lubrication of the gears and bearings. Premium, inhibited circulating oils of a viscosity suitable for the gears are usually used. In some cases, oils with enhanced antiwear properties may be desirable.

Sliding Vane Compressors

Sliding vane compressors have vanes that are free to move in slots in a rotor mounted eccentrically in a casing; see Fig. 14-11. Rotation of the rotor causes the vanes to move in and out in the slots, creating pockets that increase and decrease in volume. The vanes can be held out against the case by coupling pins as in this case, by springs, or by centrifugal force alone. Where pins are used, the casing is not circular since all diameters through the axis of the rotor must be equal.

Sliding vane compressors are available in capacities up to about 6000 cfm (10,200 m³/h), and for discharge pressures up to about 100 psi (690 kPa) for single stage, and 125 psi (860 kPa) for two stage.

Two general types of cooling are used for rotary vane compressors. Large stationary machines are water cooled. Portable and plant air machines may be air or water cooled, with the addition of oil injection into the cylinder. As with screw compressors, these latter machines may be referred to as flood lubricated, or by such terms as oil injection cooled or direct oil cooled. As with flood lubricated screw compressors, an external system to remove the oil from the discharge air and cool the oil is required.

Lubricated Parts All of the sliding surfaces in the cylinder require oil lubrication to minimize friction and wear. The lubricating oil also aids in protecting internal surfaces from rust and corrosion, and helps seal the clearances between vanes, rotors, cylinder walls, and heads. Shaft bearings of either plain or rolling element type are designed for oil lubrication. Where the shaft passes through the head, packing glands or seals are used, and these are supplied with oil to minimize wear and friction and to assist in preventing leakage of high pressure air or gas.

Lubricant Recommendations In sliding vane compressors, some of the cylinder surfaces are subjected to heavy rubbing pressures as a result of the action of the vanes in their slots. The forces acting on the extended vanes tend to tilt the vanes backward, causing heavy pressure at the inner leading edges of the vanes and at the trailing edges of the slots. This results in high resistance to the inward motion of the vanes and, consequently, heavy pressure between the vane tips the cylinder wall. This tends to rupture the oil films and cause wear and roughening of the surfaces involved.

Flood lubricated rotary vane compressors are now usually lubricated with the same oils recommended for flood lubricated helical lobe compressors. These oils give good protection against wear under the conditions encountered in the cylinders.

Large stationary rotary vane compressors differ considerably in their lubrication requirements from the flood lubricated type. Water jackets are used to cool these machines. Cylinder lubricant is fed by force feed lubrica-

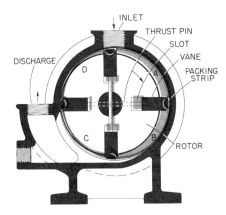

Fig. 14-11 Rotary Sliding Vane Compressor The ports are located so that air (or gas) is drawn into pockets of increasing volume A and discharged from pockets of decreasing volume B. In this design the bore of the cylinder is not circular since the total length of two vanes plus a pin, measured through the axis of the rotor, is constant.

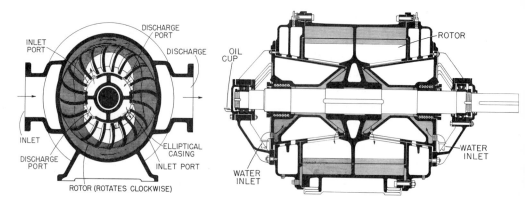

Fig. 14-12 Rotary Liquid Piston Compressor The water annulus, in combination with the oval casing, forms pockets between the rotor blades that increase in volume much like those in a sliding vane compressor. However, since there is no contact between the rotor blades and casing, no internal lubrication is required. Water is fed continuously to replace water carried out with the compressed gas.

tors. Feed rates, in relation to compressor capacity, must be much higher than for reciprocating machines. These machines tend to run cool at the inner cylinder walls, so condensation is a problem. Heavy duty engine oils of SAE 30 or 40 grade are widely used. Where the discharge temperature is above about 300 to 325°F (149–163°C), a higher viscosity oil is usually recommended. Compounded oils of comparable viscosity are also used.

Liquid Piston Compressors   In a liquid piston compressor (Fig. 14-12), rotation of the multibladed rotor causes a liquid, usually water, to be thrown outward forming an annulus that rotates at rotor speed and maintains a seal between the rotor blades and the housing. As the annulus rotates, the pockets formed between the blades and the annulus increase and decrease in volume in the same manner as in a sliding vane compressor. Because of the annulus, the suction and discharge ports must be located in the cylinder heads, rather than on the outer perimeter of the cylinder. The liquid annulus eliminates the need for sliding, sealing surfaces so there are essentially no wearing surfaces inside the cylinder.

Liquid piston compressors are available in capacities up to about 16,000 cfm (27,300 m³/h) and for pressures up to about 100 psi (690 kPa).

Lubricated Parts   The only parts of liquid piston compressors requiring lubrication are the shaft bearings. All models have rolling element bearings which may be lubricated by oil bath or with grease.

Lubricant Recommendations   Oil lubricated bearings of liquid piston compressors are usually lubricated with permium, inhibited circulating oils. Viscosities are selected according to the bearing requirements, as discussed in Chap. 5. Grease lubricated bearings are lubricated with premium, ball, and roller bearing greases.

Dynamic Compressors   In contrast to positive displacement compressors where successive volumes of gas are confined in an enclosed space, which is then decreased in size to accomplish compression, dynamic compressors first convert energy from the prime mover into kinetic energy in the gas. This kinetic energy is then converted into pressure. The rotors of dynamic compressors do not form a tight seal with the casings, thus, some leakage occurs past the rotor blades.

Centrifugal Compressors

In a centrifugal compressor, a multibladed impeller rotates at high speed in a casing; see Fig. 14-13. Air or gas trapped between the impeller blades is accelerated and thrown outward and forward in the direction of rotation. The air leaves the blade tips with increased pressure and high velocity and enters a diffuser ring. In the diffuser ring, due to increasing area in the direction of flow, a reduction in velocity and a substantial increase in pressure take place. The air then enters a volute casing where, again due to increasing area in the direction of flow, a further reduction in velocity and increase in pressure take place. The outward flow of air through the impeller creates a reduced pressure at the inlet, or eye, or the impeller, causing air to be drawn into the compressor.

Centrifugal compressors are particularly adapted to supplying large volumes of air (or gas) at relatively small increase in pressure. They are inherently suited to high speed operation—up to 20 000 rpm or higher—and can, therefore, be direct connected or geared to high speed driving motors or tur-

Fig. 14-13 Single Stage Centrifugal Compressor This is what is called a "pedestal" machine. The impeller is overhung on a short shaft supported by two bearings mounted on a pedestal. The thrust bearing is of the fixed shoe type.

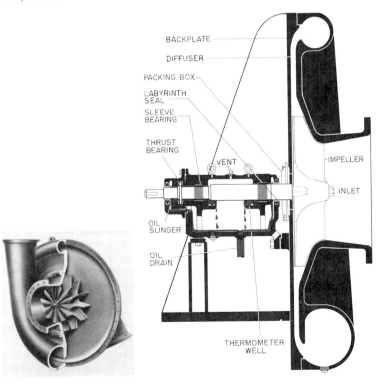

BACKPLATE

DIFFUSER

PACKING BOX

LABYRINTH SEAL

SLEEVE BEARING

THRUST BEARING

VENT

IMPELLER

INLET

OIL SLINGER

OIL DRAIN

THERMOMETER WELL

bines. Speed increasing drives may also be used. Centrifugal compressors are built with capacities up to about 650,000 cfm (1,100,000 m³/h).

Compression ratios range up to about 1.7 with one impeller or stage. Thus, single stage machines with atmospheric intake may be designed for discharge pressures up to about 10 psi (69 kPa). Multistage compressors with up to 10 stages are built. Discharge pressures may be 150 psi (105 kPa) or higher.

Centrifugal compressors are used extensively as boosters in petroleum refineries, chemical process industries, and natural gas pipelines. In booster service, air or gas is taken in at a pressure above atmospheric and discharged at a still higher pressure. Where suction pressures are high, casings and shaft seals must be designed to withstand the high pressures involved.

Centrifugal compressors are being built for plant air applications. These units typically have capacities from 100 to 15,000 cfm (1700–15,500 m³/h) at pressures of up to 125 psi (860 kPa). From two to five stages are used, with three or four stages being the most common. The impellers are driven by pinions mounted around a large bull gear. In this way each impeller is driven at its optimum speed. Where a turbine drive is used, the bull gear may be driven directly from the turbine. With an electric motor drive, the bull gear may be driven through a speed increaser. Intercooling between stages is normally used.

Centrifugal compressors discharging at low pressures usually have no provision for cooling. Higher pressure machines are cooled by water passages in the diaphragms between stages by intercooling between stages, or by injecting a volatile liquid into the gas stream.

Where the shaft passes through the casing, seals are provided. Several types are used: soft braided packing, labyrinth seals, carbon ring seals, oil film seals, and mechanical (end face) seals. The type used depends on such factors as pressure, speed, and the amount of leakage that can be tolerated.

Lubricated Parts The bearings of centrifugal compressors require lubrication, but there is no need for lubrication within the impeller casing. Oil film seals used on high pressure compressors, and contact seals, must be supplied with oil.

With a single impeller, or a group all facing in the same direction, there is an axial thrust due to the backward force exerted by the discharging gas on the rim of the impeller. This thrust force can be balanced internally by means of a balancing drum or dummy piston, but a thrust bearing must be used to support all or part of the thrust load and keep the impeller accurately located in the casing. Thrust bearings are commonly of annular contact ball, collar, fixed shoe, or tilting shoe type.

Shaft bearings may be of plain or rolling element type. Plain bearings, including collar, fixed shoe, and tilting shoe thrust bearings, are oil lubricated. Rolling element bearings may be either oil or grease lubricated.

Where a compressor is driven by a step-up gear set, the drive gears and one or more of the bearings may be lubricated from the same system. In the case of the packaged plant air units, the bull gear and bearings are contained in an oil tight casing, and oil is circulated to them under pressure from a shaft driven oil pump. Where a speed increaser is used, it is lubricated from the system that supplies the bearings.

Fig. 14-14 Six Stage Axial Flow Compressor This compressor is equipped with plain journal bearings and a tilting shoe thrust bearing. Oil is supplied to the bearings by a circulation system.

Axial Flow Compressors

An axial flow compressor contains alternating rows of moving and fixed blades. High velocity is imparted by the moving blades to the air (or gas) being compressed. As the air flows through the expanding passages between the fixed and moving blades, the velocity is reduced and transformed into static head, that is, pressure. A stage consists of one row of moving and one row of stationary blades, and a relatively large number of stages — up to 20 in one casing — are usually used; see Fig. 14-14. Axial flow compressors are compact, high speed machines suitable for handling large volumes of air at compression ratios up to about 6:1 in a single cylinder or casing. Machines capable of handling 1,000,000 cfm (1,700,000 m³/h) or more have been built. They may be driven by electric motors, steam turbines, or gas turbines, the latter fairly common in chemical process industries where large volumes of hot gas are available to drive the turbine, and large volumes of compressed air or gas are required. Axial flow compressors are also used in all except the smaller sizes of gas turbines—whether of industrial or aircraft type—that are used for power generation and ship propulsion.

Lubricated Parts As with centrifugal compressors, the parts of axial flow compressors requiring lubrication are the shaft bearings, the thrust bearings that are used to take axial thrusts and maintain axial rotor position, and any oil lubricated seals that may be used.

Lubricant Recommendations Since the lubricants used in centrifugal and axial flow compressors are not exposed to the conditions inside the cylinder or casing, lubricant selection is based on the requirements of the bearings. Where a speed increaser supplied from the same lubricating oil system is used, the oil must be selected to meet the gear requirements, which are generally more severe than those of the bearings. Premium, inhibited circulating oils are used most commonly. Where only bearings require lubrication, oils of ISO Viscosity Grade 32 are usually used. Where a speed increaser is used, somewhat higher viscosity oils, typically ISO Viscosity Grade 68, are usually used.

In packaged plant air units, there is considerable tendency for air to be whipped into the oil and foaming to result. Oils with good air release properties and foam resistance are preferred. The premium, long life turbine oils in ISO Viscosity Grade 32 are usually recommended.

Where a dynamic compressor is driven by a gas turbine, both machines may be on the same system. The lubricant selection is then based on the requirements of the gas turbine. Both specially developed mineral oil based and synthesized hydrocarbon based lubricants are used. Generally, the viscosity used is ISO Viscosity Grade 32.

Some difficulties have been experienced with dynamic compressors handling ammonia or "syngas" for ammonia production. As pointed out under reciprocating compressors, ammonia can react with some inhibitors and oil oxidation products to form insoluble soaps and sludges. These soaps and sludges can be thrown out in the coupling connecting the speed increaser to the compressor. This can cause coupling imbalance with vibration, and interfere with lubrication of the coupling. Deposits may also form in the speed increaser. Highly stable oils with inhibitors that are not affected by ammonia have been effective in reducing these problems. Both mineral oil and synthesized hydrocarbon base products are used. The viscosity used is usually ISO Viscosity Grade 32.

REFRIGERATION AND AIR CONDITIONING COMPRESSORS

The basic principles of the refrigeration compression cycle are shown in Fig. 14-15. The five essential parts basic to every such system are shown: evaporator, compressor, condenser, receiver, and expansion valve (or capillary). Liquid refrigerant flows from the receiver under pressure through the expansion valve to the evaporator coils where it evaporates, absorbing heat and cooling the room (or other space where cooling is desired). The vapor is then drawn into the compressor where its pressure and temperature are raised. At the higher pressure at the discharge of the compressor, the con-

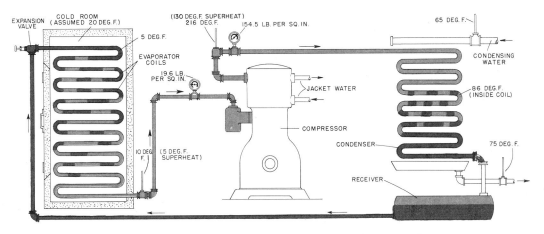

Fig. 14-15 Basic Single Stage Compression Refrigeration System The elements shown here are common to all compression refrigeration systems, whether air refrigeration or air conditioning.

densing temperature of the refrigerant is higher than it would be at atmospheric pressure, so when the hot, high pressure vapor flows from the compressor to the condenser, the cooling water removes enough heat from it to condense it. This liquid refrigerant then flows to the receiver ready for another cycle. The heat removed from the refrigerant in the condenser is equal to the amount of heat removed from the cold room plus the heat resulting from the mechanical work done on the refrigerant in the compressor that is not removed by the jacket cooling of the compressor. In many cases in commercial installations, the evaporator cools a heat transfer fluid such as brine which is then pumped through the area to be cooled. Smaller units such as home refrigerators and freezers, room air conditioners, and automotive air conditioners have air cooled conditioners rather than water cooled.

In commercial installations, two or three stages of compression may also be used. Two stage compressors are used, or combinations of separate single stage compressors. Rotary sliding vane and screw compressors are sometimes used as the low pressure or booster stage. Multistage centrifugal compressors are used for large air conditioning installations. Sliding vane compressors are now being introduced for automobile air conditioning systems. Some very small units such as dehumidifiers may be equipped with diaphragm type compressors. In nearly all other cases, reciprocating compressors are used.

Most reciprocating compressors for commercial installations are of the single acting, trunk piston type and have closed crankcases. Due to refrigerant leakage past the pistons, the crankcases are then filled with a refrigerant atmosphere. The same is true of the axial piston units used for automobile air conditioning. Crosshead and double acting compressors have open crankcases. The majority of small to medium size electric motor driven refrigeration and air conditioning units are hermetically sealed, with all of the operating parts, including the electric motor, inside the sealed unit. Evaporators may operate either dry or flooded. In dry evaporators, only refrigerant vapor is present, flooded evaporators have both liquid and vapor present.

Factors in the Compressor Affecting Lubrication

Cylinder Conditions The oil film on the cylinder walls of a reciprocating refrigeration compressor is subjected to low temperatures at the suction ports and to moderately high temperatures near the cylinder head. Since viscosity decreases with temperature, the oil near the suction ports will have considerably higher viscosity than the oil near the cylinder head. Nevertheless, the oil must spread in a thin film over the entire working surfaces. This is accomplished by the piston rings (or the piston itself in small compressors without piston rings) as the pistons move back and forth. The oil must distribute rapidly, but to do this it must not be too high in viscosity. On the other hand, too low a viscosity oil will not protect against wear.

Oil carried out of the cylinders to the valves and discharge piping is subjected to the temperature of the discharging refrigerant. Ordinarily the temperature of the discharging refrigerant is not high; for example, the discharge temperature of a single stage ammonia compressor with a compression ratio of about 5:1 should not be much in excess of 250°F (121°C). Some single stage units operate at higher ratios, and higher discharge temperatures, but in most small compressors the valve temperatures remain moderate because of the relatively large cooling area in proportion to cylinder volume. The discharge temperatures of compressors operating on fluorocarbon refrigerants are lower than equivalent machines operating on ammonia, although the compressors of automobile air conditioning systems may operate at quite high discharge temperatures.

When two or more stages of compression are used, the operating temperature in each stage will usually be lower than in single stage machines. In rotary compressors, discharge temperatures are also usually moderate because of low compression ratios.

Oxidation In compressors with closed crankcases, temperatures are normally moderate and the entire machine is filled with refrigerant vapor, only a small amount of air being present. Under these conditions oxidation, in the usual sense, does not occur, although it is doubtful that it can be avoided entirely. Limited oxidation does not impair the lubricating value of an oil since the initial oxidation products formed are soluble in oil. If oxidation progresses too far, a condition is eventually reached where some of the soluble oxidation products become insoluble when the oil is cooled.

Bearing System Conditions The general requirements of the bearing systems of refrigeration compressors are similar to those of other comparable compressors. However, some special factors must be considered.

In compressors with closed crankcases, there is very little exposure to oxygen so the oxidation stability of the oil is not a major concern. However, where the same oil is used for both the bearings and cylinders, as in many small units, the oil must have adequate oxidation stability for the cylinder conditions.

Where ammonia is used as the refrigerant in compressors with closed crankcases, any additives used in the oil must be types that are not affected by ammonia. If the refrigerant is soluble or partially soluble in the oil, such as the majority of the fluorocarbon refrigerants, this will dilute the oil and reduce its viscosity, which must be considered in the selection of the oil viscosity.

The motor in hermetically sealed units is completely surrounded by a mixture of refrigerant and oil. As a result, the oil must have good dielectric properties, must not affect the motor insulation, and must not react with the copper motor windings, and other metals at elevated temperatures. Since most of these units are operated on fluorocarbon refrigerants, the dilution effect of the refrigerant on the viscosity must also be considered.

In compressors where the crankcase and cylinders are completely isolated from each other, as in units having crosshead construction, the oil in the crankcase is exposed to air and there is intimate mixing of the warm oil with air. These conditions are favorable to oxidation and require a chemically stable oil to resist oxidation.

Factors in the Refrigeration System Affecting Lubrication

If the oil carried out of the compressor cylinder forms gummy deposits in the condenser, or if it congeals or forms waxlike deposits in the evaporator, it may seriously obstruct the flow of heat. Heat insulating deposits in the evaporator make it necessary to carry a lower evaporator temperature in order to produce the required refrigeration effect. This in turn requires a lower evaporator pressure and increases the power required by the compressor for a given refrigeration duty, due to the increased pressure range through which the gas must be compressed. In addition, at the lower suction pressure the vapor density is lower so the compressor will have to handle a greater volume of vapor, reducing refrigerating capacity. Heat insulating deposits in the condenser increase the temperature difference between the cooling medium (water or air) and the condensing refrigerant. The resulting higher condenser temperature makes higher compression necessary and increases power consumption.

Whether or not heat insulating deposits will be formed depends on the properties of the lubricating oil, the refrigerant in use, the evaporator temperature, and the equipment used in the system. The effects of some of the more common refrigerants are considered separately.

Fluorocarbons This class of refrigerants can be divided into three groups according to relative miscibility with oil:

1. R-13* and R-14 are not miscible with oil in any proportions under the conditions found in refrigeration systems
2. R-11, R-12, R-113 and R-500 are miscible with oil in all proportions at all temperatures and pressures.

*R = refrigerant. R-13 and other standard American Society of Heating, Refrigerating and Air Conditioning Engineers (ASHRAE) refrigerant designations are also widely known by trade names such as Freon 13, Isotron 13, Genetron 13.

3. R-22, R-114 and R-502 are generally miscible with oil under the condition found in the high side of the system, that is the condenser, but are only partly miscible or immiscible in the evaporator.

R-13 and R-14 are generally used for extremely low temperature refrigeration systems. Under these conditions, oils suitable to cylinder lubrication will solidify in the evaporator, so it is necessary to arrange the system so that no oil carryover reaches the evaporator.

With the refrigerants that are miscible, or partly miscible with oil, enough of the refrigerant dissolves in the oil to depress the pour point of oil sufficiently that congealing of oil on evaporator surfaces does not usually occur. However, there is a temperature at which waxlike materials separate out of these mixtures. The temperature at which this occurs depends on the refrigerant, the percentage of oil in the refrigerant, and on the oil. Refrigeration compressor oils are normally evaluated for this property in terms of the Freon 12 floc point. This is the temperature at which a heavy, flocculent precipitate first appears when a mixture of Freon 12 and 10 percent of the oil is chilled. Refrigeration systems using fluorocarbon refrigerants are often arranged so that approximately 10 percent oil is present in the evaporator. In some cases, the evaporator is actually charged with this amount of oil. Under these conditions, the floc point of the oil represents the lowest temperature that can be used with that oil.

The waxy materials that precipitate from these oil refrigerant mixtures can also clog expansion valves and capillary control tubes, preventing their proper functioning. However, the concentration of oil in the refrigerant at the expansion valve is usually lower than in the evaporator, so the floc point is depressed below what it would be at a 10 percent concentration. As a result, if the oil selected has a low enough floc point for conditions in the evaporator, it will usually not cause difficulties in the expansion valve or capillaries. Difficulties in these areas attributed to oil are frequently the result of minute quantities of water in the system forming ice crystals.

Ammonia, Carbon Dioxide Oil is not soluble in anhydrous ammonia or carbon dioxide, and not enough of either dissolves in the oil to have a significant effect on the pour point of the oil. Thus, if the pour point of the oil is above the evaporator temperature, oil will congeal on evaporator surfaces and form an insulating film that interferes with heat flow and the efficient performance of the system. To remove the oil, the evaporator must be periodically warmed in order to liquify the oil so that it will drain from the surfaces to where it can be removed. With flooded evaporators, refrigerant flow may be so rapid that there is little or no opportunity for the oil to collect on evaporator surfaces, and the pour point of the oil may not be a major concern.

Sulfur Dioxide Sulfur dioxide has a selective solvent action which with conventional lubricating oils results in sludge. It, therefore, requires the use of specially refined white oils.

Methyl and Methylene Chloride These refrigerants form acids in the presence of moisture which tend to form sludges with oils. These refrigerants are also soluble in oil. Where moisture can be kept from the systems, it is possible to use conventional refrigerating oils. Otherwise, water white oils must be used.

**Lubricating Oil
Recommendations** The requirements of oils for refrigeration systems can be summarized as follows:

1. The oil should be of proper viscosity to distribute readily at the low temperatures near the suction ports and yet protect against wear at the head end of the cylinder and in the crankcase.
2. The oil should have adequate chemical stability to resist oxidation and the formation of deposits in crankcases open to the atmosphere, and to resist the deteriorating influence of high temperatures at the compressor discharge.

In the latter respect, some modern units operating on volatile refrigerants, such as automotive air conditioning compressors, require inhibited oils to withstand the high temperatures developed.

In hermetically sealed units, the oil should have high dielectric strength, which involves exceptional freedom from moisture and contaminants. Freedom from moisture is also important in many other systems where it reduces the amount of moisture introduced into the system with the oil.

Where the refrigerant is not soluble in oil, the pour point of the oil should be below the evaporator temperature in order to avoid congealing of oil on the evaporator surfaces.

With refrigerants that are miscible or partly miscible in oil, the Freon 12 floc point of the oil should generally be below the evaporator temperature. With these refrigerants also, the viscosity reduction of the oil caused by the dilution with refrigerant must be considered in selecting the oil viscosity.

Most refrigeration compressors are operated on naphthenic straight mineral oils. Oxidation inhibitors are used where compressor discharge temperatures are high. These oils have low natural pour points, low Freon 12 floc points, and are relatively free of waxes that might congeal and cause difficulties. Deeply dewaxed paraffinic oils are also used. Oil viscosities recommended vary from as low as ISO Viscosity Grade 15 to as high as ISO Viscosity Grade 100.

Synthesized hydrocarbon base lubricating oils are now being used in refrigeration compressor applications, particularly ammonia and R22 systems. These oils have very low solubility for these refrigerants so the dilution effect is minimized, and separation of the oil from the refrigerant is improved. They are particularly applicable in reciprocating compressors operating under severe conditions with ammonia as the refrigerant, and in screw compressors with R22 as the refrigerant. The extremely low pour points and complete freedom from wax of the synthesized hydrocarbons are also beneficial in many applications.

Polyglycol based lubricating oils are used for flood lubricated screw type refrigeration compressors with either R12 or a hydrocarbon gas as the refrigerant. Since R12 and hydrocarbon gases are relatively insoluble in polyglycols, oil carried from the compressor in the refrigerant can be separated readily. The polyglycols also offer very low pour points, freedom from wax, good high temperature stability, and increased operating efficiency.

Bibliography

Mobil Technical Books
Compressors and Their Lubrication
Refrigeration Compressors and Their Lubrication

Mobil Technical Bulletin
Air Compressor Lubrication — Single- and Two-Stage Reciprocating Type

15 Handling, Storing, and Dispensing Lubricants

Mobil's lubricants are quality products made to exacting standards. Their quality levels are designed to provide effective and economical performance when they are used as recommended in the applications for which they are intended. Every practical precaution in product storage and handling is taken by Mobil to ensure that the products are maintained at refinery specification quality until they are delivered to the customer. After delivery, it becomes the responsibility of the customer to exercise proper care in handling, storing, and application to protect the quality built into each lubricant so that it can deliver optimum performance in the use for which it is intended, and to maintain product identification and any precautionary labelling that may exist.

The first steps in obtaining optimum performance from correctly selected, proper quality oils and greases are proper handling, storing, and dispensing. These are necessary for two primary reasons: First—to preserve the integrity of the products; and second—to preserve identification and any precautionary labelling. It is poor practice to buy high quality lubricants and then permit degradation of them through contamination of deterioration; or to run the risk of misapplication because the identification on the containers has become illegible through improper handling, or the products have been transferred to inadequately or improperly marked containers. Proper handling, storing, and dispensing are also important for plant and personnel safety, to protect against health hazards, and to minimize the risk of environmental contamination. Most petroleum products are combustible and require protection against sources of ignition. They are not generally health hazards, but it must be recognized that excessive exposure to them can be undesirable and should be avoided. Finally, contamination or leakage produces waste that must be disposed of, aggravating the disposal problems.

Contamination of lubricants — with subsequent damage to machines — can occur during storage or during transfer of oil or grease from the original container to the dispensing equipment, or to the equipment to be lubricated unless simple precautions are observed. Pumps, oil cans, measures, funnels, and other dispensing equipment must be kept clean at all times and covered when not in use. Where operating conditions justify them, centralized dispensing or lubrication systems which keep the lubricants in closed systems and, therefore, protected against contamination, are highly recommended. Systems of this type are available to handle many types of oils and greases. There are other advantages; lubrication servicing generally can be performed faster, with less waste; integral metering devices can supply important consumption data; and bulk deliveries with economies in the purchase cost of lubricants frequently can be obtained.

Deterioration of lubricants can result from exposure to heat or cold, intermixing of brands or types, oxidation, prolonged storage, chemical reaction with fumes or vapor, and water contamination.

The observance of relatively simple precautions and procedures in the handling, storing, and dispensing of lubricants and associated petroleum products can achieve significant economic and operating benefits.

Economic benefits that can be obtained are based mainly on the elimination of waste due to the following preventable factors.

1. Leakage or spills from damaged or improperly closed containers
2. Contamination due to exposure of lubricants to dust, metal particles, fumes, and moisture
3. Deterioration caused by storage in excessively hot or cold locations
4. Deterioration due to prolonged storage
5. Residual oil or grease left in containers at the time of disposal or return
6. Mixing different brands or types of lubricants that are incompatible
7. Leaks, spills, and drips when charging a reservoir or lubricating a machine

Operating benefits, which also are reflected in dollar savings, include the following:

1. Reduction of lubricant caused machine problems, and, therefore, reduced downtime for repairs.
2. Reduced material handling time. It has been estimated that labor costs for lubricant application can be as high as eight times the cost of the lubricant applied.
3. Better housekeeping. Oil or grease spilled on floors is a major safety and fire hazard, and is unsightly.

While the information contained in this Chapter provides general suggestions on good practices, it is the responsibility of the purchaser to identify and adhere to Federal, state, and local regulations in such areas as plant safety, handling of flammable or combustible materials, fire prevention and protection, ventilation and disposal of wastes. The recommendations and suggestions in this Chapter are believed to be consistent with the standards of the Federal Occupational Safety and Health Administration (OSHA) that have been issued as of the date of publication, but it must be recognized that in many cases more stringent state or local standards may apply. These should always be checked so that conformity with them can be established and maintained.

HANDLING

In the sense used here, *handling* includes those operations involved in the receipt of supplies of lubricants and their transfer to inplant storage. The type of handling involved depends on the form of receipt of the lubricants — either in packages or in bulk.

Packaged Products All shipments of oils, greases, and associated petroleum products in containers up to and including 55 gal (U.S.) oil drums and 400 lb grease drums are considered to be packaged products.

The size, shape, and weight of the most popular standard lubricant containers are shown in Fig. 15-1. To aid in planning storage space, racks, or shelves, the dimensions of the containers illustrated and cartons of smaller packages are shown. Most containers are made of sheet steel in thicknesses and construction suitable for their contents, the intended service, and applicable freight regulations. Exceptions are grease gun cartridges, and composite 1 qt and 1 gal food type, motor oil cans which are made of spirally wound fiber board with an oil or grease proof liner and metal ends.

As delivered, each container is closed with appropriate covers, gaskets, lids, bungs, caps, and seals. In addition, each container is stamped, embossed, stencilled, lithographed, or labelled with the brand name of the lubricant that it contains.

Most packaged lubricants are delivered to the user by truck or freight car. The 55 gal oil drum or 400 lb grease drum is the most common container, since it is usually the most economical way to purchase packaged lubricating oil, greases, or other petroleum products. In the case of medium or large users, bulk deliveries, discussed later, may offer still greater savings.

Unloading Drums can be unloaded without damage from trucks or freight cars by sliding them down wood or metal skids (Fig. 15-2). These are commercially available in the 16 in. (0.41 m) widths and in lengths from 6 to 12 ft. (1.8–3.6 m). Before unloading, the brakes of the truck should be set firmly, and if any possibility of movement exists, the wheels should be blocked. The skid should be securely attached to the truck or freight car bed.

Caution.   A full oil or grease drum weighs about 450 lb (204 kg). To minimize the risk of bodily injury, it is good practice to have drums handled by two people, unless mechanical aids such as fork lift trucks, drum handlers, or chain hoists are used.

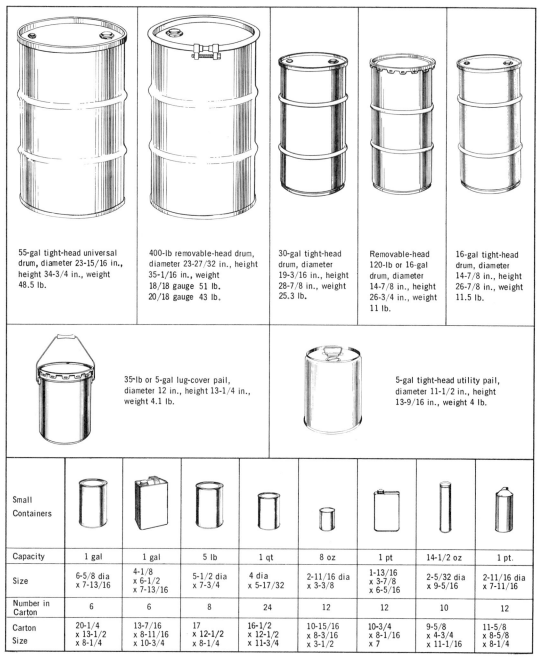

Small Containers								
Capacity	1 gal	1 gal	5 lb	1 qt	8 oz	1 pt	14-1/2 oz	1 pt.
Size	6-5/8 dia x 7-13/16	4-1/8 x 6-1/2 x 7-13/16	5-1/2 dia x 7-3/4	4 dia x 5-17/32	2-11/16 dia x 3-3/8	1-13/16 x 3-7/8 x 6-5/16	2-5/32 dia x 9-5/16	2-11/16 dia x 7-11/16
Number in Carton	6	6	8	24	12	12	10	12
Carton Size	20-1/4 x 13-1/2 x 8-1/4	13-7/16 x 8-11/16 x 10-3/4	17 x 12-1/2 x 8-1/4	16-1/2 x 12-1/2 x 11-3/4	10-15/16 x 8-3/16 x 3-1/2	10-3/4 x 8-1/16 x 7	9-5/8 x 4-3/4 x 11-1/16	11-5/8 x 8-5/8 x 8-1/4

The drum descriptions within the figure:

55-gal tight-head universal drum, diameter 23-15/16 in., height 34-3/4 in., weight 48.5 lb.

400-lb removable-head drum, diameter 23-27/32 in., height 35-1/16 in., weight 18/18 gauge 51 lb. 20/18 gauge 43 lb.

30-gal tight-head drum, diameter 19-3/16 in., height 28-7/8 in., weight 25.3 lb.

Removable-head 120-lb or 16-gal drum, diameter 14-7/8 in., height 26-3/4 in., weight 11 lb.

16-gal tight-head drum, diameter 14-7/8 in., height 26-7/8 in., weight 11.5 lb.

35-lb or 5-gal lug-cover pail, diameter 12 in., height 13-1/4 in., weight 4.1 lb.

5-gal tight-head utility pail, diameter 11-1/2 in., height 13-9/16 in., weight 4 lb.

Fig. 15-1 Mobil Lubricant Containers Standard containers that are commonly used for Mobil products, with their outside dimensions in inches are shown. Carton sizes for smaller packages are also given. Note that a 55 gal or 400 lb drum on end requires about 4 ft² of floor space, and on its side, about 6½ ft². The 14½ oz package shown is a grease cartridge for loading hand guns. The 1 pt container at the lower right is a spray can used for Mobiltac E.

Fig. 15-2 Drum Skid Drums can be unloaded from trucks or freight cars by sliding them down a wooden or metal skid. The skid must be securely fastened to the bed of the truck or freight car.

Most trucks used in the regular delivery of drums and similar heavy packages are now equipped with hydraulic tail gates (Fig. 15-3) which can be used to lower drums from the truck bed level to the ground or loading platform level. Hand winch hoisters (Fig. 15-4) can do the same job. Special jaws are available which fit the forks or replace the forks of lift trucks (Fig. 15-5) and grip drums about the diameter under the upper rolling hoops for direct movement from the truck. Under no conditions should drums or pails be dropped to the ground or onto a cushion such as a rubber tire. Doing so may burst seams, with consequent leakage losses, and possible contamination of the contents. Leakage also may constitute a safety or fire hazard. In most

Fig. 15-3 Hydraulic Lift Gate Many trucks used for delivering packaged products are now equipped with hydraulic lift gates which can be used to lower the packages to the ground or a loading dock.

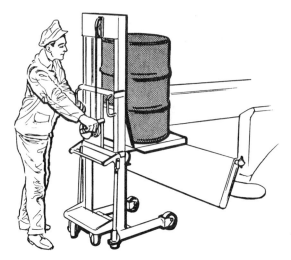

Fig. 15-4 Hand Winch Hoister A simple hoister of this type can be used to lower a drum to ground level from a truck bed. A floor lock holds the hoister in position while the drum is being loaded or lowered, and the casters permit moving for short distances.

areas also, the empty drums have a substantial resale value if they are not damaged.

Increasing amounts of lubricants in both drums and smaller packages are being delivered to customers on pallets. These can be unloaded with a fork lift truck and transported directly to storage. Generally, this unloading must be accomplished at a loading dock since it is usually necessary to drive

Fig. 15-5 Lift Truck for Handling Drums The hydraulically actuated arms on this truck will clamp and lift four drums. Other models are available that will clamp one or two drums at a time, and rotators can be added to permit inverting the drums.

Fig. 15-6 Manual Drum Handler Hand operated hydraulic systems clamp the drum then lift it for transporting.

the lift truck onto the truck or freight car to complete the unloading. Where a lift truck is to be driven onto a truck or freight car, applicable OSHA, state or local regulations relative to such operations must be adhered to. These include regulations regarding the method of blocking the truck or freight car wheels, the type and method of attaching any bridging required to reach the loading dock, and may include other requirements.

Moving to Storage After unloading, drums (55, 30, and 16 gal oil, and 400 and 120 lb grease) can be moved safely to the storage area by properly equipped fork lift trucks, either on pallets or held in specially equipped fork jaws (Fig. 15-5). If fork trucks are not available, the drums can be handled and moved with barrel trucks or drum handlers (Fig. 15-6). Drums should be strapped or hooked to the frame of a barrel truck and unloaded carefully. In some cases, it will be possible to roll drums directly on the plant floor to the storage area, provided it is only a short distance. The rolling hoops will protect the shell from damage. For long distance transfer by rolling, metal tracks should be laid on the floor.

When rolling drums, care should be taken to avoid hard objects that are higher than the drum body clearance and which might puncture the shell and result in leakage or contamination of the contents.

Caution. When rolling drums, maintain firm control to prevent the possibility of a drum running away and injuring personnel, or damaging machinery or itself.

Hand drawn or powered lift truck and skids can be used to move drums or pails along hard surfaces for long distances. An inclined ramp simplifies the task of getting drums onto pallets or skids.

Five gallon oil, or 35 lb grease pails are usually shipped on pallets. They should be handled with the same care as the larger containers.

Smaller containers of lubricants are usually packed in fiberboard cartons. These should be unloaded and moved to storage with care. The cartons should be left sealed until the product is required.

Bulk Products The advantages of bulk delivery and storage of lubricants have resulted in continuing increases in this method of operation in those locations where the quantities of lubricants used are sufficient to justify the installation costs. The term "bulk" in this context refers not only to deliveries in tank cars, tank trucks, tank wagons, and special grease transporters, but also to deliveries in any containers substantially larger than the conventional 55 gal or 400 lb grease drums. In this latter category are a considerable number of special bulk bins or jumbo drums that will transport on the order of 375 gal (1420 L) or more of oil, or 3000 lb (1360 kg) or more of grease. The bin in Fig. 15-7 which is used in a number of locations, has a nominal capacity of 485 gal (1836 L) of oil or 3680 lb (1670 kg) of grease. Bulk bins usually are offloaded and left on the customer's premises while the product in them is being used, but the lubricant also may be pumped off into bulk storage tanks, or supply tanks for central lubrication or dispensing systems.

Bulk handling and storage systems offer both economic and operating benefits to the lubricant user. Economic benefits include the following:

Fig. 15-7 Bin for Grease or Oil This type of bin, shown here stacked ready to be moved into a plant, is used most widely by Mobil, but other designs with different capacities may be used in some areas.

1. Reduced handling costs—When delivery is made directly into permanent storage tanks, handling costs are reduced markedly compared to those for drums or smaller packages.
2. Reduced floor space requirements—Permanent tanks occupy much less floor space than is required for an equal volume of oil in drums.
3. Reduced contamination hazards—Bulk storage tanks usually are filled directly from the tank car or tank truck through tight fill connections. The exposure of the lubricant to contamination is greatly reduced compared to handling and dispensing the same volume of packaged product.
4. Reduced residual waste—The user is charged only for the amount of product delivered into a bulk tank. Practice varies with bulk bins, but in many cases the user is allowed credit for residual material in the bin, or charged for only the amount added to the bin when it is refilled. Since the amount of residual oil left in drums is usually on the order of about one gallon, and the amount of grease left in drums may range up to 20 lb (9 kg) the amount of lost lubricant attributable to residual waste with these packages is significant.
5. Simplification of inventory control—The use of bulk storage facilities sharply reduces the number of individual items to be ordered, tallied, routed, and processed for return or salvage.

The operating benefits to be achieved through the use of bulk facilities include the following:

1. Simplification of handling—Less handling time means that personnel are freed for other tasks.
2. Constant availability of needed lubricants—Time spent waiting for placement and opening of new drums and installation of pumps is minimized.
3. Reduced contamination hazards mean fewer lubrication problems on the manufacturing floor.
4. Lubricants can be made available readily at strategic points—Pumps and pipe lines can be used to move lubricants from permanent tanks to locations where repetitive lubrication operations are performed.
5. Improved personnel and facility safety—Any reduction in handling operations reduces exposure to conditions that could result in injuries to personnel, plant, or environment.

These benefits can be achieved to their fullest extent only when the bulk system is properly designed and installed, and cooperation between the supplier and user results in uncontaminated products being delivered to the plant at the time that they are required.

Unloading Prior to the receipt of bulk deliveries of lubricants, whether they are in tank cars, tank wagons, or special grease transporters, certain precautions should always be observed. The storage tanks should be gaged to ensure that there is sufficient capacity available for the scheduled delivery. Empty tanks should be inspected and flushed or cleaned if necessary. Where large tanks must be entered for manual cleaning, applicable safety rules should always be observed. One person should never enter a tank alone, a second person should be available outside of the tank to provide assistance if an emergency arises. Breathing apparatus suitable for oxygen deficient atmospheres or other toxic atmospheres may be necessary, and suitable protective clothing should be worn.

Before the delivery is unloaded, a check should be made to be sure that the correct fill pipe is being used, the valves are set correctly, and any cross-over valves between storage tanks are locked shut.

Each type of bulk delivery requires some additional special precautions. Some of the more important of these are outlined in the following sections.

Tank Cars and Tank Wagons   Only trained employees should unload tank cars of flammable or combustible liquids. The car brakes should be set, wheels blocked, and "stop" signs set out. Before attaching the unloading connection, the dome cover should be opened and the bottom outlet valve checked for leakage. If the valve is not leaking its cap will usually be loose. After unloading is complete, the hose should be disconnected, the dome cover closed, and the valve closed immediately.

Special Bulk Grease Vehicles   While tank trucks have been used successfully to carry bulk grease, cleaning problems are considerable and tend to discourage the use of this type of vehicle. A more satisfactory approach has been the use of specially designed and constructed bulk grease vehicles. The current vehicles (Fig. 15-8) are capable of carrying 38,900 lb (17,645 kg) of grease in two 19,000 lb (8618 kg) compartments. The tanks are fully insulated for heat retention in long hauls. Two power pods, each equipped with its own engine and pumping unit, can unload the grease at a rate of up to 1000 lb (454 kg) per minute. The two power pods are interchangeable to provide maximum dependability.

These units are designed primarily for use in refilling bulk bins or the main storage tanks in centralized lubricant dispensing systems in plants having large volume requirements for specific lubricants. They usually are used for transporting pumpable greases in the consistency range up to and including NLGI No. 1. Other greases can be handled in bulk; however, each application requires careful study and engineering to assure satisfactory operation.

Fig. 15-8 Bulk Grease Transporter

Obviously not all marketing areas can be served economically by bulk grease vehicles. In some areas, excessively long hauls might be required, while in others, the number of customers requiring such large volume deliveries may be insufficient to provide adequate utilization of the equipment.

STORING

Proper storage of lubricants and associated products requires that they be protected from sources of contamination, from degradation due to excessive heat or cold, and that the identification be maintained. Also important are the ease with which the products can be moved in and out of storage, and the ability to operate on a "first in, first out" basis. Increasingly important in the selection, location, and operation of petroleum product storage facilities are applicable fire, safety, and insurance requirements.

The National Fire Protection Association in its flammable liquids code NFPA 30,* defines a *flammable* liquid as any liquid having a flash point below 100°F (37.8°C), except any mixture having components with flash points of 100°F (37.8°C) or higher, the total of which make up 99 percent or more of the total volume of the mixture. Flammable liquids are known as Class I liquids and are further subdivided as follows:

Class IA includes those liquids having flash points below 73°F (22.8°C) and having a boiling point below 100°F (37.8°C).

Class IB includes liquids having flash points below 73°F (22.8°C) and having a boiling point at or above 100°F (37.8°C).

Class IC includes liquids having flash points at or above 73°F (22.8°C) and below 100°F (37.8°C).

Combustible liquid is defined as any liquid having a flash point at or above 100°F (37.8°C). Combustible liquids are divided into two classes as follows:

Class II liquids include those with flash points at or above 100°F (37.8°C) and below 140°F (60°C), except any mixture having components with flash points of 200°F (93.3°C) or higher, the total volume of which make up 99 percent or more of the total volume of the mixture.

Class III liquids include those with flash points at or above 140°F (60°C). They are divided into two subclasses, as follows:

Class IIIA liquids include those with flash points at or above 140°F (60°C) and below 200°F (93.3°C), except any mixture having components with flash points of 200°F (93.3°C) or higher, the total volume of which make up 99 percent or more of the total volume of the mixture.

Class IIIB liquids include those with flash points at or above 200°F (93.3°C). This class of liquids, because of their lower flammability, does not require the special handling precautions that apply to the more flammable materials.

When a combustible liquid is heated for use to its flash point it must be handled in accordance with the next lower class of liquids.†

*The definitions contained here are those of NFPA 30 as revised in 1973.

†Recently issued OSHA (Department of Labor) regulations, which apply to any work place having employees, require that when a combustible liquid is heated for use to within 30 Fahrenheit degrees (16.7 Celsius degrees) of its flash point, it must be handled in accordance with the requirements of the next lower class of liquids.

Most lubricants, because of their relatively high flash points, under these regulations are Class IIIB combustible liquids. Petroleum solvents that are commonly used for cleaning parts and equipment are Class II combustible liquids. However, depending on the conditions of storage and use, they may require handling as Class IC flammable liquids.

Suggestions included in the section on Oil Houses, Size and Arrangement are believed to be consistent with the Occupational Safety and Health Standards of the Occupational Safety and Health Administration (OSHA) of the Department of Labor in effect as of the date of publication, which are similar to those of the National Fire Protection Association, but it must be remembered that more stringent state or local regulations could apply.

Packaged Products Packaged lubricants and associated products can be stored: (1) outdoors, (2) in a warehouse, or (3) in an oil house. Outdoor storage should be avoided whenever possible. The best storage area for lubricants is a well arranged, properly constructed, and conveniently located oil house. Warehouse storage is desirable when the oil house lacks the space needed to stock the complete lubricant inventory required.

Outdoor Storage Storing lubricant drums or other containers out-of-doors is a poor practice. The hazards of this type of storage include the following:

1. Identifying drum markings may fade and become unreadable under the combined action of rain, sun, wind, and airborne dirt. Rusting can obliterate drum markings. When the markings become so deteriorated that the drum contents cannot be identified, it may be necessary to take a sample and perform a laboratory analysis to determine the nature of the contents. This is a costly and time consuming procedure. On the other hand, accidental use of the wrong lubricant as a result of incorrect identification can cause severe damage to machines or other equipment.

2. Seams of containers may be weakened due to alternating periods of heat and cold which cause the metal to expand and contract cyclically. The net result may be loss of the contents by leakage or contamination if the drums are subsequently subjected to rough handling or improper storage.

3. Water may get into the drum around the bungs and contaminate or destroy the contents. A drum standing on end with the bungs up can collect rain water or condensed atmospheric moisture inside the chime (Fig. 15-9). This water can gradually be drawn in around the bungs by the breathing of the drum as the ambient temperature rises and falls. This can occur even with the bungs drawn tight and the tamperproof seals in place.

4. Dirt and rust that accumulate inside the chime and around the bungs may contaminate the contents when the drum is finally opened for use.

5. Extremes of heat or cold can change the physical properties of some products and render them useless. Water may separate from soluble oils, or fatty oils congeal in compounded oils. Emulsifiers in wax emulsions may separate and cause irreversible degradation. Often cold lubricants must be warmed up before use, with the possibility of further damage due to overheating.

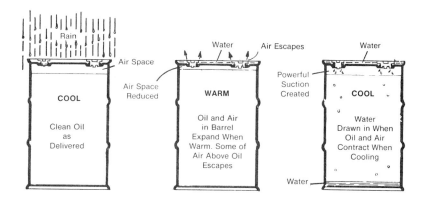

Fig. 15-9 Breathing of Moisture by Upright Drum

6. Contaminating rust can develop inside a container if water leaks in from any source.

Of course, any of these conditions that result in irreparable damage to the drums can mean that the resale value of the drums is lost and the drums become a scrap disposal problem.

To minimize the harmful effects of unavoidable outdoor storage, a few simple precautions and procedures can be very helpful.

As a general rule, lubricants in containers smaller than drums (550, 390, and 16 gal oil drums and 400, and 120 lb grease drums) should never be stored out-of-doors. When drums must be stored outside, a temporary shelter (Fig. 15-10) or leanto, or a water proof tarpaulin, will protect them from rain or snow. Drums should be laid on their sides with the bungs approximately horizontal as shown. In this position, the bungs are below the level of the contents so that breathing of water or moisture is greatly reduced, and water cannot collect inside the chime. For maximum protection, the drums should be stood on end with the bung ends down on a well-drained surface. Where all of these approaches are impractical, drum covers such as those shown in Fig. 15-11 can be used. These are available in both metal and plastic. Drums that have a bung on the side should be stored either on end, or on

Fig. 15-10 Outdoor Drum Shelter

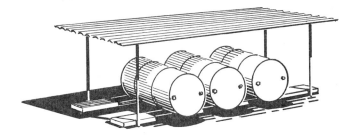

Fig. 15-11 Metal Drum Covers Weights have been placed on these covers to reduce the possibility that the covers will be blown off by high winds.

the side with the bung down. Regardless of the position, drums are stored in, they should always be placed on blocks (Fig. 15-12) or racks several inches above the ground to avoid moisture damage.

When drums must be stored outdoors with the bung end up, they should be cleaned carefully to eliminate the hazard of collected rust, scale, or dirt falling into the drum contents.

The drum storage area should be kept clean and free of debris that might present a fire hazard.

Fig. 15-12 Correct Method of Blocking Up a Drum

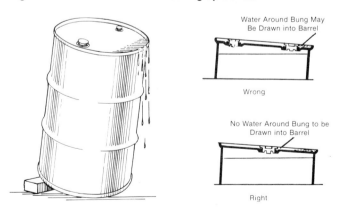

Water Around Bung May Be Drawn into Barrel

Wrong

No Water Around Bung to be Drawn into Barrel

Right

Fig. 15-13 Drum Stacker The lifting arms slip under a drum resting on the floor. The drum can then be lifted and moved into the storage rack. This unit is battery powered, but manual models are also available.

Warehouse Storage Mechanical handling equipment is needed for efficient movement of drums into and out of a plant warehouse. Hand or power operated fork lift trucks or stackers (Fig. 15-13) are widely used for this purpose. They offer the advantage of a single handling operation from warehouse storage to the oil house or point of use. A chain block or trolley with a proper drum sling mounted on an I-beam bridge (Fig. 15-14) can be used to move drums in and out of storage. This type of equipment does not require the aisle space needed to maneuver a fork lift. Mechanical handling equipment is closely regulated by the Occupational Safety and Health Administration. A thorough study is required to assure that the system and equipment meet all requirements.

Racks and shelving should provide adequate protection for all containers. Aisle space should be adequate for manuevering whatever type of mechanical handling equipment is used. Each type of lubricant stocked should not be allowed to block access to the older stacks, so that a "first in, first out" basis can be maintained. This will minimize the hazard of deterioration due to excessive storage time. A wide variety of racks and shelving (Fig. 15-15), both assembled and in components for assembly on site, is commercially available.

When lubricants and other petroleum products are to be stored in a plant warehouse, location and rack construction should be considered with respect to applicable insurance, fire, and safety regulations. The material storage location should be considered also on the basis of receiving and dispensing convenience.

Fig. 15-14 Chain Block and Trolley A hand or power operated chain block mounted on an I-beam bridge can be used to move drums into and out of storage racks. The racks shown are constructed of 2 in. pipe and standard fittings. The pairs of flat strips supporting the upper drums can be removed to permit access to the drums below.

Fig. 15-15 Storage Racks Racks of this type can be shop built or obtained commercially in various heights and width. Drums can be conveniently moved into and out of this rack with the stacker shown in Fig. 13.

Oil Houses

Function The function of an oil house in an industrial plant is to provide a central point for the intermediate storage and day-to-day dispensing of lubricants, cutting fluids, and other related materials needed to lubricate and maintain the plant's production equipment. Where soluble cutting fluids are used, the oil house generally is the location where emulsions are prepared. In many cases, also, it is equipped to purify or recondition used or contaminated oils.

The oil house should be equipped to properly maintain and clean all the equipment used in daily lubrication work. This includes pumps, grease guns, oil cans, portable dispensing units, strainer and filter elements, oiler bottles, and grease gun fillers. In addition, storage space should be provided for small containers of products, guns and cans, etc.

The administrative function of the oil house includes the maintaining of machine lubrication charts and inventory records, conformance with established lubrication service schedules, and good housekeeping practices.

Facilities The simplicity or elaborateness of oil house facilities will depend largely on the size of the plant and the complexity of its lubrication program. Speaking generally, a well equipped oil house will contain adequate stocks of the following items:

1. Drum racks, either of the rocker type or tiered
2. Oil and grease transfer pumps
3. Drum faucets
4. Grease gun fillers
5. Grease guns
6. Oil cans
7. Portable equipment such as oil wagons, lubrication carts, sump drainers, and power grease guns
8. Maintenance supplies such as wiping rags, wicks, oiler bottles, replacement screens and filters, and spare grease fittings
9. Containers of sawdust or other absorbent materials for cleaning up oil spills

In addition, batch purification equipment to recondition used oils may be desirable, soluble oil mixing equipment may be required, and solvent tanks for cleaning parts and lubrication equipment generally are required.

When plant usage of a specific oil, grease, or fluid warrants, a permanent bulk tank may be installed in the oil house.

Size and Arrangement These factors are also determined largely by plant size and lubrication service requirements. The total rack storage needed can be calculated by determining the quantity of each type of oil, grease, solvent, etc., that will need to be stocked in the oil house and then converting this to rack dimensions by means of the container dimensions shown in Fig. 15-1. The amount of space needed for miscellaneous equipment (solvent tanks, mixing equipment, cabinets for lubrication service personnel, portable dispensing equipment etc.) must also be considered.

Individual lockers for lubrication service personnel can contribute to good housekeeping and general oil house efficiency. Lockers of steel construction are durable and easy to keep clean. They should be of sufficient

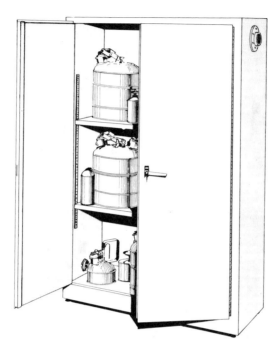

Fig. 15-16 Individual Locker Adequate space is provided for the lubrication service-man's equipment. The locker should be provided with top and bottom ventilation to dissipate fumes and vapors. Where flammables are to be stored in the locker, a safety locker of approved design must be used.

size to store normal individual equipment, with adjustable shelving to permit maximum space utilization (Fig. 15-16). This type of storage cabinet should be ventilated to prevent accumulation of vapors or fumes. If safety cans of cleaning solvents are to be stored in them, the lockers must be of an approved design.* Used wiping rags or waste should never be stored in these lockers, but should be placed in an approved disposal container.

An adequate cleaning facility is an oil house essential. Dispensing equipment (grease guns and the like) must be cleaned regularly for proper functioning. Airborne dirt and dust collect quickly on oil wetted surfaces, so oil containers, faucets, and similar equipment should be cleaned regularly to remove these contaminants. Solvent tanks should be large enough to handle the largest equipment used. In general, baths of nontoxic safety solvents are permitted by safety regulations provided that the tanks have properly designed covers of the automatic selfclosing type, and ventilation is adequate. Two tanks, one for cleaning and one for rinsing, will assure proper cleaning of equipment. As shown in Fig. 15-17, a metal grating over a drain leading to the central trap prevents spilled solvent from collecting on the floor to form a safety or fire hazard, and from finding its way to the sewers to contaminate

*Where a reference is made to an "approved design" or "approved type" of container, locker, etc., the intent is an item that has been tested by and carries the label of one of the recognized testing organizations such as Underwriter's Laboratories (UL) or Factory Mutual Laboratory (F/M).

the plant effluent water. The approved safety drip cans under the drain faucets collect any leakage from these sources.

One oil house layout (Fig. 15-18) is offered as a guide to planning. Note that the sliding doors (1 and 6) are wide enough to admit the largest expected fork lift load. The doors shown in this layout are selfclosing fire doors which may not always be required by local fire regulations, but generally offer the advantage of increased safety. The portable equipment storage area (13) provides ample maneuvering space for lift trucks to remove and replace the oil and grease drums (2 and 12). The open layout permits easy, unimpeded access from any part of the oil house to the door (6) leading to the plant.

Location For optimum utilization of manpower, an oil house should be located as centrally as possible with respect to the lubrication service activity. A study of service requirements based on total travel of lubrication service personnel to their work from the oil house can help to determine the most economical and efficient location for the oil house. Travel distances from the warehouse, unloading dock, or other storage facilities should also be taken into account. In multibuilding plants it may be advantageous to build a separate oil house in a central location within the plant area. In such cases, an

Fig. 15-17 Cleaning Tanks While the overhead hood with forced ventilation is not mandatory in all cases, it is generally good practice. Self-closing covers, with a fusible link to permit automatic closing in the event of fire, are now required by OSHA regulations.

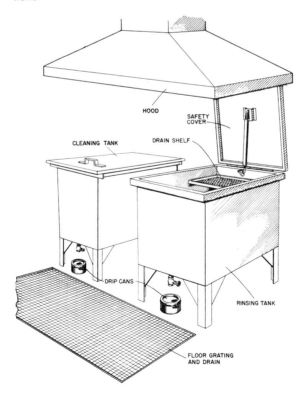

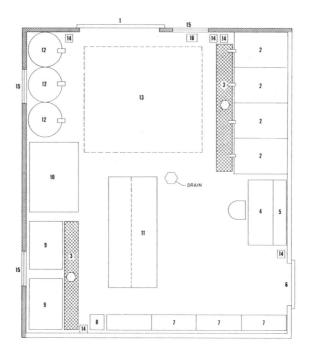

Fig. 15-18 Oil House Layout Some of the features of this layout are as follows: 1 and 6, self-closing fire doors wide enough for passage of lift trucks or other material handling equipment; 2, drum racks; 3, grating and drain; 4, desk; 5, filing and record racks; 7, individual lockers or storage cabinets; 8, waste disposal container; 9, solvent cleaning tanks; 10, purification equipment, soluble oil mixing equipment, or other special equipment; 11, cabinets and racks for equipment, supplies and small containers; 12, grease drums with pumps; 13, parking area for oil wagons, etc.; 14, fire extinguishers; 15, ventilators; 16, container of sawdust or other absorbent.

oil warehouse and the oil house may be combined in one building, with consequent savings in handling and storage costs.

Housekeeping   Orderliness and cleanliness in the oil house are essential. In an orderly storage arrangement, the chances of mistaken identification of products or applications are greatly reduced. Regular cleaning schedules should be set up and maintained. Each container or piece of equipment should bear a label showing clearly the product for which it is used. This label should provide enough information to enable personnel to correctly identify the desired product in the original containers. For example, identifying an oil container as "Mobil D.T.E. Oil" would not be adequate identification since the same plant might use several Mobil D.T.E. Oils, such as Mobil D.T.E. 24 and 25 for hydraulic systems, Mobile D.T.E. Oil 105 for compressors, and Mobil D.T.E. Oil Heavy Medium for bearing lubrication. Color coding is frequently suggested for this identification, but it must be remembered that a considerable number of people are colorblind, which can defeat a color coding system. Labels should be renewed as frequently as necessary to maintain legibility.

Note: While comparatively few lubricants and associated products require precautionary labelling, in those few cases where it is required, similar precautionary labels should be affixed to all equipment used for those products. Dispensing and application equipment used for products requiring precautionary labelling should not be used for any other products unless thoroughly cleaned first.

Every piece of equipment used in the oil house should have a space reserved for it — and should be in its place when not actually in use. A chart or list of these locations posted near the product storage racks will facilitate location of needed items.

Observance of cleanliness and an orderly routine will be reflected in the attitude and efficiency of the lubrication personnel. Their increased sense of pride and responsibility will have a direct bearing on the lubrication service in any plant. It will also be a contributing factor to optimum safety against fire and personal injuries. In this respect, it is considered to be of such importance that it receives major emphasis from compliance officers of some of the regulatory bodies.

Safety and Fire Prevention   Warning signs should be posted in every oil house to alert personnel to the presence of combustible materials. If flammables are used or stored, or Class II combustibles are used in such a way that they must be treated as flammables, then the applicable signs warning of the hazards associated with these materials should also be posted. Standard warning placards complying with OSHA or other applicable safety regulations relative to flammables or combustibles are generally available. "No Smoking" signs, in red, should be prominently displayed and the no smoking rule rigorously enforced.

Hand fire extinguishers and automatic systems are essential for oil house safety. Hand operated fire extinguishers should be mounted at strategic points throughout the oil house, particularly near those areas where cleaning tanks are located. They should be inspected periodically (at least as frequently as required by applicable fire regulations) to be sure that they are in satisfactory operating condition.

Fire extinguishing methods for flammable or combustible liquid fires include the following:

1. Suppression of vapor by foam
2. Cooling below the flash point by water spray or fog (not a direct stream, which could spread the fire!)
3. Excluding oxygen, or reducing it with carbon dioxide (CO_2), to a level insufficient to support combustion
4. Interrupting the chemical chain reaction of the flame with dry chemical agents or a liquified gas agent

All personnel employed in the oil house should be thoroughly instructed in the location and use of the fire extinguishing equipment.

Rags, paper, or other solid materials that have been soaked in flammable or combustible liquids should be placed in an approved type of disposal container with a selfclosing cover. The container should be emptied at the end of each shift and the contents removed to a safe location for reconditioning or incineration. All spills should be cleaned up promptly. If an absorbent is used, it should be swept up promptly, placed in an approved disposal container, and removed to a suitable disposal area.

Ventilation equipment should be kept in first class operating condition at all times. Without good ventilation, vapors could collect within the oil house and be a hazard to personnel safety and health, as well as a fire hazard. The solvents used for cleaning parts and equipment in an oil house may present fire and health hazards. A few simple precautions can minimize these hazards.

1. Use only nontoxic solvents with flash points above 100°F (38°C). Mobil products suitable for the purpose include Sovasol No. 5 (flash point 105°F, 41°C), Sovasol No. 35 (flash point 120°F, 49°C), and kerosene.

2. Static electricity generated by the flow of products such as solvents can cause sparking which can be a source of ignition. This is not generally a problem with products with flash points above 100°F (38°C) but simple precautions will eliminate the hazard. Drums containing solvents should preferably be grounded during opening and while installing faucets or pumps. If the drum is not resting on a grounded, conducting surface, an adequate ground can be obtained by clipping a wire between the drum chime and a water pipe. When filling small containers from a solvent drum, a bond between the drum and container should be made. If both the drum and container are resting on the same conducting surface, adequate bonding will be obtained if the faucet or nozzle is kept in contact with the container, otherwise a bond wire clipped between the container and the drum should be used (Fig. 15-19).

3. Use only approved type safety containers (Fig. 15-20) for transferring solvents from drums or bulk storage to the cleaning tanks or point of use.

4. Prolonged inhalation of solvent fumes can cause headaches, dizziness, or nausea. Good ventilation around areas where solvents are used should be provided, preferably in the form of an exhaust system with hoods.

5. Keep covers on the solvent tanks when they are not in use. If an exhaust system with hoods is used over the solvent tanks, make sure that the exhaust system is operating before the covers of the tanks are opened.

In cases where inadequate ventilation conflicts with safety regulations, or problems of disposing of dirty solvents make it impossible to clean with petroleum solvents inside the oil house, arrangements should be made for

Fig. 15-19 Bonding for Control of Static Electricity At the left, if either container is on an insulating surface, a bond wire should be used. At the right, if both are on the same conducting surface, an adequate bond can be obtained by keeping the faucet or nozzle in contact with the container being filled.

INSULATING SURFACE

CONDUCTING SURFACE

Fig. 15-20 Safety Container The container is self-closing with a liquid tight seal. The container must have a U/L and/or F/M label.

cleaning in an area where the requirements can be met—in a painting department or general cleaning area, for example.

Security Experience indicates that the oil house should be kept closed, and sometimes locked, to other than lubrication personnel. This is necessary to prevent the confusion which may arise when unauthorized or uninstructed personnel are able to obtain their own choice of lubricants from an area of open containers. Inadvertent choice of an improper lubricant may lead to damage to machinery or manufactured products.

Lubricant Deterioration in Storage Lubricants can deteriorate in storage, usually as a result of one of the following causes:

1. Contamination, most frequently with water
2. Exposure to excessively high temperatures
3. Exposure to low temperatures
4. Long term storage

Some contaminated or deteriorated products can be reconditioned for use, while others must be degraded to inferior uses, or destroyed or otherwise disposed of. In addition, portions of some contaminated products can be salvaged for use. The decision as to which course of action to follow is dependent on such factors as the amount of product involved and its value compared to the cost of reconditioning or salvaging, the type and amount of contaminants present, the degree of deterioration that has occurred, and the effect of the contamination or deterioration on the functional characteristics of the product. Some of these considerations are discussed at more length in the following paragraphs.

Water Contamination In many cases, water contamination is relatively easy to identify, although a quantitative indication of the amount present usually requires laboratory analysis. Frequently a quantitative determination is not required, and the presence of water can be detected by haze or suspended water. Where doubt exists a simple "crackle" test in which a small amount of the oil is placed in a shallow dish and heated to the boiling point of water on a hot plate, provides a fairly positive identification of water. When water contamination has been identified, the steps that can be taken depend on the type of product, and its intended use.

High Temperature Deterioration **Greases** Oil separation (bleeding) may occur in some greases when they are stored for prolonged periods in a high temperature environment. The separated oil will accumulate on the surface of the product. The amount of oil separated should be considered in deciding whether or not the grease can be used satisfactorily. Where the amount of oil separated is relatively small, it can be poured or skimmed off and the remaining grease used satisfactorily. Where excessive oil separation has occurred, it is probable that the bulk grease has changed sufficiently to be unsuitable for its intended use, so must be disposed of.

Evaporation of the original water content of a waterstabilized calcium soap grease can cause separation of the product into oil and soap phases. However, this usually occurs only at quite high storage temperatures. Where it occurs, the grease usually will be rendered unfit for service.

Lubricating Oils Most premium grades of hydraulic, process, circulation, and engine oils are not affected by even prolonged storage at temperatures below 150°F (66°C). Prolonged storage at higher temperatures—direct container contact with furnaces, steam lines, etc.—may cause some darkening due to oxidation. When in doubt as to the ability of the oil to perform satisfactorily, a sample should be sent to the laboratory for analysis and evaluation—assuming there is sufficient product involved to justify the cost of the analysis—before using the product.

Products with Volatile or Aqueous Components This group of products includes diluent type open gear lubricants, solvents, naphthas, conventional soluble oils, wax emulsions, emulsion type fire resistant fluids, and water-glycol fire resistant fluids.

Long storage at elevated temperatures of diluent type open gear lubricants can cause evaporation of the diluent. If a significant proportion of the diluent is lost by evaporation, the product will be difficult to apply.

Caution The diluent used in some diluent type open gear lubricants is classed as a health hazard on inhalation and precautionary labelling is required. Containers of such materials should be kept tightly closed when not in use, and stored in a cool place.

Volatile products, such as solvents and naphthas, may suffer loss by evaporation. Usually the quality of the remaining product is not affected. If the products are in sealed containers, high temperatures can cause distortion or bursting.

Some products that contain significant amounts of water, such as fire resistant water-in-oil emulsions and water-glycol fluids, are usually unharmed by small amounts of water evaporation. Losses can be replaced by adding water as needed. Agitation usually is necessary when water additions are made to these products.

Wax emulsions stored at over 100°F (43°C) will lose water content due to evaporation. Minor losses will cause the formation of a surface skin on the product. This skin can be removed, leaving the remainder of the emulsion fit for use. However, before use the product should be passed through a 40 mesh screen to remove any remaining undesirable materials. Excessive water loss may upset the chemical balance of the product and cause separation of the wax and water, producing a curd-like appearance. Under these conditions the product cannot be reconditioned readily and usually must be disposed of.

Low Temperature Deterioration

Below freezing temperatures normally will not affect the quality of most fuels, solvents, naphthas, and conventional lubricating oils and greases. The major difficulty associated with storage outdoors or in unheated areas during cold weather is with dispensing of products that are not intended for low temperature service. When possible, containers should be brought indoors and warmed before attempting to dispense the product.

Products that contain significant amounts of water should not be exposed to temperatures below 40°F (4°C). Freezing of wax emulsions will cause separation of the wax and water phases giving the product a lumpy, curd-like appearance. Under these circumstances, wax emulsions normally must be disposed of, since the products usually cannot be restored to their original condition.

Repeated freezing or long term exposure to freezing temperatures may destroy the emulsifying properties of conventional soluble oils. Usually there is no change in the appearance of the product, although it may have a cloudy cast. However, it may not emulsify properly with water. Under these conditions, the product must be disposed of.

Some oils, when subjected to repeated fluctuations of a few degrees above and below the pour point, may under go an increase in their pour point (poor point reversion) of 15 to 30 Fahrenheit degrees (8 to 17 Celsius degrees). As a result, dispensing may be extremely difficult even when the ambient temperature is above the specified pour point of the product. Oils that contain pour point depressants or that have a relatively high wax content (e.g., steam cylinder oils) are the most prone to pour point reversion. Although this problem occurs relatively infrequently, products that may exhibit the phenomenon should be stored at temperatures above their pour points.

A product that has undergone pour point reversion may return to its normal pour point when stored for a time at normal room temperature. Cylinder oils may require 100°F (38°C) storage for reconditioning.

Table 15-1 Product Storage Life

Product	Maximum Recommended Storage Time, Months
Diesel fuels	6
Defoamants (Mobil Par Lettered series)	6
Gasoline	6
Greases (inc. Mobilplex 40 series)	12
Calcium complex greases (except Mobilplex 40 series)	6
Emulsion type fire resistant fluids	6
Soluble oils	6
Custom blended soluble oils	3
Wax emulsions	6

Long Term Storage

Long term storage at moderate temperatures has little effect on most premium lubricating oils, hydraulic fluids, process oils, and waxes. However, some products may deteriorate and become unsuitable for use if stored longer than three months to a year from date of manufacture. An approximate guide to the maximum recommended storage times for these products is shown in Table 15-1. Where storage time and conditions for a product are extremely critical, these facts are generally indicated on the package.

DISPENSING

Dispensing in the sense used here includes withdrawal of the products from the oil house or other storage, transfer to the point of use, and application at the point of use. The point of use is defined as the bearing, sump, reservoir, friction surface, system, etc., where the product will perform its required function. Packaged products may be dispensed directly from the original containers, or transferred to various types of dispensing or application equipment. Bulk products may be transferred by means of portable equipment to the point of use, but frequently are dispensed directly from the tanks or bins through permanently installed piping systems to automatic application devices, or automatic or manually operated dispensing points. Where lubricants and related products are dispensed by methods other than completely closed systems, the basic requirements for effective dispensing are as follows:

1. Containers or devices used to move lubricants and related products should be kept clean at all times.
2. Each container or device should be clearly labelled for a particular product and used only for that product.
3. The device used for introduction of a product to the point of final use should be carefully cleaned before the filling operation is started; this includes grease fittings, filler pipes, filler holes, etc. Sumps and reservoirs should be thoroughly cleaned and flushed before filling the first

time, and should be checked when being refilled and cleaned as necessary.

The most common dispensing equipment used in industrial manufacturing or process plants includes the following, in sizes, types, and quantities suitable to specific needs:

In the oil house
    Faucets
    Grease gun fillers
    Grease paddles
    Highboys
From oil house to machine
    Lubrication carts and wagons
    Oil cans
    Safety cans
    Grease guns
    Portable greasing equipment
    Portable grease gun fillers
    Oil wagons
    Sump pumps
    Portable fluid purifiers

In the Oil House

Faucets Faucets (Fig. 15-21) are used to dispense oils, solvents and other fluids from storage tanks or drums to containers for use in the manufacturing area. They are available in different sizes for fast or slow flowing fluids, and to fit either the 2 in. or ¾ in. drum opening. The faucet can be inserted in the opening, and the drum lifted by crane, lift truck, or chain block to a rack which will hold it in a horizontal position while the fluid is dispensed. An alternate method is to use one of the many types of rocker drum racks (Fig. 15-22) to tip the drum onto its side and support it during dispensing. Racks of this type with casters permit easy positioning of the drum after it is on the rack.

Drum faucets are available commercially in steel, brass, and stainless steel to suit specific fluid needs (noncorrosive, nonsparking, etc.). Some types provide a padlock hasp to prevent inadvertent or unauthorized opening. Faucets with a selfclosing feature are generally preferable since they minimize spillage if the faucet is accidentally opened. Faucets equipped with flame arresters are available for use with flammable liquids.

Flexible metal hose extensions may be attached to some faucets to contact the lip or edge of the container being filled to reduce the risk of static electricity discharge.

Transfer Pumps Pumps are used to transfer oils, greases, solvents, coolants, etc., from drums or tanks to other containers. In the manufacturing area they may be used to fill reservoirs or lubrication equipment directly from drums or lubrication wagons.

A pump that can be inserted into the bung opening of a drum is desirable for dispensing oil. These pumps are of the positive delivery type, and can be obtained to deliver measured quantities of oil. The simplest types are hand operated. An excellent type of hand operated pump has a closable return (Fig. 15-23) actuated by a spring, to return drippage to the drum without the danger of the contamination that would occur through an exposed drain. Air

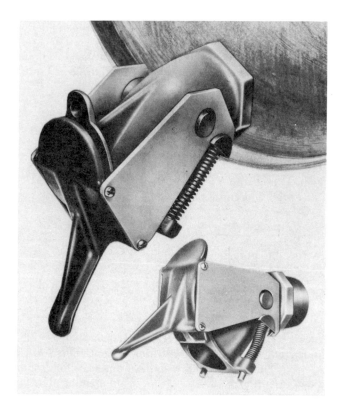

Fig. 15-21 **Faucet** This faucet has the desirable self-closing feature.

Fig. 15-22 **Rocker Type Drum Rack** The hook on the handle engages the top drum chime while the pads slip under the bottom chime.

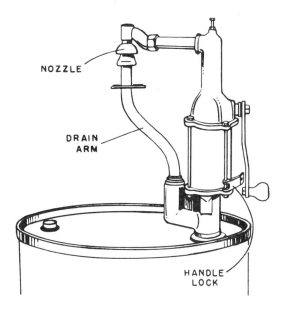

NOZZLE

DRAIN
ARM

HANDLE
LOCK

Fig. 15-23 Drum Pump with Return

operated drum pumps (Fig. 15-24) and electrically operated pumps are also available.

Drum pumps also are available for dispensing semifluid or very soft greases that are packed in closed head drums. These pumps can be fitted into the 2 in. bung opening, and deliver up to 20 lb (9 kg) of grease per minute. They are used for transferring grease to smaller servicing equipment such as grease gun fillers and portable greasing equipment.

Stiffer greases are packed in open head drums, and pumps for these products usually are mounted in the center of a drum cover that clamps on the top of the drum (Fig. 15-25). The usual pumps can handle greases of No. 2 or No. 3 NLGI consistency, depending on the type of grease. However, with some of the harder greases or greases with poor slumpability, a follower plate to force the grease down to the pump inlet may be required as shown in Fig. 15-26 which also illustrates a hoist that can be used to remove or install pumps rapidly and easily with a minimum of lubricant contamination.

Grease Gun Fillers   These are used to fill grease guns, either directly from the original container or from a reservoir in the gun filler. Models are available to fit pails, on 400 or 120 lb open head drums. They may be hand (Fig. 15-27), air, or electrically operated.

Where grease guns, fillers, and other grease dispensing equipment are to be filled with a grease that is too stiff to pump with a transfer pump, a grease paddle normally must be used. Preferably, the paddle should be made of metal to prevent contamination with splinters that might occur with a wooden paddle. After filling is completed, the cover of the grease container should be replaced securely to prevent contamination.

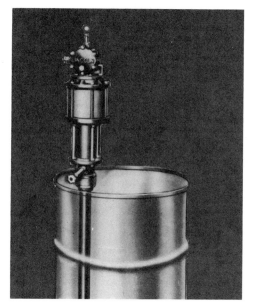

Fig. 15-24   Air Operated Oil Pump

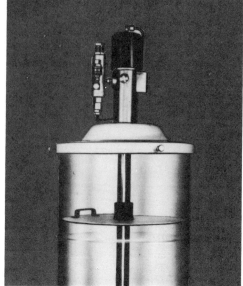

Fig. 15-25   Air Operated Grease Pump

Highboys   At one time, highboys with permanently mounted pumps were used to some extent for oil storage and dispensing. They have the advantages of neat appearance, compactness, and sturdiness, but have the disadvantages that they are difficult to clean and require an additional product transfer—with its risk of contamination—to fill them. The usual highboy design has no drain for cleaning, so there is a tendency for foreign matter to collect in the bottom of the tank. When new oil is added to the tank, the resulting agitation may cause this foreign material to be suspended in the oil. As a result, oil drawn off before this can settle may contain harmful contaminants.

Where highboys must be used because of space limitations, they should be equipped to permit easy draining and cleaning.

From Oil House to Machine   Moving lubricants from the oil house to the machinery where they are to be used is a critical phase of dispensing that justifies the same care as handling in the oil house. The problem again is one of preventing contamination and confusion of products. It is usually further complicated, moreover, by the necessity for transporting a variety of lubricants in different types of containers or lubrication equipment. This phase of the dispensing problem is essentially a matter of choosing containers or equipment that can be handled economically with minimum risk of contamination or confusion. As a general rule, the most desirable containers or pieces of equipment are those that can be filled in the oil house and then emptied directly into a machine, or used to lubricate a machine. Each product should have its own container or piece of dispensing equipment which should be clearly marked for that product. Such equipment should not be considered interchangeable, unless it is emptied and thoroughly cleaned before being used for another product.

Caution: Galvanized containers should never be used for transporting oil. Many of the industrial oils used today contain additives which react with zinc to form metallic soaps. These soaps may thicken the fluid and clog small oil passages, wicks, etc.

Oil Cans Although being outmoded rapidly and outnumbered by automatic oiling devices ranging from simple bottle oilers to complex centralized oiling systems, the common oil can is still widely used to carry oil to machines. Its chief advantages are traditional acceptance and low cost, combined with the need for an easy method of applying small quantities of oil to open bearings. Its chief disadvantages are high labor cost per unit of oil dispensed, the excessive hazard of lubricant contamination as compared with closed automatic systems, and the generally inferior performance of bearings designed for hand oiling. When hand oil cans must be used, thorough precautions should be taken to maintain lubricant cleanliness.

The simplest hand oil can is the diaphragm type. A more practical and desirable oil can is the positive delivery type which delivers a definite quan-

Fig. 15-26 Air Operated Pump Hoist
The pump is equipped with a follower plate in order to ensure feeding of relatively stiff greases.

Fig. 15-27 Hand Operated Grease Gun Filler
This unit fits a 120 lb open head drum and is equipped with a follower plate. Similar units fitting 25 lb or 36 lb pails or 400 lb drums are available.

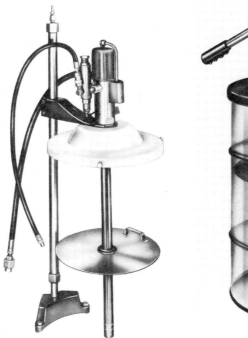

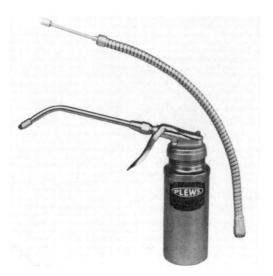

Fig. 15-28   Pistol Oiler   The flexible spout can be used to reach hard to get at oil holes.

tity of oil from any position (Fig. 15-28). The trigger actuates a simple pump incorporated in the can and an adjustment permits varying the quantity of oil delivered.

Where larger quantities of oil than can be delivered with an oil can are required, special containers are used. Open pails and cans invite contamination. A practical container for quantities of oil that can be hand carried is the safety can (Fig. 15-29). Various styles are available. They can be obtained in many capacities with selfclosing spouts and fill covers. Some are provided with removable spouts of flexible metal hose to permit filling of less accessible reservoirs. Such cans are sturdy, easily cleaned, and closed against con-

Fig. 15-29   Larger Oil Containers

tamination. Selfclosing safety cans of approved design should always be used for transporting flammable materials.

Plastic containers for oils in 2 qt to 5 gal sizes are being used in some plants. Lightness, low cost, and freedom from rust or corrosion are advantages. However, some plastics are affected by oils or by the additives in some oils, so it is always necessary to check with the manufacturer to be sure that the plastic used is compatible with the fluid to be carried.

Portable Oil Dispensers A wide variety of equipment is available to transfer quantities of oil from the oil house to the point of use. Included are bucket pumps (Fig. 15-30) which can be used to transport a few gallons of a liquid lubricant and discharge it directly into a machine sump or reservoir. Where larger quantities must be transferred, wheeled dollies (Fig. 15-31) which will transport a standard 16 gal or 55 gal drum are available. The drum can be equipped with a manual, air, or electrically operated transfer pump so that the lubricant can be pumped directly into the machine. Even larger quantities can be transported in special oil carts (Fig. 15-32), which also may be equipped with a manual or power operated pump.

Portable Grease Equipment Small quantities of grease (3 to 24 oz) (85 to 680 g) can be dispensed from hand grease guns. These are available in push, screw, lever (Fig. 15-33), and air operated types. Some of the lever and air operated guns can be loaded by suction, with a gun filler, by cartridge, or by any of the three. Where small quantities of grease at high pressure are required, special lever operated guns, sometimes called booster guns, are available.

Couplers and coupling adapters can be fitted to grease guns to permit their use with the various types of fittings. Extension hoses can be attached to many guns to facilitate lubrication of hard to reach fittings.

Before loading a grease gun, it should be checked for cleanliness, and cleaned if necessary. When using a grease gun filler, the fittings on both the filler and the gun should be wiped clean with a lint free cloth. Hand loading should be avoided whenever possible. If it is necessary to load from an open grease container, check the surface of the grease for dirt and if necessary scrape off the top layer to remove any dirt or hardened grease. Use a clean metal paddle to transfer the grease from container to gun, or load the gun by suction filling. Before suction filling, carefully wipe the surfaces of the gun that will be submerged in the grease. Grease cartridges should be wiped to remove dust before being loaded into a gun. The metal top of the cartridge should be pierced or torn off carefully to avoid the possibility of metal particles falling into the grease.

Grease guns should be marked clearly as to the type of grease they are used for. A gun should be used for only one type of grease to avoid problems that might arise due to incompatibility of different brands or types of grease. If it is necessary to change the type of grease to be used, the gun should be thoroughly cleaned in solvent and dried before refilling. After filling, grease should be pumped through the nozzle to flush out any remainder of the previously used product and any traces of solvent.

Where more of a particular grease is required than can be provided by a single filling of a gun, several alternatives are available. A gun filler on a pail

Fig. 15-30 Lever Operated Bucket Pump This unit holds 30 lb of a lubricant such as a gear lubricant. The extension on the base is to step on to hold the unit firmly while it is being operated.

Fig. 15-31 Drum Dolly With the wheels raised, a drum can be rolled onto the platform of the dolly. The wheels can then be lowered and the drum moved to the point of use where the product can be dispensed with a transfer pump.

or drum of grease, or a portable gun filler (Fig. 15-34) can be taken to the location where lubrication is being performed. Manual bucket pumps on standard 35 lb pails can also be used. A variety of air and electric power guns are also available. Most of this equipment will handle semifluid and soft greases without difficulty. Some fibrous and harder greases may require a follower plate to ensure that the grease slumps to the pump inlet. Very hard greases may require an arrangement that provides positive priming, as shown in Fig. 15-35, to ensure feeding to the pump inlet.

Lubrication Carts and Wagons The lubrication program of a plant may be so organized that personnel are regularly assigned to service comparatively distant parts of the plant where they must apply a number of different oils and/or greases to various types of machines. It is important in these cases to supply each person with a practical means for transporting the necessary supplies. Such equipment may be either purchased or built. It may be elaborate or simple, depending on the variety of lubricants used and on the machines to which the lubricants are to be applied. A simple, practical lubrication wagon may be nothing more than a shop cart with sufficient room for oil containers, grease guns, hand oil cans, grease pails or portable gun fillers,

Fig. 15-32 Oil Cart The drain on the right facilitates cleaning the tank.

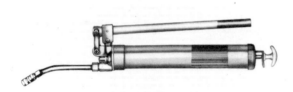

Fig. 15-33 Lever Operated Grease Gun Models are available to deliver various pressures. Capacity of this model is 24 oz, and it can be loaded by suction, by paddle, or with a grease gun filler. Similar guns of smaller size are available for cartridge loading.

and a supply of miscellaneous necessities such as spare grease fittings, clean wicks for wick feed cups, filled bottles for bottle oilers, replacement cartridges, clean lint free cloths for cleaning application points, and lubrication instructions for the machinery to be serviced. Such a cart is flexible and permits a quick change in the type of lubricants carried.

More elaborate lubrication carts can be obtained from suppliers of lubrication equipment. The type shown in Fig. 15-36 provides space for portable grease gun fillers, grease guns, and a number of containers of fluid lubricants. The types shown in Fig. 15-36 handle mainly fluid lubricants, but space is available to carry miscellaneous supplies. If there are provisions for practical and easy cleaning of the tanks, these carts are very satisfactory.

Closed System Dispensing A number of types of lubricant dispensing and application systems can be considered as essentially closed systems in which the lubricant is exposed to

Fig. 15-34 Portable Gun Filler The 30 lb tank on this unit can be filled in the oil house. The unit can then be taken to the plant to refill guns as required.

Fig. 15-35 Electric Power Gun Cutaway The cutaway view shows the helix arm and worm gear which ensure feeding to the pump inlet of stiff greases.

Fig. 15-36 Commercial Lubrication Cart Space is provided for gun fillers, fluid lubricants, grease guns, and miscellaneous supplies.

the minimum possibility of contamination. These systems are not truly closed in that the supply tank or reservoir must be charged with lubricant periodically, or the lubricant may be dispensed at the point of use through a nozzle or metering device, but the risk of contamination is greatly reduced compared to the dispensing methods already discussed. Some of the more common types of these systems are disucssed in the following sections.

Central Dispensing Systems
While lubricants can be dispensed from bulk bins or bulk tanks into containers, oil wagons, grease gun fillers, etc., for transfer to points of use, at locations where usage is high, transfer by some form of piping system often is justified. Such systems are usually custom designed and built where large volumes are to be handled, but smaller systems may be assembled from equipment modules available from lubrication equipment suppliers. Electric, hydraulic, or air operated transfer pumps, similar to those discussed earlier for dispensing from drums that will handle greases or oils are available to fit tanks or bulk bins. Where lines extend beyond the capacity of the tank mounted pump, booster pumps can be installed at suitable points along the lines leading to the dispensing stations. When distribution lines must be run outdoors or through cold areas, steam or electric tracers on the lines may be required to maintain pumpability of the lubricant.

At the remote dispensing stations, delivery can be made to grease gun fillers, power grease guns, the reservoirs of centralized lubrication systems, to any type of oil dispensing unit, or directly to the point of final use such as the reservoirs of large lubrication systems or units being lubricated on an assembly line. An example of lubricant being dispensed directly to the point of use is shown in Fig. 15-37. The plant in which this diesel engine is in-

Fig. 15-37 Dispensing System for Large Diesel Engine The overhead hose reel (arrow) is used to dispense oil directly to the cylinder lubricators through a metering nozzle. Crankcase oil is brought from the main storage tank to a secondary tank indoors, then pumped directly to the crankcase.

Fig. 15-38 Metering Control Valve The dial can be preset to deliver 1 to 60 qt of oil and then shut off flow. A counter provides a record of the total amount of lubricant dispensed.

stalled is equipped with a 5000 gal (19,000 L) underground storage tank for cylinder oil, and a 3,000 gal (11,400 L) tank for crankcase oil. Cylinder oil is pumped to the overhead hose reel and then dispensed through a metering nozzle to the cylinder lubricators. Crankcase oil is pumped to a 275 gal (1040 L) secondary tank indoors for preuse warming, then is pumped directly to the crankcase. The pump for the latter operation is controlled from the engine location so that the correct amount of makeup can be added. Dispensing nozzles of the type shown in Fig. 15-38 can be used to control dispensing in systems of this type, or in assembly line filling of equipment such as gear units. The dial can be preset to deliver from 1 to 60 quarts and to shut off automatically. The meter also records the total amount of oil dispensed, up to 9999 gal.

Maintenance and Service   Comparatively little maintenance is required in central dispensing systems. Proper cleanliness precautions should be observed when filling storage tanks or reservoirs. An adequate quantity of lubricant should always be maintained in the tank so the transfer pump will not suck air which might cause binding. Pump pressures should be checked periodically, and lines should be inspected for leaks or damage.

16 In-Plant Handling for Lubricant Conservation

Conservation of our natural resources and improvement of our environment have become commonplace goals of society today. No one will disagree with the desirability of the goals, only possibly with the means to achieve them. When a single operation by modification can contribute to both environmental improvement and conservation of natural resources, its achievement is to be highly recommended. Control of lubricants within the plant to prevent the generation of oily wastes is just such an operation. Any lubricant that becomes a waste becomes a cost item from the standpoint of material purchases and waste disposal. A plant lubricant may become a waste product and as such may require reclamation or disposal, under proper control to prevent pollution, for the following reasons:

1. Contamination with water, foreign matter, dust, process materials, wear metals from the lubricated machine, and fuel dilution
2. Degradation during use due to depletion of additives, increase in total acid number, formation of oxidation products, change in viscosity, or loss of lubricity
3. Escape from system through leaks, spills, or drips; excess lubricant in all loss systems; carry off on products in process

The key to preventing any lubricant from becoming a waste product lies in the selection, storage, and handling of these products in the plant from the receipt as new materials to the disposal as used materials.

With increasing emphasis on the subject of pollution of our ground and surface waters, stricter regulations are being legislated and enforced on the composition of industrial plant effluent to both surface waters and municipal sewage plants. Regulations covering plant effluent or stream quality vary from area to area and, depending on the jurisidictional control of the particular body of water or sewage plant authority, may be under the control of Federal, state, municipal or regional groups. The first step for anyone concerned with the problem of preventing pollution by control of plant effluent is to know the effluent, sewage, or stream standards governing his particular situation. It is impossible to cover the details of the myriad of regulations in this Chapter. Suffice it to say that the basic Federal law concerned with the abatement of both industrial and municipal water pollution is the Water Quality Act of 1965, Public Law 89-234. This Act was further strengthened in 1966, followed by the enactment of the Water Quality Improvement Act of 1970. Basically, these legislative actions prohibit the discharge of any waste effluents which may result in deterioration of the quality of these waters to levels below established standards. The law applies to interstate waters, but states have also developed standards for intrastate waters. These standards are based on economics, health, and esthetic considerations, and are set to protect not only present, but also anticipated future uses of waters. A municipality or industry which discharges liquid wastes directly or indirectly to receiving waters is obligated to conform to the specific water quality standards set by local and state agencies and approved by the Federal Water Quality Office. Plans for implementing and enforcing water quality standards are different in the various states and also vary with the use of a particular body of water, i.e., whether it is for municipal water supply, recreation and fish propagation, agricultural or industrial use, etc.

Water quality standards may be established either for the receiving waters into which wastes are discharged, "stream standards," or for the waste itself, "effluent standards." Standards established for industrial waste effluents depend on whether disposal is to receiving waters or to a municipal sewage treatment plant. With respect to the latter, many municipalities have, or are developing, sewer use ordinances to control and regulate the volume, strength, and nature of the waste discharged to the sewers. Generally, such regulations prohibit the discharge of gasoline, fuel oil, spent lubricating oils, phenols, as well as toxic substances such as heavy metals and other materials which may interfere with the sewage treatment process. Also, sewage discharges may be limited in respect to their Biological Oxygen Demand (BOD), perhaps to no greater than 300 mg/L for a 5 day BOD test. Current restrictions in many localities, in line with the increasing efficiency of sanitation plants, are lowering the limit to 25 mg/L and Total Organic Content (TOC) is being evaluated as the criterion in place of BOD.

As previously indicated, before discharging any fluid waste into a waterway, it is essential to comply with the existing regulatory requirements at the particular location. In general, as far as oil content is concerned, the effluent should be substantially free of visible floating oil (not in excess of about 15 mg/L). This level assumes a dilution effect in the receiving stream. For example, if the stream is to be used for municipal water supply, its oil content needs to be kept below 0.2 mg/L. If, on the other hand, the receiving body of water is being used for recreational, agricultural, or industrial uses, as much as 10 mg/L may be acceptable. By the same token, in order not to interfere with the planned use of the stream, effluent limits generally must be met for color, odor, turbidity (should be removed to below obvious levels), dissolved oxygen, heavy metals such as lead and mercury, phenols, phosphates, suspended solids, etc.

OVERVIEW OF IN-PLANT HANDLING

The objective of proper in-plant handling, then, may be defined as efficiently utilizing petroleum lubricants to prevent them from prematurely becoming waste products and, further, disposing of them when they do become waste products in such a way that they do not become part of the plant effluent to ground or surface waters.

Annually, worldwide,* approximately 5.3 billion gallons of petroleum lubricating and process oils are produced and used as automotive and industrial products. About half of this, 2.7 billion gallons, is used industrially, with some 920 million gallons as process and other oils consumed in use and 1.78 billion gallons as lubricants, hydraulic fluids, and cutting fluids. The last quantity requires reclamation for continued beneficial use to conserve our natural resources or disposal in an approved manner to prevent pollution.

Mobil, as a member of the petroleum industry, is keenly aware of its responsibility to utilize, with due regard for conservation and environmental quality, one of nature's primary resources. It realizes this responsibility does not end with ensuring that its own operations use petroleum resources in the best interest of society, but also desires to assist its customers in realizing the ultimate benefits from petroleum products and the proper disposal of them once they have served their primary purpose. Therefore, Mobil will continue to carry on its operations with full concern for the protection of public health and property from harmful pollution of air and water and will cooperate with industry and government in identifying sources of harmful pollution. Where the weight of scientific evidence establishes the need, Mobil will take prompt, sound, and realistic corrective action.

In addition, Mobil's Lubrication Program should be an initial step toward efficient lubricant usage and waste oil disposal. Mobil's Engineering Service, part of this program, helps to insure proper product selection, maximum product life in service, minimization of leaks, beneficial machine maintenance, optimum drain intervals, and improved handling and storage to prevent contamination and spills. It is better not to generate a waste oil in the first place than to dispose of it later.

Although each industrial operation and individual plant limits are sufficiently different to make the minimization of waste oil generation and disposal to control pollution unique problems requiring specific solutions, experience shows that the following concrete suggestions and overall

*Free world total.

recommendations can be made concerning the handling of lubricants within the plants, which may be adapted to each individual case:

1. Select lubricants, including hydraulic fluids, gear lubricants, cutting fluids, coolants, and crankcase oils, to obtain long service life.
2. Establish a program of good preventive maintenance and efficient housekeeping.
3. Where feasible, utilize a plantwide multimachine circulation system to replace small, single machine reservoirs.
4. Provide purification equipment for circulation systems to ensure optimum recycling where feasible.
5. Survey to ascertain the nature and source of waste generation and disposal problems.
6. Attack the problem at the source, not at the plant effluent stage.
7. Know local regulations and regulatory bodies.
8. Keep cutting fluid and lubricant streams separated and distinct from each other to prevent complexity in mixed streams.
9. Maintain separate sewer systems for sanitary waste, process water, and storm drainage.
10. Don't use dilution as the solution to pollution control.
11. Concentrate before you treat for final disposal — usually volume increase causes increased costs and space requirements.

PRODUCT SELECTION

Since the purchase and receipt of lubricants into a plant initiate the cycle leading to eventual consumption or disposal, it is at this point we must start to examine what can be accomplished to prevent the new product from becoming a premature waste oil and a possible pollutant.

Many factors enter into the selection of the proper product depending on its use as a lubricant, hydraulic fluid, or cutting fluid. Speed, load, and temperature must be considered in selection of gear and bearing lubricants; type of metals and severity of operation in machining; and the type of engine, speed, and fuel, in crankcase and cylinder lubrication of stationary diesels, gas engines, and gas turbines. These factors and their effects on product selection are well known and usually adhered to. There are additional factors which will minimize the waste oil disposal problems in a plant. These factors, which should be considered in the selection of a product, are long service life in any application, the ability to help control leakage, the ability to minimize contamination, compatibility with other products, value as byproduct waste oil, and ease of disposal. One general rule that should be a guide in product selection to achieve the optimization of in-plant handling is the use of the smallest number of multipurpose premium products as possible.

Long Service Life One important characteristic of a lubricant that will do much to minimize the waste oil disposal problem is long service life. It is apparent that the longer the interval between drains in any machine using a closed circulation system for lubrication, the less waste oil will be generated and require disposal. Long service life is a result of many factors including machine condition, proper selection of lubricant and product quality. Product characteristics inherent in the base oil or additive package which insure long life are chemical stability to insure minimum buildup of oxidation products; resistance to changes in viscosity, increase in acidity and deposit formation; additive sufficiency to insure the needed film strength, detergency, and viscosity-temperature characteristics; and resistance to depletion of these additive qualities. All of these items lead to selection of the highest quality product to achieve the longest service life. In addition, premium product selection is usually the most economical, for although the initial material cost is higher, the cost of application, maintenance and disposal in manhours is sharply decreased and the overall cost of lubrication is lower.

Compatibility with Other Products

Where the possibility exists of one product mixing with another during use, such as the hydraulic fluid in a machine tool with the cutting fluid, the products used for both fluids should be selected so that cross contamination will be avoided. For example, in the instance mentioned, if a chemical coolant type of cutting fluid could be used as the hydraulic fluid as well, disposal will not be required because of admixture. Other instances will become apparent in a survey of the types of lubricants used in the plant and the product availability from lubricant suppliers. Even at the disposal stage compatibility of waste oils will permit consolidation without forming complex mixtures with attendant difficulties of disposal.

Value as Byproduct

Since eventually most lubricants become waste oils and require disposal, consideration should be given at the selection stage to the value of the waste oil as a byproduct. Many lubricants after satisfying their primary function can be used for less demanding service, be used as a fuel, or serve as a feed material for rerefining or reclamation. The ability of a lubricant to serve as a byproduct after initial use should be considered in the selection of a new product, both from the point of view of conservation of natural resources and that of waste disposal.

Ease of Disposal

The last factor to be considered in lubricant selection is the ease of disposal once it becomes a waste product. This factor is related to the type of products for example, oil-in-water with water-in-oil emulsions differ in the ease with which the emulsion can be broken. Also products have been developed as cutting fluids which may be disposed of in municipal sewage systems. In addition, the value as a byproduct, as noted previously, will increase the ease of disposal.

Following the selection of products to fulfill the lubricating and other plant requirements, the next area in which to prevent premature generation of waste oils and subsequent disposal or pollution control problems is the handling, storing, and dispensing of lubricants and associated petroleum products after receipt from the vendor. This subject is covered in Chap. 15.

IN-PROCESS HANDLING

Once a lubricant is charged to a reuse system and until it requires drainage and replacement, much can be done to prevent escape of the lubricant from the system and its premature degradation requiring early draining. The prevention of leaks, spills, and drips, which complicate the disposal of lubricants and may lead to accidental pollution of plant effluent, will be considered in this section. The elimination of unnecessary contamination through proper system operation and preventive maintenance will also be treated along with in-process purification to prolong the service life of lubricants, and hydraulic and cutting fluids.

Reuse versus All Loss Systems

There are many ways of applying lubricants to bearings, gears, machine ways, cutting tools, etc., varying from hand oilers, grease guns, through bottle and wick oilers and splash and mist feed, to the most sophisticated circulation systems and centralized bulkfed systems. Some of these types of application are all loss systems in which the lubricant is used in a once through operation. In some cases, the lubricant is consumed in the process of lubrication, such as in a two cycle engine where the lubricating oil is mixed with the fuel; is carried off on the product or in exhaust gases; or is collected for disposal. A closed cycle system such as a gear case or a circulation system, as used in many bearing systems for large engines, turbines, or machine tools, continually reuses the same lubricant until its lubricating characteristics change or are degraded. The latter systems are recommended wherever possible both from the standpoint of conservation of product and prevention of pollution. Circulation systems are by far more economical in terms of costs of material, application labor, housekeeping, and disposal. Further, these systems when coupled with purification systems actually furnish better conditioned lubricant throughout the use period to the machine's bearings, gears, and other lubricated parts.

Closed cycle and circulation systems can be more readily adjusted to prevent excessive lubrication and can be sealed to prevent contamination of the lubricating oil or cross contamination of the lubricant with cutting fluids, process fluids, and materials being fabricated.

Although an excellent method for specific application and economical in lubricant consumption, mist lubrication systems unless carefully controlled may cause fogging and condensation on floors and machine exteriors, causing unsafe and possibly polluting conditions. The same is true of many hand controlled oilers which need constant attention to insure proper feed without excess lubrication which may cause drips and leaks.

Prevention of Leaks, Spills, and Drips

Any loss of lubricant from a system or machine components, not called for in the design and for which a collection system is not provided, complicates the disposal problem. It has been estimated that leakage from circulating systems, including hydraulic systems, approximates 100,000,000 gallons a year and this would require 5,500,000 manhours to provide makeup. With proper leakage control, over 75 percent of this waste of material and manpower could be eliminated. In addition to the saving in material and labor the benefits that accrue from proper leakage control include increased production, decreased machine downtime, prevention of product spoilage, cross contamination of other lubricant or coolant systems; elimination of safety hazards to plant and personnel; and prevention of accidental pollution of the plant's storm, process, or sanitary waters. In addition, leaks, spills, and drips seriously complicate the disposal problem. Product thus disposed of is lower in byproduct value, most often it must be incinerated to be disposed of in a nonpolluting manner. Further, by cross contamination, it often forms emulsions or complex mixtures with other liquids which are not easily separated. The impact of even small drips can be seen from the data in Table 16-1.

System leakage control on any machine or complex of machines requires attention to two general classes of joints through which fluid may be lost, namely, moving joints and static joints. Moving joints include rod or ram packings, seals for value stems, pumps and fluid motor shafts and in some instances, piston seals. Static joints include transmission lines, pipes, tubing, and hoses, fittings and couplings, and seals and packing for manifolds, flanges, cylinder heads, and equipment ports.

Little or no leakage will occur past newly packed moving joints if correct material and procedures have been used. Some of the causes of leakage in moving joints include improper installation, causing seal and packing damage, and rough or scarred finishes on rods or shafts. However, both internal leakage past pistons and valves and external leakage past rod, shaft and valve-stem packings may be expected to develop in time even under normal service conditions. Only a properly planned and executed preventive maintenance program will find and correct such leakage. Internal leakage may cause problems in machine operation and loss of lubricant through consumption in the machine. External leakage may cause problems of loss and pollution noted previously.

Leakage of oil from static joints may be the result of one or more of several common causes:

1. Use of unsuitable type of joint or transmission lines
2. Lack of care in preparing joint — machining, threading, cutting, etc.

Table 16-1   Losses by Oil Leaks

	Loss in 1 day		Loss in 1 month		Loss in 1 year	
Leakage	Gallons	Value[b] dollars	Barrels[c]	Value[b] dollars	Barrels[c]	Value[b] dollars
One drop[a] in 10 sec	0.112	0.10	0.06	3.30	0.72	29.60
One drop[a] in 5 sec	0.225	0.20	0.12	7.80	1.44	79.20
One drop[a] per sec	1⅛	1.10	0.62	33.80	7.44	409.20
Three drops[a] per sec	3¾	3.70	2.05	112.80	22.60	1243.00
Stream breaks into drops	24	24.40	13.1	720.50	157.2	8646.00

[a]Drops approximately ¹¹/₆₄ in. diameter. One drop in 10 sec = 200% makeup per year; one drop in 5 sec = 400% makeup per year; one drop in 100 sec = 20% makeup per year.
[b]Based on $1 per gallon.
[c]Based on 55 gallon bl.

3. Lack of care in making up the joint
4. Faulty installation such that joints are loosened by excessive vibration or ruptured by excessive strain
5. Severe system characteristics which subject the joints and lines to peak surges of pressure due to "water hammer"

In addition to slow leakage, a combination of these items may contribute to fatigue failure, line breakage, and loss of large quantities of lubricant or hydraulic fluid.

Several general measures to reduce leakage are as follows:

1. Use joint types and packing material for installation and maintenance work that have proved satisfactory in service.
2. Train maintenance personnel in principles of proper installation of joints and packing, and maintain surveillance to ensure proper execution of accepted procedures.
3. Minimize the number of connections and make all lines and connections accessible for checking and maintenance.
4. Design installations to avoid excess vibration, mechanical strain, twisting, and bending.
5. Locate and protect tubing, valves, etc., from damage by shop trucks, heavy work pieces, or materials handling equipment.
6. Protect finely finished surfaces in contact with packings from abrasives and mechanical damage.

One other area which should be given consideration to prevent spills is in dispensing and draining of lubricant from machine reservoirs. The dispensing of new lubricants has already been discussed and the drainage of waste oils will be treated later in this chapter.

Elimination of Contamination   As outlined in the introduction, lubricants become waste products by contamination with foreign matter, by degradation products formed from the lubricant under the action of the lubricating process, or by cross contamination through admixture with other process fluids. The result of any one of these contaminating actions is a loss of primary function of the lubricant with consequent need for reclamation or disposal as a waste oil, a downgrading in value as a byproduct, or an increase in complexity of disposal procedures. Therefore, in the interest of minimizing waste and maximizing ease of disposal, contamination of any kind should be eliminated or controlled to the highest degree possible.

Central Reservoir Maintenance   Reservoirs for circulating lubricating systems, which include hydraulic and cutting fluid systems, may be any one or combination of more than one of three types: integral with machine and located in base, separate from machine but individually associated with it, and centrally located reservoirs serving multiple machines. Whether a central system is economically justified depends on many factors, among which are the size of the shop, the number of different grades or brands of fluid in use, and general shop layout. Obviously in any shop with multimachine lubrication requirements, a central system can be better justified if only one or at most two multipurpose lubricants are handled.

Table 16-2   Circulating Oil System Contaminants

Cleaning solvents or compounds	Packing and gasket fragments
Coal dust	Paint flakes
Core sand	Persistent emulsions
Cutting fluids	Pipe scale
Dirt	Pipe-thread compound
Drawing compounds	Rust particles
Dust	Rust preventives
Grease from pump bearings	Water
Lint from cleaning rags or waste	Way lubricants
Metal chips	Weld spatter
Metal wear particles	Wrong oil

All reservoir types can become contaminated in similar fashion, and preventive methods apply equally to each. Dirt, water, dust, lint, and other foreign substances contribute to the formation of emulsions, sludges, deposits, and rust when present in a circulating lubrication system and must be removed or the bulk lubricant drained and replaced. Such contaminants, present in the waste oil, limits value as a byproduct for use as a fuel or for reclamation. Table 16-2 lists numerous contaminants sometimes present in circulating systems. Considering each of these individually may suggest sources of contamination and means to prevent entry into the system. Contamination prevention of foreign matter involves both good system design and good housekeeping practices. The following general recommendations will help in controlling reservoir contamination by foreign matter; see Fig. 16-1:

Fig. 16-1   Contamination Control in Central Reservoir

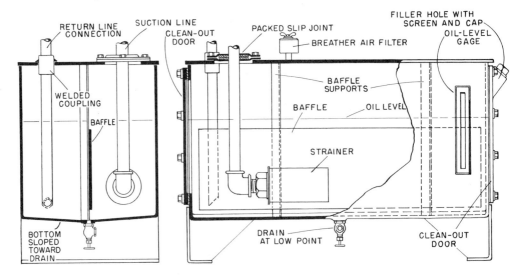

1. The reservoir cover, if of removable type, should fit well, be gasketed and tightly bolted on.
2. Clearance holes in the reservoir cover for suction and drain lines should be sealed, preferably by a compressible gasket and bolted flange type retainer.
3. The oiler filler hole should have a fine mesh screen and a dusttight cover.
4. The breather hole should be provided with an air filter and checked regularly.
5. The suction should be equipped with a strainer to prevent dirt and other foreign matter from entering the system and should be inspected regularly.

Cross Contamination   Where a machine utilizes separate systems and products for hydraulic operation, bearing and gear lubrication, and machine tool cooling, every effort of design and operation must be exerted to prevent cross contamination. For example, hydraulic systems of machine tools are subject to contamination with water-soluble chemical coolants or oil emulsions which can, due to poor oxidation resistance, chemically active agents, and fatty acids, cause deposits on valves, emulsion formation, and foaming in the hydraulic system. Similarly, in the case of machine tool operation, chips may be allowed to pile up around cutting fluid drains. Such piles act as dams and may cause pools of cutting fluid to overflow onto the ways and contaminate the way lubricating system. Further, as mentioned in leaks and drips, lubricating systems and packing on rotating shafts may allow cross contamination of the cutting fluid with tramp oil.

Proper System Operation   Another type of contamination which will cause a lubricant to become a waste oil is caused by degradation products formed in the lubricant as a result of system operating conditions. For example, excessively high operating temperatures can cause oxidation of the base oil and the oxidation products formed could result in changes in viscosity, an increase in acidity, and the formation of varnish and deposits. The excess temperature may be the result of improper selection of system capacity, system malfunction, or poor original design. While premium lubricants have high oxidation stability, proper system design and operation should help this inherent characteristic. Adequate capacity of the lubricant system or the use of an oil cooler will help control the bulk oil temperature during recycling. Also, proper oilfeed rate will assist in holding temperatures to a reasonable limit, and properly sized components will assist by carrying the load applied. Soot, unburned fuel, water vapor, combustion, and ash and wear metals in piston blowby can also contaminate a lubricating oil. Proper maintenance of the engine will minimize these contaminants.

IN-PROCESS PURIFICATION

In circulating systems, the lubricant, hydraulic oil, or cutting fluid should be kept as free of contaminants as possible, first, as noted previously by preventing entry of contaminants into the system, and second by in-process purification during operation. Purification, depending on the nature and extent

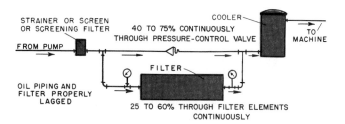

Fig. 16-2   Continuous Bypass Purification

of the contaminants, may consist of continuous bypass, full-flow treatment, or a combination of the two incorporated into the circulating system, portable units for periodic purification, independent purification units connected permanently to large reservoirs, and separate batch units used to purify or reclaim drained lubricants for reuse.

Continuous Bypass Purification

In the continuous bypass system (Fig. 16-2), a portion of the oil or coolant delivered by the pump is diverted continuously from the main line for purification. The cleaned oil or coolant is then returned to the system. The remaining unpurified oil or coolant is delivered through the main line to the system. This system of purification is reasonably effective when the rate of contamination is not too high.

The bypass is limited to applications where continuous purification of 5 to 10 percent of the total pump discharge is sufficient to keep the entire charge in good condition for a satisfactory period.

Flow to the bypass filter is limited by a throttling orifice to prevent too great a withdrawal of bypass fluid. Gradual clogging of the elements reduces the flow of dirty fluid through the filter. This condition causes an increase in pressure on a gage on the filter between the orifice and the filter elements. When this gage shows a predetermined increase in pressure, the filter elements should be replaced.

Continuous Full-Flow Purification

In a continuous full-flow system (Fig. 16-3), the entire volume of lubricating oils or coolant is forced through a filter before passing through the cooler to the lubricating system. The oil may pass through the cooler ahead of the filter if the cooled oil is still at a good filtering viscosity.

Clogging of the filter decreases the flow of fluid through elements. This condition is indicated when gauges on the inlet and outlet of the filter show

Fig. 16-3   Continuous Full-Flow Purification

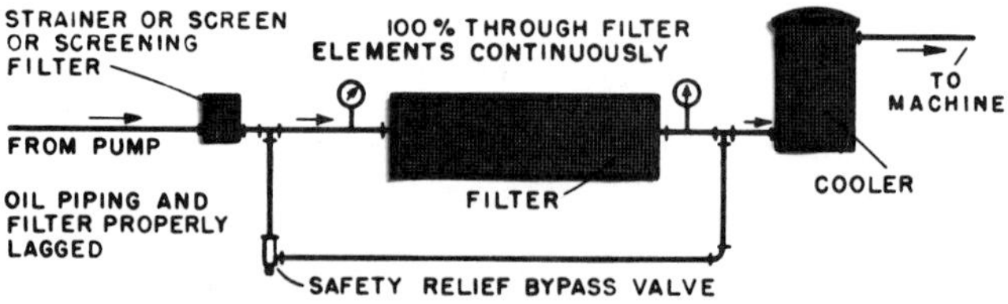

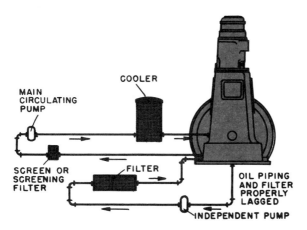

Fig. 16-4 Continuous Independent Purification

an increasing drop in pressure across the filter. Before this drop becomes excessive, a relief valve in the filter opens and bypasses unpurified fluid around the elements to maintain a full supply of oil or coolant to the system. Before this happens, the filtering elements should be replaced. Filters installed in parallel with the required valves permit replacement of elements in one filter at a time, without shutdown. This should be done whenever the pressure differential across a filter reaches the value recommended by the filter or machine manufacturer.

Continuous Independent Purification Lubricating oil or coolant is sometimes purified continuously by means of a system entirely independent of the main circulating system. Used oil is drawn from the sump by an independent pump, delivered to a centrifuge (purifier) or filter, or both, and after purification, returned to the system (Fig. 16-4).

Although this system would operate normally during a system operation, it can be used when the system is shut down, and can be shut down, without system interruption, at any time to replace filter elements or clean the purifier whenever pressure drop across the unit dictates.

Periodic Batch Purification In this method, the entire charge of lubricating oil or coolant is removed from the system for purification. The oil or coolant may be allowed to settle, then reheated and passed through a centrifuge, reclaimer or other type of purification equipment.

Unless this oil is changed quite frequently, batch purification alone cannot be expected to keep an engine as clean as effective continuous purification.

Full-Flow and Bypass In such a purification system, the dirty fluid from the machine passes through the purification system at a full flow rate and a portion of the stream delivered to the machine is diverted and returned for further purification. The bypass stream is valved so that when the main stream is shut down the

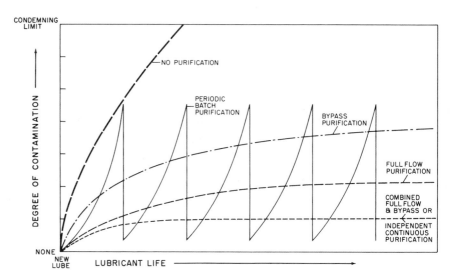

Fig. 16-5   Effectiveness of Purification Methods

purification system may continue to run. Such a system when utilizing filtration can provide for gross filtration of the full-flow stream and fine filtration of the bypass stream at a slower flow rate with the combination yielding a cleaner fluid.

A comparison of the four methods (Fig. 16-5), full-flow, continuous bypass, combination full-flow and bypass, and batch purification, shows that the batch method yields the cleanest fluid at the least cost but requires downtime and has a widely fluctuating efficiency. The combination method, which can function when the system is either operating or shut down and can use more stringent purification method on the bypass portion without sacrificing operating flow rates, yields the highest overall efficiency.

Many industrial oils can be retained in service for long periods before draining by use of in-system or portable purification systems. Other contaminated oils can be reused after independent purification; they can be returned to the system provided they are clean, retain their original characteristics, and if additives have not been removed by the purification process. If not serviceable for the original uses, purified lubricants may be used in a less critical application (e.g., a purified hydraulic oil might be used in a noncritical gear case, an all loss system such as hand oiling of chains, or for flushing purposes). Lubricants cross contaminated with cutting fluids or vice versa cannot be adequately purified for their original purpose and, thus, must be drained and used for heating, rerefined, or disposed of in another way.

PURIFICATION METHODS

The purification methods most commonly used in industrial plants are settling, size filtration, and clay or depth filtration. The contaminants removed and the various purification methods are outlined in Table 16-3.

Table 16-3   Materials Removed by Purification[a]

Materials	Settling	Size Filtration	Clay Filtration	Centrifuging	Reclaiming
Water	Yes	*	*	Yes	Yes
Solids	Yes	Yes	Yes	Yes	Yes
Soluble oil-oxidation	No	No	Yes	No	Yes
Fuel, solvents	No	No	No	No	Yes
Oil Additives	No	No	Yes	**	Yes

*A little water can be adsorbed or held back by the filter material.
**Some additive is removed along with water.

Settling, flotation and size filtration are often used on the machine or in the circulating system. A central reservoir of proper design will permit settling and may be equipped with size filtration. Clay filtration and centrifuging may be incorporated into the system as separate units. Reclaiming systems are usually a batch operation conducted on the drained lubricant.

Settling   Water and heavy solid contaminants will separate by gravity whenever oil is held in a quiescent state for a suitable period. For separation of solids, water, and oil the system as shown in Fig. 16-6 can be fabricated. Moderate heating (160–180°F, 71–82°C) of the oil by a steam coil, electric immersion,

Fig. 16-6   Settling Tanks for Batch Purification

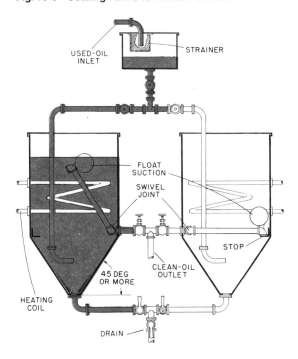

or belt type heater will lower the oil viscosity and thereby accelerate the settling process. In the dual tank for use in batch purification (Fig. 16-6), preliminary straining of gross particulate is achieved in the upper tank and used oil is introduced into one of the duplicate settling tanks well below any clarified or partially clarified oil present. The used oil is added slowly to prevent agitation of the settled contaminants. The floating suction permits withdrawing clean oil from the surface above a minimum level of water and sludge concentrate in the conical bottom.

Filtration Filtration or straining may be used either as full-flow or continuous bypass methods. Straining by metal screen or woven wire in cartridges or plates removes only the largest particles, is low in initial cost, and allows maximum flow rates indendent of fluid viscosity. It will not remove water and other soluble contaminants. Area, edge, and depth type filters allow a variation in size of particle removal with the depth and area types suitable for the finer particles. The area, edge, and depth type filters may be used on full-flow or continuous bypass. Certain types will also remove small amounts of water and additives. While high in both initial and operating costs, they are adaptable to single machines or multimachine systems. When the system is properly designed, little downtime is required.

Size Filtration This method uses wire cloth, metal edge, paper edge, surface type strainers, or depth type filters to remove solid particles from oil.

Wire cloth strainers are used to remove only the larger particles—50 μm and larger. Metal-edge type filters (Fig. 16-7) clean by forcing the oil to pass through a series of edge openings formed by a stack of metal wheels or between the turns of metal ribbons. The majority of these units will filter out

Fig. 16-7   Edge Type Oil Filter

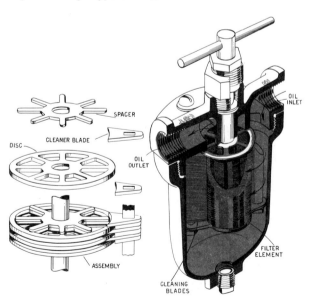

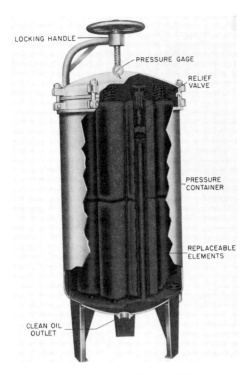

Fig. 16-8 **Typical Depth Type Oil Filter**

particles of 90 μm; some will filter particles as small as 40 μm. Paper-edge filters use elements of impregnated cellulose discs or ribbons. Impregnated cellulose discs will filter particles from 0.5 to 10 μm, ribbon elements 40 μm size particles. Depth-type filters (Fig. 16-8) use Fuller's earth, felt waste, or cellulose fibers in replaceable cartridges and can filter out particles of 1 to 75 μm size. Surface type filters use replaceable resin treated paper or closely woven cloth. The paper elements are good to 1 μm particle size, the cloth to 40 μm size.

Depth Type Filters These are probably the most versatile of industrial type filters available. The depth type filters can be equipped with cartridges using a variety of filter media. Commercially available cartridges include the following.

1. An adsorbent type filter containing Fuller's earth in a woven cloth container. This material is recommended for removal of soluble contaminants such as acids, asphaltene, gums, resins, colloidal particles, and fine solids. It will also remove polar type oil additives and is, therefore, not recommended for this type of lubricant unless the additive is replaced following purification. Although Fuller's earth will remove small amounts of water (approximately 1 qt (0.95 L) of water per element), it is not recommended where gross water contamination is prevalent. Typical applications for this type filter medium are quench oils, hydraulic fluids, engine oils, and vacuum

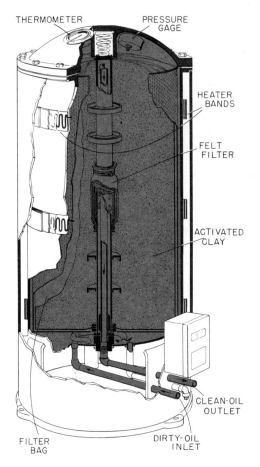

Fig. 16-9 Activated Clay Type Oil Filter

pump lubricants. It is also recommended for filtering certain synthetic fluids such as phosphate esters and silicones.

2. A cellulose type filter element, usually a combination of cotton waste and wood fibers such as redwood, excelsior, etc., is usually recommended for removal of large amounts of gross solids contaminants. It will not remove additives or water; the cellulose type retains its resiliency in the presence of water and water emulsion and usually permits higher flow rates because of its greater porosity. Both types are recommended for filtering engine oils, hydraulic fluids, turbine oils, fuel oils, and lubricating oils.

3. Resin impregnated, accordian pleated paper filter cartridge is recommended for medium contaminant loads, high flow rates, and low pressure drops such as required by full-flow, engine oil filtration. The cartridges are available for particle removal in the 5 to 20 micron size. It is also recommended for use in filtering fuel oils, metal working coolants, and washing solvents.

Clay Filtration   Certain clays such as Fuller's earth will adsorb oil-oxidation products and the depth of clay will screen out fine dirt particles. The Fuller's earth may be in cartridge type (Fig. 16-9) units or may be mixed with the oil and filtered through size filtration units. The major disadvantage of clay filtration is that certain oil additives may be removed with the contaminants.

Multipurpose Purifiers   Filtration, settling, and free water removal are often combined in a single unit as shown in Fig. 16-10. These units have capacities for treating from 90 to 2600 gph (340 to 9840 L/hr). The oil conditioner is a three compartment unit that provides dry clean oil by:

1. Removing free water and coarse solids via horizontal wire screen plates in the precipitation compartment (A)
2. Removing suspended solids via vertical, cloth covered, leaf type, filter elements in the gravity filtration compartment (B)
3. Polishing the oil and stripping it of all moisture vapor via cellulose filter cartridges in the storage and polishing compartment (C).

The leaf type filter elements may be removed individually without shutting the system down. It has a micronic selectivity down to 2 micron particle size and will not remove rust or oxidation inhibitors.

Centrifugation   Waste lubricating oils can also be purified by centrifugation, a process for separating materials of different densities whether two liquids, a liquid and a solid, or two liquids and a solid. A centrifugal force is produced by any mov-

Fig. 16-10   Combined Filtration and Settling Purifier

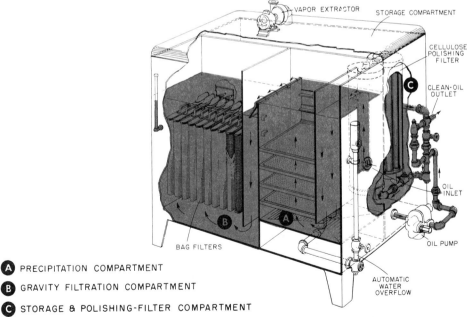

Ⓐ PRECIPITATION COMPARTMENT

Ⓑ GRAVITY FILTRATION COMPARTMENT

Ⓒ STORAGE & POLISHING-FILTER COMPARTMENT

ing mass that is compelled to depart from the rectilinear path which it tends to follow; the force being exerted in the direction away from the center of curvature of its path. The centrifugal force is directly proportional to the mass or specific gravity of a material and the higher the specific gravity the further it will travel from the center of rotation. The centrifuge is a machine designed to subject material held in it or being passed through it to centrifugal force, with means for collecting the separated materials. A batch centrifuge holds a given quantity of material and must be stopped periodically to discharge the solids, and clarified liquid, and be recharged, while a continuous centrifuge accepts a steady stream of material, applies a centrifugal force, and continuously discharges the separated components. A centrifugal separator or purifier handles a continuous stream of mixed liquids and effects their separation. A basket centrifuge is designed to hold a mass of material, usually a mixed solid and liquid, and by subjecting the mass to centrifugal force separates the liquid which passes through the basket walls leaving the solids behind.

Centrifugation by disc and tubular centrifuge, operating at high gravities, is excellent for removal, at high rates, of relatively small particles and separation of fluids of close but varying densities. The initial and operating costs are relatively high and fairly large floor space is required.

RECLAMATION OF LUBRICATING OILS

In addition to settling and filtration, vacuum distillation may be used to eliminate water and light ends usually present due to fuel dilution. This processing is referred to as reclaiming and may be either a batch or a continuous operation. Reclamation units are available in various sizes to handle up to 100 gph (378.5 L/hr.) and may be custom made as high as 3000 gph (11,355 L/hr.).

Fig. 16-11   Batch Reclaimer

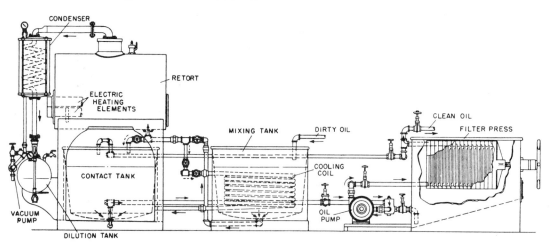

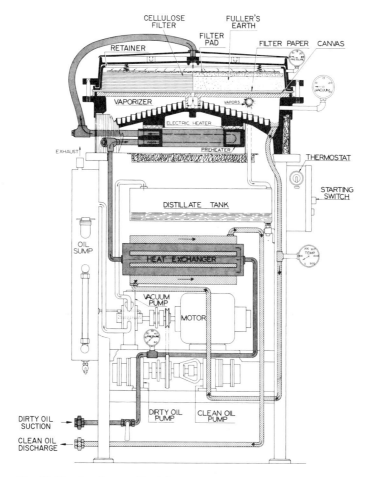

Fig. 16-12   Continuous Reclaimer

In a batch reclaimer (Fig. 16-11), the dirty used oil is mixed with activated clay in the mixing tank and is drawn by vacuum into an electrically heated retort. Under controlled vacuum and high temperature, all moisture and the light ends of fuel dilution are removed. The high temperature lowers the oil viscosity, making it easier to filter, and increases the activity of the clay. After treatment in the retort for a sufficient length of time, the heating elements and vacuum pumps are shut off and the oil is allowed to drain through heat transfer coils in the mixing tank to the contact tank where a filter aid is added. It is then pumped to a filter press containing both cloth and paper filtering media. Vapors are drawn from the retort through a condenser. The condensed vapors are collected in the dilution tank. These reclamation units are custom made to meet customer specifications.

A similar process on a continuous basis is conducted in the oil reclaimer shown in Fig. 16-12. Dirty oil is pumped through a heat exchanger and a preheater which raises the oil to proper filtering temperature. The oil passes through a filter bed composed of a filter pad, a bed of Fuller's earth, filter paper, and canvas. The filtered oil passes to a vaporizer where the oil temperature is increased. A vacuum is maintained in the vaporizer which aids in the evaporation of water and the distillation of the more volatile portions, if any. The purified oil is cooled in the heat exchanger by the incoming oil.

The unit shown in Fig. 16-12 can be obtained as a portable unit handling as low as 6 gph (23 L/hr) of oil which may be hooked up to a single machine reservoir. They are also available as custom made stationary units to handle up to 3000 gph (11,355 L/hr). The unit shown is an integral filter and reclaimer, other units have separate filter and reclaimer units to handle large volumes of solids. When activated clay (Fuller's earth) is used in these reclamation units to treat oils containing additives, the clay will usually remove the majority of the additives. The resulting clarified oil will have to be either reconstituted with a new additive package or used in a less demanding function. The unit shown in Fig. 16-12 can be equipped with cellulose discs in place of Fuller's earth when handling additive based oils, phosphate based synthetics containing additives, or other fluids adversely affected by activated clay. The cellulose material will not remove the additive package.

WASTE COLLECTION AND ROUTING

So far we have discussed the selection, storage, and handling of lubricating oils as new products and in-service materials to insure long life, minimize contamination and provide purification in process. Eventually all lubricants will become waste oil and must be drained from the system, collected into waste oil storage tanks, and prepared for one of several methods for final disposal. All this must be accomplished with two general considerations in mind:

Fig. 16-13 Multimachine Collection System

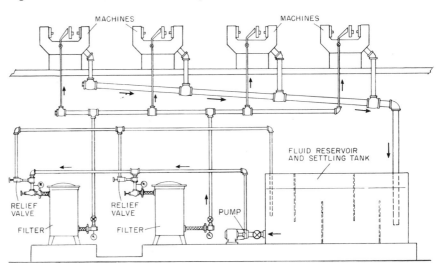

Fig. 16-14 Sump Tank Cleaner and Centrifuge

1. Handling the waste oil so that it will have the highest value as a by-product and the optimum stability for disposal
2. That it be prevented from polluting the process, storm and sewer systems in the plant and the ground and surface waters outside the plant

All this can be accomplished by proper collection and routing systems. An intermediate size multimachine collection system is shown in Fig. 16-13. Or the collection system may be simply a combination sump tank cleaner and centrifuge (Fig. 16-14) for renewing and purifying used oil containing sludge, chips, and other solid contaminants. Clean oil is retruned to the sumps by reversing a valve so that air pressure instead of vacuum is exerted on the storage tank. Waste oil is carted back to be put in storage drums for final disposal. This system can handle individual machine needs for reclamation or collection of waste oil.

Whatever the size of the systems for collection and routing of the waste oil, certain principles should be followed:

1. Separate collection systems should be used for each type of lubricant and separate storage provided until final disposal. Neat oil and lubricants should be kept separate from emulsions, chemical coolants, etc.
2. Lubricating oils should be kept separate on the basis of degree of contamination even though the type of oil may be the same. A relatively clean hydraulic oil should not be mixed with waste cutting oils or lubricants. The value of the waste hydraulic oil as a byproduct will be lowered by admixture with the solids contaminated, cutting fluid.

3. Reservoirs should be drained immediately after shutdown while the oil is still at operating temperature and solids are in suspension. Certain oxidation products become insoluble in cold oil and would precipitate with the solids, remain in the reservoir, and contaminate a new charge.
4. Prevent contamination of process-water storm drains and sewage systems by waste oil streams. This is particularly a consideration where floor drains exist and spills of lubricating oil occur. It is also possible in oil coolers or heaters where leaks in the exchangers can cause contamination of process steam or cooling water or the oil being treated.
5. Open troughs or floor gratings and drains to catch leaks, spills, and drips or to route waste oil to storage are not desirable because the waste oil will be contaminated with dirt, water, lint, etc.

The question of individual machine collection systems versus plant wide systems will be a matter of economics and feasibility; however, the plant wide system is more desirable. Individual machine systems can be serviced with oil carts such as shown in Fig. 16-15, which are commercially available. The commercial unit shown has a capacity of 80 gal (246 L) and is equipped with either a single flow or reverse suction rotary hand pump.

Where it is not feasible to connect the individual machines to a waste oil collection system, consideration should be given to a central piping system to carry waste oil to storage, with conveniently located floor drains into

Fig. 16-15   Oil Cart

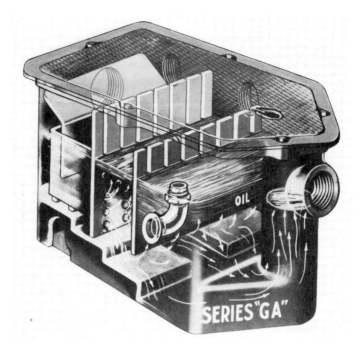

Fig. 16-16 Oil Separator for Drain System

which the oil carts could be drained. If such a system is used, it must be designed so that it can be isolated from ordinary floor drainage of water or washing solution. Where it is impossible to have separate oil drainage systems, the floor drains in areas liable to waste oil spillage should be equipped with oil traps as shown in Fig. 16-16.

FINAL DISPOSAL

Lubricants, including cutting fluids, hydraulic oils, gear lubes, etc., eventually reach a state where they must be disposed of as waste oil. The method of final disposal will depend on the type and condition of the waste oil. There are several generally accepted methods which permit disposal with adequate safeguards to prevent water or air pollution, including the following:

1. Burning as fuel
2. Rerefining and reuse
3. Incineration
4. Plowing into the soil

One of the presently most feasible and generally acceptable methods of disposing of used lubricating oils is burning as fuel. Burning waste oils as fuel is highly recommended as an approved method of final disposal for the following reasons:

1. It is the most widely applicable method from the standpoint of type of waste oil and location of generating source.
2. There are practically no volume limitations on the amount of waste oil that may be handled by this method. It is feasible for the large, medium, and smaller user of petroleum lubricants.
3. It turns what could be a liability (disposal of waste oil) into an asset. Since some contract haulers are charging to remove waste oils, and heavy fuel costs are rising, a substansial net value can be realized by using waste oil as fuel.

Further, there is an ever increasing demand for petroleum products (gaseous and liquid) as an energy source. Therefore, any secondary use of waste oil, subsequent to its primary use as lubricant, will recover additional value as a petroleum product and conserve virgin petroleum resources. Burning of used lubricants as fuel is the simplest of the disposal methods and has been widely proven.

While practically all liquid petroleum products may be oxidized into carbon dioxide and water by burning, not all are suitable for use as fuel or for mixing with conventional fuel. Some, due to contamination with large amounts of water, low flash petroleum products, sediment, or contaminants of mineral origin, lose their value as fuel and must be incinerated using supplemental fuels. This is the reason for proper in-plant handling to ensure a reasonable waste oil byproduct whether the final disposal is by burning as fuel, re-refining, or contract hauling by waste collectors.

Grossly contaminated liquid petroleum products, both low and high flash, as well as certain aqueous wastes, may be disposed of by high temperature incineration. This method will also handle greases, waxes, oil saturated solids such as floor drying compounds, and some other miscellaneous materials.

Index